International Series of Monographs in Analytical Chemistry, Volume 22

Newer Redox Titrants

A. Berka
Charles University,

J. Vulterin
Institute of Chemical Technology and

J. Zýka
Charles University, Prague

Oxidation-reduction reactions are still of great importance in titrimetric methods of analysis. Because of their variety they offer special ways of studying the quantitative course of reactions in both inorganic and organic systems. The purpose of this book is to present a survey of newer redox titrants, their evaluation, practical use and particularly an exhaustive list of references to all papers published on this topic. This is one of the first books in which the field has been so widely treated. It is a tribute to the authors' wide experience in their field, which goes back more than fifteen years; material contained in this book is based on this experience as well as on their own publications on the experimental evaluation of nearly all the reagents which are reviewed. This book should be of great interest to all scientific workers in research and service laboratories of analytical chemistry, in pharmaceutical chemistry, in the food industry and in medicine. It should be of value to scientific research experts and students of analytical chemistry and some of the related branches should be of interest to universities and colleges of technology.

INTERNATIONAL SERIES OF MONOGRAPHS IN
ANALYTICAL CHEMISTRY
GENERAL EDITORS: R. BELCHER AND L. GORDON

VOLUME 22

NEWER REDOX TITRANTS

NEWER REDOX TITRANTS

by

A. BERKA, J. VULTERIN and J. ZÝKA

Translated by

H. WEISZ

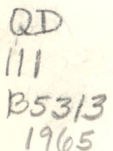

PERGAMON PRESS

OXFORD · LONDON · EDINBURGH · NEW YORK
PARIS · FRANKFURT

Pergamon Press Ltd., Headington Hill Hall, Oxford
4 & 5 Fitzroy Square, London W.1

Pergamon Press (Scotland) Ltd., 2 & 3 Teviot Place, Edinburgh 1

Pergamon Press Inc., 122 East 55th St., New York 22, N.Y.

Pergamon Press GmbH, Kaiserstrasse 75, Frankfurt-am-Main

Copyright © 1965
Pergamon Press Ltd.

First edition 1965

Library of Congress Catalog Card No. 64–7811

PRINTED IN GREAT BRITAIN BY BELL AND BAIN LTD., GLASGOW

CONTENTS

	PREFACE	vii
1.	POTASSIUM PERMANGANATE IN ALKALINE SOLUTION	1
2.	COMPOUNDS OF TRIVALENT MANGANESE	10
3.	COMPOUNDS OF TRIVALENT COPPER	14
4.	POTASSIUM HEXACYANOFERRATE (III)	18
5.	HYPOHALITES (HYPOCHLORITE, HYPOBROMITE)	29
6.	CHLORAMINE-T	37
7.	BROMINE	48
8.	N-BROMOSUCCINIMIDE	52
9.	IODINE MONOCHLORIDE	55
10.	PERIODIC ACID AND ITS SALTS	66
11.	LEAD (IV) ACETATE	76
12.	COMPOUNDS OF PENTAVALENT VANADIUM	83
13.	IRON (III) SALTS	94
14.	COMPOUNDS OF TRIVALENT COBALT	99
15.	HYDROGEN PEROXIDE	101
16.	CHROMIUM (II) SALTS	104
17.	TIN (II) CHLORIDE	120
18.	SODIUM ARSENITE	127
19.	MERCURY (I) NITRATE AND MERCURY (I) PERCHLORATE	131
20.	COMPOUNDS OF MONOVALENT COPPER	137
21.	COMPOUNDS OF PENTAVALENT AND TRIVALENT MOLYBDENUM	140
22.	COMPOUNDS OF PENTAVALENT AND TRIVALENT TUNGSTEN	143
23.	URANIUM (IV) SULPHATE	145
24.	VANADIUM (II) SULPHATE	147
25.	VANADIUM (IV) SULPHATE AND VANADIUM (IV) ACETATE	155
26.	ASCORBIC ACID	161
27.	HYDROQUINONE AND SIMILAR REDUCING AGENTS	172
28.	HYDRAZINE SULPHATE	180
29.	SODIUM NITRITE	185
30.	SOME OTHER OXIDIZING AND REDUCING TITRANTS	194
	AUTHOR INDEX	201
	SUBJECT INDEX	215
	OTHER TITLES IN THE SERIES	246

PREFACE

VOLUMETRIC determinations, based on oxidation–reduction processes, still remain among the most widely used methods of chemical analysis. The increasing number of original communications indicates that continual investigation in this field has by no means exhausted all possibilities concerning the study of oxidation–reduction processes or their application to volumetric analysis.

One of the trends which may be observed in the communications published in the field under consideration, is the study of new, or of less commonly used, volumetric reagents. The work may be concerned either with the initial attempts to introduce these agents into volumetric analysis, or with their wider evaluation; extension of the application of more well-known reagents is also considered.

It is understandable that in fundamental text-books or in monographs devoted to volumetric analysis, the main attention must be given on the one hand to the fundamentals of redox processes in general, and on the other to a detailed discussion of those volumetric agents which are already well known in practice and currently used, and on which conventional oxidation–reduction determinations are based. Naturally it is impossible to devote sufficient attention to other reagents.

This objective guided us in setting out the contents of the present book. We eliminated beforehand such reagents (or volumetric determinations) which are known in sufficient detail, and whose detailed treatment, with an enumeration of all possibilities, would require in each case practically a separate monograph (for instance manganometry, cerimetry, iodimetric titrations, titrations by bromate, iodate, titanium (III) salts, etc.). As follows from the title of the book, we have confined it mainly to such reagents, which either have not been treated in review articles at all, or only briefly. In such cases we have attempted to add the fullest possible list of literature quotations concerning the particular reagent or reaction. We regard this as being the main role of the book, i.e. to provide information to all those who work in this field and who would wish to verify in practice the application of the reagents described or their suitability for given practical cases. Wherever possible, we have also made use of our own experience for a critical evaluation of the material discussed.

Some other reagents included in this book are, on the contrary, not treated fully, for they have already been discussed at length in G. Jander et al.: *Neuere massanalytische Methoden*; F. Enke, Stuttgart 1956, or in I. M. Kolthoff, R. Belcher, V. A. Stenger and G. Matsuyama: *Volumetric Analysis III, Titration Methods: Oxidation–Reduction Reactions*, Interscience Publishers, New York 1957; in these cases we have been forced to make a compromise, presenting only the main characteristics of the given reagent, and citing the two above comprehensive works, in which readers will find more detailed information (this concerns for instance the use of periodic acid and its salts, chromium (II) salts, hypohalites, etc.). In the present volume we have attempted rather to discuss the results of the most important communications published since the time when the above books appeared.

We have further limited the scope of this book by excluding from it coulometric titrations, where the reagent—although possibly less common—is generated *in situ* (an exception to this is the titration by hypobromite "*in statu nascendi*" in the chapter on hypohalites). We have similarly not included numerous communications based on addition or substitution reactions, such as the use of bromine or interhalogens, and we have retained only those titrations where the reagent mentioned acts as oxidant.

Since the Czech edition of this book (*Vybrané oxydačně redukční metody*, SNTL, Prague) published in 1961, and the German edition (*Massanalytische Oxydations–und Reduktions-methoden*, Akademische Verlagsgesellschaft, Leipzig) published in 1964, the book has been enlarged by approximately 200 new literature quotations; this shows that the theme of the book is still justified and is being currently studied. It is possible that when the latest literature was being treated, some important work might have escaped our attention. We shall be very grateful to our readers for such information.

In concluding, we wish to thank Prof. R. Belcher (University of Birmingham) and Prof. L. Gordon (Case Institute of Technology, Cleveland) for their kindness in proposing the publication of this book by Pergamon Press, and to Prof. H. Weisz (University of

Freiburg i. Br.) for the careful translation, as well as for reminding us of some details which could be made more precise for the English edition and for drawing our attention to some errors.

> Dr. ANTONÍN BERKA,
> Department of Analytical Chemistry, Charles University, Prague, Czechoslovakia
>
> Dr. JAROSLAV VULTERIN,
> Department of Analytical Chemistry, Institute of Chemical Technology, Prague, Czechoslovakia
>
> Prof. Dr. JAROSLAV ZÝKA,
> Department of Analytical Chemistry, Charles University, Prague, Czechoslovakia

CHAPTER 1

POTASSIUM PERMANGANATE IN ALKALINE SOLUTION

The reduction of permanganate in alkaline solution is a far more complicated process than that which occurs in an acid medium; it is also much more highly dependent on the conditions of reaction (e.g. on the concentration of hydroxide ions, on the temperature, etc.). For analytical purposes, the reduction of permanganate (Mn^{VII}) to manganate (Mn^{VI}) and subsequently to manganese oxide (Mn^{IV}) are the stages through which the reactions generally proceed.

The formal potential of the redox system

$$MnO_4^- + e^- \rightleftharpoons MnO_4^{2-} \tag{1}$$

in 1 N KOH solution was found to be + 0·61 V (at 18°C); for the system

$$MnO_4^{2-} + 2\,H_2O + 2e^- \rightleftharpoons MnO_2 + 4\,OH^- \tag{2}$$

the potential was + 0·51 V.[1]

Holluta,[2,3] in the course of his studies of the reaction between permanganate and formate, observed that with increasing alkalinity the reduction of permanganate proceeds more and more towards manganate. Permanganate behaves similarly towards other not too strong reducing agents.[4]

The mechanism of the permanganate oxidation in alkaline solution has been thoroughly investigated[5,6,7] and it was assumed that hydroxide radicals, which are the real oxidation agent, are intermediately formed:

$$MnO_4^- + OH^- \rightleftharpoons MnO_4^{2-} + OH \tag{3}$$

$$MnO_4^{2-} + 3\,H_2O \rightleftharpoons H_2MnO_3 + 2\,OH^- + 2\,OH \tag{4}$$

The influence of the hydroxide ion concentration on the velocity of the two reduction steps in alkaline solution is well recognizable from the two equations. If, however, there is no oxidizable substance present, the very reactive hydroxide radicals do not only react with the MnO_4^{2-} ions (equation 3), but also with each other, thus forming water and gaseous oxygen; this explains the lower stability of alkaline permanganate solutions. With increasing temperature this reaction is greatly enhanced; it is therefore necessary in carrying out a quantitative oxidation with permanganate, to avoid working at too high temperatures, and to choose the alkalinity of the solution so that the reduction of the permanganate to MnO_4^{2-} or to MnO_2, just proceeds stoichiometrically.

Nevertheless, the correct choice of the alkalinity of the solution is not sufficient to ensure that only one of the two reduction steps is reached. A definite course of reaction can, however, be ensured under suitable conditions when barium salts or telluric acid are added.

Whereas $Ba(MnO_4)_2$ is readily water-soluble, the solubility of $BaMnO_4$ is only $2\cdot46 \times 10^{-10}$ g/l. (at 25°C).[8] If barium nitrate or barium chloride is added to the solution to be titrated, the manganate formed during the reaction precipitates as the barium salt so rapidly, that (under optimum conditions) it cannot be further reduced.[4] The $BaMnO_4$ forms deposits very well and, in spite of the blue-green colour of the precipitate one can titrate visually to the colour change produced by the first drop of excess permanganate. Often a known amount of permanganate can be titrated with the reducing agent to be determined.

In most cases, however, the determinations are carried out indirectly: the unconsumed excess of permanganate is back-titrated in alkaline medium with sodium formate (in the presence of cobalt, nickel, copper or silver salt as catalyst), formic acid, thallium (I) salt, cyanide and, in acid solution, with oxalic acid or with iron (II) salt. If it is necessary to let the reagent react for a longer time, e.g. in indirect determinations of organic substances, a blank test must be done.

Alternatively, telluric acid[9] may be used; in its presence permanganate is reduced in alkaline solution fairly rapidly and quantitatively to manganese (IV). This reaction has been applied in many direct and indirect determinations.[10]

The determination of compounds which are reduced quantitatively to MnO_2 can be carried out also by filtration of the precipitated manganese dioxide and its conversion by a reaction with Mn^{2+} (in the presence of pyrophosphate) to a complex of trivalent manganese, which can be titrated with standard quinol (hydroquinone) solution using diphenylamine as indicator.[10a] Manganese dioxide can also be reduced (with ascorbic acid) to Mn^{II}, which is then determined complexometrically,[10b] thus avoiding the errors caused by the low stability of permanganate solution (owing to its partial reduction to manganate at higher temperatures and in strongly alkaline medium).

STANDARD SOLUTION

Preparation of 0·1 M* potassium permanganate:

The distilled water is boiled for about 5 min, the necessary amount of $KMnO_4$ (15·803 g $KMnO_4$ for 1 l.) is added, and the solution again boiled for 1 min. The cooled solution is set aside for 2 hr; this allows the manganese (IV) oxide formed to settle out. It is then filtered off. The solution prepared in this way is stable for several months.

Standardization

The permanganate solution is standardized exactly as in acid solution. Thus, oxalic acid, sodium oxalate, arsenic (III) oxide, iron (II) ammonium sulphate, hexacyanoferrate (II), potassium iodide, etc., are used.

INDICATOR

The end-point of the direct titration is easily recognized by the colour owing to traces of excess $KMnO_4$; the end-point of the back-titration by the decolorization. In some cases the equivalence point is determined potentiometrically.

REVIEW OF DETERMINATIONS

Arsenic and antimony. Arsenic and antimony can be determined by potentiometric titration with potassium permanganate in 0·25 N NaOH in the presence of telluric acid.[9] As little as 2 μg of arsenic,[11] 3 μg of antimony[12] can be determined in 0·1 N NaOH by titration with 10^{-4} N standard solution. The back-titration of the permanganate with an arsenic (III) solution is also possible. In both determinations the permanganate is reduced to manganese (IV).

The indirect determination in the presence of telluric acid can be done over a wider range of hydroxide ion concentration. In 0·1–2 N NaOH, the excess of reagent should react for 5 min. Then, an excess of a precisely standardized iron (II) solution in 3 N H_2SO_4 is added and the unconsumed amount of it (after addition of phosphoric acid) is titrated with standardized permanganate.[13]

According to another publication,[14] the permanganate is reduced quantitatively to manganese (IV) (as the indirect As^{3+} determination) in 1·5–3 N NaOH, even without addition of H_6TeO_6, whereas the reduction

---

* Here and in all subsequent examples M = molar.

proceeds quantitatively to manganese (VI) in 1 N NaOH in the presence of Ba^{2+} ions. The excess of permanganate is titrated potentiometrically with standardized thallium (I) solution.

Selenium and tellurium. In the potentiometric titration of tellurium (IV) with permanganate in 0·4–0·08 N NaOH (depending on the concentration of Te^{4+})[15] the reagent is reduced to MnO_2; very precise results can be obtained in the presence of telluric acid.[9] Amperometric titration gives also good results,[15a] even in the presence of selenium (+ 0·4 V, 1 N NaOH).[15b]

The direct titration of selenium (IV) is effected in 2·5 N NaOH and 10 per cent NaCl solution (MnO_4^- is reduced to MnO_4^{2-}). At 90°C in 0·1 N NaOH and 10 per cent NaCl solution and in the presence of $AuCl_3$, permanganate is reduced to MnO_2. The reduction of permanganate with standardized selenite solution proceeds to manganate in 1–3 N NaOH. All these reactions can be done potentiometrically.[16]

The indirect determination of selenium (IV) and tellurium (IV) solutions can be carried out equally well in the presence or absence of barium ions in 1–2 N NaOH; the excess of permanganate is then back-titrated potentiometrically with a standardized thallium (I) sulphate solution in the presence of barium ions in 1–1·5 N NaOH ($MnO_4^- \to MnO_4^{2-}$). Instead of barium ions, telluric acid (H_6TeO_6) can be used; in this case the permanganate is reduced to the tetravalent state ($MnO_4^- \to Mn^{4+}$).[17] The excess of permanganate can also be determined by potentiometric titration using a standard solution of arsenic (III) or formate.[16,18]

In the determination of selenium (IV) and tellurium (IV) compounds in the presence of each other, first the tellurium (IV) is titrated potentiometrically ($TeO_3^{2-} \to TeO_4^{2-}$) and then the selenium (IV) is determined indirectly.[9]

Chromium. The direct titration of chromium (III) ions with alkaline permanganate yields inaccurate results. Good results can be obtained, however, by titration of a known amount of $KMnO_4$ with the solution of chromium (III) to be determined, in the presence of barium salts in 0·8–1·5 N NaOH ($MnO_4^- \to MnO_4^{2-}$) or, in absence of barium ions, in 0·5–2 N NaOH ($MnO_4^- \to MnO_2$).[19]

In the indirect determination, chromium (III) is oxidized with excess permanganate in 2·5 N NaOH or (in the presence of barium ions) in 1 N NaOH; the excess of reagent is back-titrated potentiometrically with formic acid[19] or with standard thallium (I) solution.[17]

Vanadium. The oxidation of tetravalent vanadium with an alkaline permanganate solution,[19a] proceeds best using an excess of reagent, and in the presence of barium ions or telluric acid. The excess of permanganate is titrated with a solution of thallium (I) or formic acid.

Cerium. Cerium (III) ions can be titrated with permanganate in an acetate-buffered solution (pH = 8). Because the manganese (III) ions formed give an intense colour to the solution, it is necessary to determine the equivalence point potentiometrically.[20]

The oxidation of cerium (III) proceeds quantitatively in 1–2 N NaOH solution in the presence of telluric acid; the permanganate is simultaneously reduced to manganese (IV). In the titrimetric determination, the excess of permanganate is back-titrated after 10 min with thallium (I) solution.[20a]

Lead. Lead (II) salts are titrated directly with permanganate in the presence of zinc and mercury (II) oxide as catalysts, using a potentiometric end-point.[21] In the indirect determination, lead (II) is oxidized in 1–2 N NaOH with excess of permanganate and the excess is determined with standard thallium (I) solution in the presence of barium ions or telluric acid.[17]

Thallium. Thallium (I) salts are titrated with permanganate in alkaline medium at 40°C.[21a]

Tin. The direct titration of stannite (II) with permanganate in alkaline solution has to be done in an inert atmosphere using a potentiometric end-point;[22] the following reaction takes place:

$$13\ SnO_2^{2-} + 6\ MnO_4^- + 3\ H_2O \rightarrow 13\ SnO_3^{2-} + 2\ Mn_3O_4 + 6\ OH^-$$

Ruthenium. The titration of di- and tetravalent ruthenium with permanganate at 70–80°C has also been described. Here, the end-point of the titration is detected by the red colour which appears, if a drop of the solution is acidified with sulphuric acid (external indication). In this titration, the permanganate is reduced quantitatively to manganese (IV) oxide and the ruthenium is oxidized to the hexavalent state. Instead of permanganate, manganate can be used; it is much more stable at higher temperatures.[23,24]

Mercury (I). Monovalent mercury can be oxidized in 0·5–1 N NaOH with excess of permanganate in the presence or absence of telluric acid. The excess of reagent is reduced in acid medium with an excess of iron (II) solution which is finally back-titrated with permanganate.[24a]

The reduction of permanganate with mercury (I) in acid solution in the presence of fluoride has been used for the indirect determination of iron (II), chromium (III) and vanadium (IV) salts.[24a, 24b, 24c] These substances are oxidized with excess of permanganate in alkaline solution. After acidifying the solution, the excess of the reagent is reduced with excess of mercury (I) solution in the presence of fluoride; the iron, chromium and vanadium now present in the higher valency state do not react. The mercury (I) not consumed by the reaction, is back-titrated with permanganate.

Hexacyanoferrate (II). Hexacyanoferrate (II) reacts with permanganate in alkaline solution in the presence of barium ions according to the reaction

$$[Fe(CN)_6]^{4-} + MnO_4^- \rightarrow [Fe(CN)_6]^{3-} + MnO_4^{2-}$$

This reaction has been applied to the potentiometric determination of permanganate (in presence of manganate) to control the technical production of permanganate.[25]

Cyanide. An excess of permanganate is added to the cyanide to be determined in 10 per cent KOH solution; Cu (II) ions have to be present in the solution as catalyst. The permanganate is back-titrated in acid medium with oxalic acid or iron (II) sulphate.
Instead of permanganate, potassium manganate can be used.[26]

Sulphide, thiosulphate, sulphite and phosphite. Sulphides can be oxidized quantitatively with excess of permanganate in alkaline medium to sulphate[27] or, in the presence of telluric acid to dithionate.[28] The excess of reagent is determined iodimetrically in acid solution[27] or, Mohr's salt (iron (II) ammonium sulphate) is added in excess and is back-titrated with permanganate.[28] Thiosulphate can be determined in the same way.[13,27] Sulphite and phosphite are determined by another method, in the presence of telluric acid.[13]

Hydrazine. Hydrazine is oxidized in about 1 N NaOH solution by excess of permanganate in the presence of telluric acid or barium ions. The excess of permanganate is titrated with thallium (I) sulphate[29] or, after acidifying with sulphuric acid, Mohr's salt is added in excess and back-titrated with permanganate.[30] Whereas the direct titration of hydrazine with permanganate in alkaline solution does not yield good results, permanganate can be determined with hydrazine in 0·75–1 N NaOH solution in the presence of barium salt ($MnO_4^- \rightarrow MnO_4^{2-}$); in the absence of barium ions, the permanganate is reduced in 0·5–2·5 N NaOH solution to MnO_2.

Hydrogen peroxide. Hydrogen peroxide is oxidized by excess of permanganate in 1 N NaOH in the presence of telluric acid and then is back-titrated with a precisely standardized thallium (I) solution.[31]

Barium peroxide and sodium peroxide. Both peroxides have likewise been determined indirectly by oxidation with excess of permanganate in 0·5 N NaOH solution in the presence of telluric acid. A potentiometric titration procedure has also been described.[32]

Formate. The reaction between permanganate and formate has been thoroughly studied in an extensive work by Stamm.[4] According to those studies, formate is best determined by titration of a known amount of permanganate with the formate solution in alkaline medium (1–3 g NaOH for each 40 ml) in the presence of barium salt. Before reaching the equivalence point, some nickel nitrate is added to the solution, to enhance the rate of reaction, and then the titration is continued until decolorization.

Potassium permanganate has also been titrated potentiometrically with formate and formic acid in the presence of barium ions (in 0·5–1·5 N NaOH solution), or in their absence (in 0·5–2·5 N NaOH solution).[33]

Hypophosphite, phosphite, iodide, iodate, cyanide, thiocyanate, methanol and formaldehyde. These substances are oxidized with excess of permanganate in alkaline solution[4] and back-titrated with formate. Hypophosphite and phosphite are oxidized to phosphate, iodide and iodate to periodate, cyanide to cyanate, thiocyanate to cyanate and sulphate, and methanol and formaldehyde to carbon dioxide and water. Iodide can be determined in the presence of bromide and chloride. Cyanide, methanol and formaldehyde can be determined by titrating directly a known amount of permanganate with the solution of the substance to be evaluated.

Acetone, fumaric acid and erythritol. In the oxidation of these substances with permanganate in alkaline solution, oxalic acid can be formed in variable amounts besides carbon dioxide and water. Accordingly, the determination is done as follows:

The substance is treated with an excess of permanganate; after the oxidation, an amount of oxalic acid equivalent to the permanganate is added and the solution is acidified with sulphuric acid. The excess of oxalic acid present in the colourless solution corresponds exactly to the amount of permanganate used for oxidizing the substance to be determined to carbon dioxide and water in alkaline solution; it is determined titrimetrically with a standard permanganate solution.

Iodide, iodate, cyanide, methanol, glycerol, phenol, salicylic acid,[34] *acetaldehyde, benzaldehyde, mandelic acid, cinnamic acid, tartaric acid, malic acid, isoamylalcohol, benzylalcohol and n-butanol.*[35] These substances have been oxidized with excess of permanganate in alkaline solution at 25°C (for inorganic substances) or at 45°C (for organic substances). After 10 min, the excess of permanganate is titrated potentiometrically with sodium formate in the presence of barium chloride.

Zirconium can be determined indirectly (as zirconium mandelate) by oxidizing mandelic acid to benzoic acid.[35a]

Citric, tartaric and salicylic acid. The influence of inorganic salts on the velocity of the oxidation of these acids with excess of alkaline permanganate has been studied.[36]

In the absence of a catalyst, tartaric acid is oxidized quantitatively within 24 hr, in the presence of calcium or copper (II) salts to the extent of 99·5 per cent within 2 hr. In the presence of chromium (III) ions, the rate of oxidation is diminished.

In the oxidation of citrates and salicylates, no catalytic influence of inorganic salts has been observed.

Formaldehyde, formic, salicylic acid, methanol, hydrazine and thiosulphate. For the determination of these compounds, procedures based on the determination of manganese dioxide formed by the reaction, have been recently described.[10a, 10b]

4-Dimethylamino-1, 5-*dimethyl*-2-*phenylpyrazolone (amidopyrine).* This substance consumes 4 equivalents of permanganate per mole when oxidized in alkaline solution. The permanganate is added in excess, and the residual amount is determined iodimetrically after acidifying the solution.[37]

Determination of reducing substances in natural and waste water. This determination is likewise carried out with an excess of alkaline permanganate solution in the presence of barium chloride; the excess of oxidizing agent is back-titrated with formate. When at least a three-fold excess of permanganate is used, the oxidation is complete after 1 min.[38]

Permanganate can oxidize in alkaline solution many other organic substances. For their quantitative determination, the reagent is, as a rule, added in excess. Moreover, the various conditions of reaction have to be considered (e.g. alkalinity of solution, temperature, duration of reaction) for the reaction to proceed stoichiometrically. Several procedures have been recommended for back-titrating the excess of permanganate. Because it is not possible to quote here all these methods, the reader is referred to the extensive and systematic study of Imhof[46] and to the publications of Drummond and Waters, who investigated the oxidizing action of alkaline solutions of MnO_4^-, MnO_4^{2-}, MnO_4^{3-}, $[Mn(H_2P_2O_7)_3]^{3-}$ and $Mn_2(SO_4)_3$ on various organic compounds.[39-45]

REFERENCES

1. SACKUR, O. and TAEGENER, W., *Z. Elektrochem.*, **18**, 718 (1912).
2. HOLLUTA, J., *Z. phys. Chem.*, **102**, 276 (1922).
3. HOLLUTA, J., *Z. phys. Chem.*, **113**, 464 (1924).
4. STAMM, H., *Angew. Chem.*, **47**, 791 (1934).
5. DUKE, F. R., *J. Amer. Chem. Soc.*, **70**, 3975 (1948).
6. SYMONS, M. C. R., *J. Chem. Soc.*, 3956 (1954).
7. SYMONS, M. C. R., *J. Chem. Soc.*, 3676 (1954).
8. SCHLESINGER, H. I. and SIEMS, H. B., *J. Amer. Chem. Soc.*, **46**, 1965 (1924).
9. TOMÍČEK, O., PROČKE, O. and PAVELKA, O., *Collection (Czech. Chem. Comm.)*, **11**, 449 (1939).
10. BERKA, A., *Chemie*, **10**, 187 (1958).
10a. BERKA, A., *Collection (Czech. Chem. Comm.)*, **29**, 2844 (1964).
10b. FLASCHKA, H. and GARETT, G., *Chemist-Analyst*, **52**, 101 (1963).
11. ISSA, I. M. and ELSHERIF, I. M., *Recu. Trav. Chim.*, **75**, 447 (1956).
12. ISSA, I. M. and ELSHERIF, I. M., *Chemist-Analyst*, **45**, 78 (1956).
13. ISSA, I. M. and ISSA, R. M., *Chemist-Analyst*, **45**, 62 (1956).
14. ISSA, I. M. and ELSHERIF, I. M., *Analyt. Chim. Acta*, **14**, 300 (1956).
15. ISSA, I. M. and AWAD, S. A., *Analyst*, **78**, 487 (1953).
15a. ISSA, I. M., ISSA, R. M. and ALLAM, M. G., *Analyt. Chim. Acta*, **23**, 196 (1960).
15b. SONGINA, O. A. and TOYBAEV, B. K., *Izv. Akad. Nauk. Kazakh. SSR., Ser. Khem.* No. 2, 53 (1962); *Ref. Zhur.*, **20**, G83 (1963).
16. ISSA, I. M., EID, S. C. and ISSA, R. M., *Analyt. Chim. Acta*, **11**, 275 (1954).
17. ISSA, I. M. and ISSA, R. M., *Analyt. Chim. Acta*, **13**, 323 (1955).
18. ISSA, I. M., ISSA, R. M. and ABDUL AZIM, A. A., *Analyt. Chim. Acta*, **11**, 512 (1954).
19. ISSA, I. M., ABDUL AZIM, A. A. and ISSA, R. M., *Analyt. Chim. Acta*, **12**, 92 (1955).

19a. Issa, I. M. and Daess, A. M., *Recu. Trav. Chim. Pays-Bas*, **75**, 51 (1956).
20. Weiss, L. and Sieger, H., *Z. anal. Chem.*, **113**, 305 (1938).
20a. Issa, I. M. and Allam, M. G. E., *Z. anal. Chem.*, **182**, 244 (1961).
21. Issa, I. M., Issa, R. M. and Abdul Azim, A. A., *Analyt. Chim. Acta*, **10**, 474 (1954).
21a. Issa, I. M. and Issa, R. M., *Analyst*, **79**, 771 (1954); *Chem. Abstr.*, **49**, 2940 (1955).
22. Issa, I. M. and Awad, S. A., *J. Indian Chem. Soc.*, **32**, 23 (1955).
23. Gall, H. and Lehmann, G., *Ber. dt. chem. Ges.*, **60**, 2491 (1927).
24. Ruff, O. and Vidic, E., *Z. anorg. Chem.*, **136**, 58 (1924).
24a. Issa, I. M., Hamdy, M. and Hadidy, A. E., *Egypt. J. Chem.*, **2**, 59 (1959); *Z. anal. Chem.*, **176**, 364 (1960).
24b. Issa, I. M. and Hamdy, M., *Z. anal. Chem.*, **175**, 110 (1960).
24c. Issa, I. M. and Hamdy, M., *Z. anal. Chem.*, **174**, 418 (1960).
25. Řezáč, Z. and Kadič, K., *Chem. Průmysl*, **6**, 192 (1956).
26. Gall, H. and Lehmann, G., *Ber. dt. chem. Ges.*, **61**, 670 (1928).
27. Kolthoff, I. M., *Pharm. Weekbl.*, **61**, 841 (1924).
28. Issa, I. M. and Hamdy, M., *Z. anal. Chem.*, **169**, 334 (1959).
29. Issa, I. M. and Issa, R. M., *Analyt. Chim. Acta*, **14**, 578 (1956).
30. Issa, I. M. and Issa, R. M., *Chemist-Analyst*, **45**, 40 (1956).
31. Issa, I. M., Fathalla, A. H. and Issa, R. M., *Analyt. Chim. Acta*, **14**, 573 (1956).
32. Issa, I. M. and Khalifa, H., *J. Indian Chem. Soc.*, **33**, 778 (1956).
33. Issa, I. M. and Issa, R. M., *Analyt. Chim. Acta*, **11**, 192 (1954).
34. Singh, B., Singh, A. and Singh, G., *J. Indian Chem. Soc.*, **30**, 488 (1953).
35. Singh, B., Singh, A. and Nahan, R. K., *Res. Bull. East Panjab Univ.*, Nr. 33, **93** (1953).
35a. Schneer, A. and Hartmann, H., *Magyar Kém. Folyóirat*, **65**, 31 (1959).
36. Zolotukhin, V. K. and Molotkova, A. S., *Trudy Komis. Anal. Khim. Akad. Nauk USSR*, **5** (8) 179 (1954); *Ref. Zhur. Khim.*, **13**, 270 (1955).
37. Schulek, E. and Menyhárth, P., *Z. anal. Chem.*, **89**, 426 (1932).
38. Stamm, H., *Angew. Chem.*, **48**, 150 (1935).
39. Drummond, A. J. and Waters, W. A., *J. Chem. Soc.*, 435 (1953).
40. Drummond, A. J. and Waters, W. A., *J. Chem. Soc.*, 3119 (1953).
41. Drummond, A. J. and Waters, W. A., *J. Chem. Soc.*, 2456 (1954).
42. Levesley, P. and Waters, W. A., *J. Chem. Soc.*, 217 (1955).
43. Drummond, A. J. and Waters, W. A., *J. Chem. Soc.*, 497 (1955).
44. Pode, J. S. F. and Waters, W. A., *J. Chem. Soc.*, 717 (1956).
45. Land, H. and Waters, W. A., *J. Chem. Soc.*, 4312 (1957).
46. Imhof, J. G., Doctoral Thesis, Utrecht 1932; Kolthoff, I. M., Belcher, R., Stenger, V. A. and Matsuyama, G., *Volumetric Analysis*, III, Titration Methods: Oxidation–Reduction Reactions, p. 113. Interscience, New York 1957.

CHAPTER 2

COMPOUNDS OF TRIVALENT MANGANESE

ALL manganese (III) salts are strong oxidizing agents. The normal potential of the redox system

$$Mn^{3+} + e^- \rightleftharpoons Mn^{2+}$$

is + 1·4 V. Hence trivalent manganese is nearly as strong an oxidizing agent as permanganate. Unlike permanganate, manganese (III) salts have the valuable property of oxidizing iron (II) quantitatively even in the presence of chloride ions. The comparatively low stability of manganese (III) solutions is however, disadvantageous, but the stability of the solution can be enhanced by suitable complexing agents. One such complexing agent is diphosphate, but of course, the above mentioned redox potential is diminished to + 1·22 V owing to complex formation.[1]

The application of manganese (III) compounds in titrimetric analysis (apart from the determination of iron (II) in presence of chloride ions) is therefore limited to systems with low redox potentials, where other oxidizing agents are equally satisfactory. Of greater significance is the behaviour with organic compounds, which is similar to the action of periodate and lead (IV) salts.[2]

STANDARD SOLUTION

Since Ubbelohde[3] used manganese (III) sulphate for the first time as a reagent in titrimetric analysis, various authors have tried to stabilize this standard solution,[4,4a,29] for example, by the use of double sulphates of trivalent manganese[5,6] or by introducing complexing agents such as fluoride.[1,27] Diphosphate is the most suitable for stabilizing manganese (III).[1,7-14,18,19,25] The less familiar diphosphate can be replaced by phosphate,[4a] the standard solution prepared with phosphate contains the complex ion $[Mn(PO_4)_2]^{3-}$ as the active component[15-17,23] and is stable for about 10 days, a somewhat shorter time than a solution prepared with diphosphate. This standard solution is nevertheless well suited

COMPOUNDS OF TRIVALENT MANGANESE 11

for titrations. It is prepared by oxidizing a manganese (II) solution with bromate in phosphoric acid medium at elevated temperature; care should be taken that potassium ions are present in sufficient amount. After the bromine has been removed by boiling, an intensive violet solution remains which is suitably diluted (to 0·05 N or 0·01 N). Bismuth (V) or chromate (VI) at higher temperature can likewise be used for the oxidation.[30] The manganese (III) solution can also be prepared by reducing a permanganate solution with hydrogen peroxide.[20]

Standardization

The normality of the manganese (III) solution is determined by potentiometric titration with iron (II) or with hydroquinone.[15-17]

INDICATOR

Saito and Sato[8] state that the first drop of standard solution in excess gives a violet colour and so indicates the equivalence point. In other publications, the use of diphenylamine is described as indicator for the titration of iron (II),[18] and barium diphenylamine sulphonate for the determination of arsenic (III).[1] Other conventional redox indicators can be used.[19] All these determinations can be carried out potentiometrically. In one of the more recent publications[31] photometric titration has been applied to the determination of hexacyanoferrate (II), tin (II) and iron (II).

REVIEW OF DETERMINATIONS

Inorganic compounds. Arsenic *(III)*, antimony *(III)* and tin *(II)* can be titrated directly in acid medium with the above described solution of the diphosphate or phosphate complex of trivalent manganese. More details of the potentiometric determination of a mixture of these ions can be found in the original publications.[1,8,15,16,29,31]

Iron *(II)* can even be titrated in a 4·8 N chloride ion solution.[16,18,26] For the determination of iron in steel, an excess of reagent is added.[28]

Methods for the analysis of inorganic compounds mentioned later have no essential advantage, e.g. the indirect determinations of *peroxides*,[1,8] *nitrites*,[1] *vanadyl*[1] and also the determination of *thiocyanates, iodides*, etc., in presence of catalysts.[29] *Chromium, titanium, molybdenum, tungsten* and *uranium* compounds in their lower valency states can be titrated directly.[16] In the determination of *vanadium (II)* compounds, two oxidation steps can be reached quantitatively, i.e. $V^{2+} \rightarrow V^{4+}$ and $V^{4+} \rightarrow V^{5+}$.[25]

Organic compounds. The reaction of manganese (III) compounds with organic substances has been applied mainly for preparative work or for studies on the velocity of oxidation of some α-hydroxy acids, aldehydes and alcohols, or for comparing the action of permanganate and manganese (III) salts.[21,22] From some recent publications it is evident, however, that various organic compounds can be determined volumetrically by oxidation with manganese (III). The direct potentiometric determination of *oxalic acid*[1,8] (at higher temperature), of *hydroquinone* and of some p-*substituted phenols* and *amines*, of *hydrazine* and its derivatives, of the *hydrazide* of the *isonicotinic acid*, of *ascorbic acid*, etc., have been described so far.[17,29] The indirect titration has proved useful so far only for following up the oxidation, and in the determination of *tartaric acid, malonic acid, citric acid* and *salicylic acid* (oxidation to CO_2).[24] Formic and acetic acids do not interfere with these determinations. With potassium diphosphato manganate (III), experiments have also been carried out on the glycol splitting of mannitol and related compounds.[17]

REFERENCES

1. BELCHER, R. and WEST, T. S., *Analyt. Chim. Acta*, **6**, 322 (1952).
2. LEVESLEY, P. and WATERS, W. A., *J. Chem. Soc.*, 3119 (1953); 2456 (1954); 217 (1955); 4312 (1957); 2129 (1958).
3. UBBELOHDE, A. R. J. P., *J. Chem. Soc.*, 1605 (1935).
4. BELCHER, R. and TOWNEND, J., unpublished studies, see ref. [1].
4a. TANINO, K., *Rep. sci. Res. Inst.*, **32**, 20 (1956); *Chem. Zbl.*, **131**, 8287 (1960).
5. ETARD, A., *Compt. Rend.*, **86**, 1400 (1878).
6. MEYER, R. J. and BEST, H., *Z. anorg. Chem.*, **22**, 22 (1899).
7. IKEGAMI, I. H., *J. Chem. Soc. Japan, Ind. Chem. Sect.*, **52**, 173 (1949); *Chem. Abstr.*, **49**, 4171 (1954).
8. SAITO, K. and SATO, N., *J. Chem. Soc. Japan, Ind. Chem. Sect.*, **55**, 59 (1952); *Chem. Abstr.*, **48**, 9848 (1953).
9. ISHIBASHI, M. and SHIGEMATSU, T., *Japan Analyst*, **7**, 646 (1958).
10. KITAHARA, S., *Rep. Sci. Res. Inst. Japan*, **32**, 20, 129 (1956); *Chem. Abstr.*, **51**, 10595 (1956).
11. TANINO, K., *Rep. Sci. Res. Inst. Japan*, **32**, 24 (1956); *Chem. Abstr.*, **51**, 10595 (1956).
12. WEST, T. S., *Metallurgia*, **47**, 45 (1953).
13. ISHIBASHI, M., SHIGEMATSU, T. and SHIBATA, S., *Japan Analyst*, **7**, 644 (1958); 647 (1958).
14. KITAHARA, S. and TANINO, K., Japan Patent, **30**, 4474 (1959); *Chem. Abstr.*, **53**, 19708 (1959).
15. KLUH, I. DOLEŽAL, J. and ZÝKA, J., *Z. anal. Chem.*, **177**, 14 (1960).
16. DUŠEK, O., Diploma Thesis, Charles University, Prague 1960.
17. KŘÍŽOVÁ, J., Diploma Thesis, Charles University, Prague 1960.
18. ISHIBASHI, M., SHIGEMATSU, T. and SHIBATA, S., *Japan Analyst*, **8**, 377 (1959).
19. ISHIBASHI, M., SHIGEMATSU, T. and SHIBATA, S., *Japan Analyst*, **8**, 380 (1959).
20. TSUBAKI, I., *Japan Analyst*, **8**, 318 (1959).
21. DRUMMOND, A. Y. and WATERS, W. A., *J. Chem. Soc.*, 3119 (1953); 2456 (1954); 217 (1955); 4312 (1957); 2129 (1958).
22. LHOTKA, J. F., *Stain. Technol.*, **28**, 245 (1953); *Chem. Zbl.*, 11060 (1955).
23. ISHIBASHI, M., SHIGEMATSU, T. and SHIBATA, S., *Japan Analyst*, **7**, 644 (1958).
24. ISHIBASHI, M., SHIGEMATSU, T. and SHIBATA, S., *Japan Analyst*, **8**, 380 (1959).

25. IKEGAMI, H., *J. Chem. Soc. Japan, Ind. Chem. Sect.*, **52**, 173 (1949).
26. TSUBAKI, I., *Japan Analyst*, **3**, 253 (1944).
27. TARAYAN, V. M., EKIMYAN, G. M. and KOKHORYAN, A. T., *Izvest. Akad. Nauk S.S.S.R., Khim. Nauk*, **10** (2) 105 (1957).
28. KITAHARA, S. and TANINO, K., *Rep. Sci. Res. Inst.*, **32**, 24 (1956); *Chem. Zbl.*, **131**, 8287 (1960).
29. SINGH, B. SAHOTA, S. S. and VERMA, B. Ch., *Res. Bull. Panjab Univ.*, **10**, 261 (1959); *Chem. Abstr.*, **54**, 24096 (1960).
30. KITAGAWA, H. and SHIBATA, N., *Japan Analyst*, **9**, 597 (1960).
31. MALIK, W. U. and AJMAL, M., *Anal. Chemistry*, **34**, 207 (1962).

CHAPTER 3

COMPOUNDS OF TRIVALENT COPPER

THE strong oxidative power of trivalent copper has been used for the oxidimetric determination of both inorganic and organic compounds. In these oxidimetric titrations, solutions of complexes of trivalent copper with periodate, $K_7[Cu(IO_6)_2]$, or with tellurate (VI), $K_9[Cu(TeO_6)_2]$, are used; they are sufficiently stable and are easily prepared. The redox potential of the system

$$Cu^{3+} + e^- \rightleftharpoons Cu^{2+}$$

in which the copper is bound to periodate or tellurate (VI) is quoted to be 1·8 V in alkaline medium at 25°C.[1] According to Jenšovský[19] the formal potential of the periodate complex changes from 0·7 V at pH 12 to 1·1 V at pH 8, and that of the tellurate complex from 0·77 V at pH 12 to 1·12 V at pH 9.

Titrations with copper (III) standard solutions are carried out nearly without exception in alkaline medium, mostly in KOH solutions. In the presence of sodium ions, the reagent can be used to analyse only very low concentrations of the substances to be determined, owing to the low solubility of sodium periodato cuprate (III).

When some systems are titrated with complexes of trivalent copper, with periodate or tellurate (VI), the latter two compounds can act as oxidants. It is then necessary to determine an empirical factor for the standard solution (see below), as e.g. in the titration of arsenic (III), in order to obtain reproducible results. Accordingly, the low accuracy of the determination of some substances reported by Beck can be explained by the unsuitable standardization of the titrant with arsenic (III).[1a]

STANDARD SOLUTION

The standard solutions are prepared by oxidation of copper (II) sulphate with potassium peroxydisulphate[2-5] in alkaline medium and in presence of an excess of periodate or tellurate (VI).[6] The standard solution still contains an excess of periodate or tellurate, both of which cause the undesirable inaccuracy in a number of determinations, owing to their own oxidative power.

Recently it has been found that an excess of periodate or tellurate (VI) is not necessary in preparing the standard solution, so that a precisely defined solution can be made up as follows:[7]

Dissolve 12·485 g of crystallized copper (II) sulphate in 400 ml of boiling water and add 23·002 g potassium metaperiodate. Slowly pour a saturated solution of 56 g of KOH into the boiling yellow–green suspension of copper (II) periodate. Add in small portions about 20 g of solid potassium peroxydisulphate to the dark green solution which contains partially dissolved copper (II) periodate. The solid residue dissolves and the solution assumes a brown colour. Decompose the excess of peroxydisulphate by boiling for 30 min. Crystals of potassium sulphate deposit from the solution as it cools. Transfer the solution to a measuring flask and make up to 500 ml.

The solution thus prepared is 0·1 M with regard to diperiodato cuprate (III) and, at the same time, about 2 M with regard to potassium hydroxide.

Standard solutions can also be made by weighing the crystallized compound; the synthesis is, however, complicated and gives only a poor yield.[7]

The standard solutions are stable if stored in polyethylene bottles[6] or in brown glass flasks, coated with paraffin wax.[7]

Standardization

The standardization is carried out by potentiometric titration of potassium hexacyanoferrate (II) in a solution containing potassium hydrogen carbonate, potassium borate or potassium acetate. In some cases, standard solutions are used with an empirical factor, i.e. solutions, which have been standardized against a precisely known amount of the substance to be analysed, under the same conditions as in the actual determination.

INDICATOR

Because the standard solution has an intensive brown colour, the colour caused by a slight excess of standard solution can be used for visual indication of the end-point. With some compounds (especially organic substances) the solutions colour blue or green during the titration, or a yellow–green precipitate is formed before the equivalence point is reached, i.e. before an excess of reagent colours the solution brown. In such cases, the visual determination of the endpoint is not precise. Accordingly, the *electrometric* titration is preferred for determinations with copper (III). Particularly " dead-stop " titration with platinum electrodes,[6] potentiometric[7,8] or spectrophotometric[9] titrations have proved satisfactory.

REVIEW OF DETERMINATIONS

Arsenic and antimony. The potentiometric titration of arsenic (III) with diperiodato cuprate (III) (or the reverse titration) can be carried out in about 10 M KOH solution. An empirical factor has to be determined for the standard solution,[7] because the copper (III), as well as the periodate ion, oxidize the substance to be determined. Neither visual titration[2] nor " dead-stop " titration[6] give satisfactory results.

Visual titration in alkaline solution has been recommended for the determination of small amounts of antimony.[2,10]

Hexacyanoferrate (II). The potentiometric titration of hexacyanoferrate (II) (see standardization) can be done also in the presence of sodium ions. Dead-stop titration did not yield good results.[6]

Thallium. Thallium has been titrated potentiometrically with periodato cuprate (III).[10]

Amongst other inorganic substances, the compounds As_2S_3, Sb_2S_5, P_2S_5, MoS_3, SnS_2 and the ions Cr^{3+}, Pb^{2+}, CN^- and $S_2O_3^{2-}$ have been titrated visually in alkaline solution.[2]

The oxidation of *iodides*, *iodates* and *cyanides* has been investigated in 0·5 M KOH solution by " dead-stop " titration; only cyanide ions can be determined satisfactorily; the oxidation of iodide and iodate proceeds too slowly. *Thiosulphates* can also be determined in $NaHCO_3$ solution by this method, but potassium ditellurato cuprate (III) is used as titrant.[6]

Calcium. The gravimetric or colorimetric determination of calcium with naphthalhydroxamate has been modified by Beck to a volumetric determination. Calcium forms with naphthalhydroxamate a red precipitate, which, after having been washed, can be suspended in a 5 per cent KOH solution and titrated hot with potassium diperiodato cuprate (III). The end-point of the titration is recognized by the green colour of the solution.

Tartrate. Tartrate can be determined by amperometric titration in 2–10 M KOH solution with potassium diperiodato cuprate (III) using a platinum vibrating electrode.[11] The interfering action of the periodate has no influence, because the periodate is reduced at a more negative potential than the copper (III) ions.

4,4-Dihydroxydiphenyl. This compound is oxidized by potassium diperiodato cuprate (III) to 4,4′-diphenoxy quinone.[8] This determination is carried out in sodium hydrogen phosphate solution using potentiometric indication of the equivalence point.

In the determination of *sugar* in *blood* and *urine*,[15] after preparing the sample one titrates until a green colour persists for 30–60 sec.

For the analysis of *amino acids* and *proteins* the so-called *differential percuprimetric titration* (with the use of which time reaction curves are obtained) has been suggested. The added amount of reagent is plotted graphically against the time necessary for its quantitative reduction by the substance. In this way, characteristic curves are obtained from the various compounds, from which also some conclusions can be drawn about the constitution of these substances.[12,13] Curves obtained in the oxidation of proteins in blood serum in some maladies, show characteristic deviations from the normal curves.[14,17] The visual determination can be made more precise by photometric methods.[9,18]

The oxidation of other organic substances has likewise been investigated with the aid of visual titration:

Mannitol, starch, glycerol, fumaric acid, mannose, lactose, formaldehyde, tartrate, aceto-acetic acid, oleic acid, maleic acid, diacetyl, dimethylglyoxime, oxine, hydroquinone, resorcinol, pyrocatechol, quinone, aniline, acetanilide, phenyl hydrazine, pyrogallol, phloroglucinol, ethyleneglycol, ethylenediamine, vitamin B, inositol[3,16] and various *proteins*.[2] Most of these reactions proceed very slowly, the end-point of the titrations is often very difficult to see and the results are erroneous. In some cases, boiling temperatures are used or an excess of the reagent is added, which is then back-titrated with arsenic (III).

REFERENCES

1. LATIMER, W. M. and HILDEBRAND, H., *Ref. Book of Inorg. Chem.* 1947; KOVÁTS, J., *Acta Chim. Akad. Sci. Hung.*, **21**, 247 (1959).
1a. BERKA, A., *Chemie*, **10**, 543 (1958).
2. BECK, G., *Mikrochemie*, **36**, 245 (1950).
3. BECK, G., *Mikrochemie*, **38**, 152 (1951).
4. MALATESTA, L., *Gazz. Chim. Ital.*, **71**, 467, 580 (1941).
5. VOTISS, M., *Recu. Trav. Chim.*, **44**, 425 (1925).
6. KEYWORTH, D. A. and STONE, K. G., *Anal. Chemistry*, **27**, 833 (1955).
7. JENŠOVSKÝ, L., *Chem. Listy*, **50**, 1103 (1956); *Collection (Czech. Chem. Comm.)*, **22**, 1564 (1957).
8. MATRKA, M. and NAVRÁTIL, F., *Chem. Průmysl*, **7**, 136 (1957).
9. BECK, G., *Mikrochemie*, **39**, 147 (1952).
10. ALIMARIN, I. P., CHENG, J. A. and PUZDRENKOVA, I. V., *Vestnik Moskov. Univ.*, **13**, 201 (1958); *Chem. Abstr.*, **53**, 15856 (1959).
11. JENŠOVSKÝ, L., *Chem. Listy*, **50**, 1313 (1956).
12. BECK, G., *Mikrochemie*, **38**, 1 (1951).
13. BECK, G., *Analyt. Chim. Acta*, **9**, 241 (1953).
14. BECK, G., *Mikrochemie*, **39**, 22 (1952).
15. BECK, G., *Mikrochemie*, **35**, 169 (1950).
16. BECK, G., *Mikrochemie*, **40**, 258 (1953).
17. BECK, G. and SPENGLER, G. A., *Helv. Med. Acta*, **22**, 98 (1955).
18. BECK, G. and SPENGLER, G. A., *Recu. Trav. Chim.*, **74**, 543 (1955).
19. JENŠOVSKÝ, L., Doctoral Thesis, Charles University, Prague 1959.

CHAPTER 4

POTASSIUM HEXACYANOFERRATE (III)

POTASSIUM HEXACYANOFERRATE (III) is distinguished by its great oxidative power in alkaline solution. The direct and indirect determinations of numerous inorganic and organic substances with a standard potassium hexacyanoferrate (III) solution are based on this fact.

The normal potential of the redox system

$$[Fe(CN)_6]^{3-} + e^- \rightleftharpoons [Fe(CN)_6]^{4-}$$

is quoted to be + 0·45 V.[1] The potential depends not only on the concentration of alkali metal ions[1a,1b] but also on the concentration of salts present in the solution. According to Adams, Reilley and Furman,[58] the formal potential of the system $[Fe(CN)_6]^{3-}/[Fe(CN)_6]^{4-}$ in alkaline medium equals the square of ionic strength at a constant pH. The most important fact for practical purposes is that the potential does not change significantly over a wide range of hydroxyl ion concentration. The potential changes, e.g. in the range from 1 to 5 N NaOH solution by the order of 0·05 V, whereas the redox potential of most of the other systems becomes noticeably negative in alkaline solution. This makes it possible to determine these substances with standard potassium hexacyanoferrate (III) solution. The titrations are mostly carried out in an alkali hydroxide or alkali carbonate solution. The determination can also be carried out in ammoniacal medium at normal or elevated temperature. Osmium (VIII) oxide is used as catalyst in some determinations[2,3,3a]

STANDARD SOLUTION

To prepare a 0·1 N standard solution, 33·0 g of the pure dried compound are dissolved in distilled water and made up to 1 l. If the salt is impure, it has to be recrystallized several times from aqueous solution and dried at 100°C to constant weight.

The standard potassium hexacyanoferrate (III) solutions should not be prepared in large amounts, because they decompose fairly rapidly. The normality of the solution has to be determined anew each week.

POTASSIUM HEXACYANOFERRATE (III)

Standardization

The determination of the normality is mostly done *iodimetrically*.

The hexacyanoferrate (III) ions are reduced in strong acid solution with potassium iodide; the iodine thus formed is then titrated with sodium thiosulphate.

In the *argentimetric* determination, the excess of silver nitrate, added to the hexacyanoferrate (III), is titrated with ammonium thiocyanate. The hexacyanoferrate (III) can also be determined directly by potentiometric titration with silver nitrate.

Following other methods, potassium hexacyanoferrate (III) can be determined with standard mercury (II) perchlorate solution in the presence of thiocyanate and acetic acid.[4]

INDICATOR

The end-point is generally detected *potentiometrically* or *amperometrically*. Platinum electrodes (also rotating platinum electrodes) serve as indicator electrodes.

The indirect determinations are carried out visually in most cases:

The amount of potassium hexacyanoferrate (II) formed by the reduction of potassium hexacyanoferrate (III) in acid solution, is measured by titration with permanganate, ceric sulphate or iodimetrically. The end-point can be detected also by the appearance of the red colour of the iron (III) compounds with dimethylglyoxime,[4a] with diphenylamine, methylene blue or thionine.[4b,4c]

The potentiometric or amperometric indication of the end-point is, however, the most suitable and precise method.

REVIEW OF DETERMINATIONS

Arsenic, antimony and tin. Arsenic (III), antimony (III) and tin (II) compounds can be titrated with potassium hexacyanoferrate (III) in concentrated sodium hydroxide solution at 50–70°C.[5] The equivalence point is detected potentiometrically. This not too fast reaction can be accelerated catalytically by osmium (VIII) oxide.[2,3,6,6a,6b] Tin (II) compounds are titrated in an inert atmosphere.

Recently, a visual direct titration of tin (II) compounds with hexacyanoferrate (III)[6b] has been worked out. The titration is carried out at a pH of 11·0–12·5. Cacotheline is used as indicator. Arsenic (III) and antimony (III) ions are determined indirectly by visual titration. Both ions are precipitated as sulphides, redissolved in sodium hydroxide or sodium carbonate solution; the sulphide ions are titrated with hexacyanoferrate (III) using sodium nitroprusside as indicator.

Cerium. The potentiometric determination of cerium (III) compounds is carried out in a 25–50 per cent potassium carbonate solution, in a CO_2 or N_2 atmosphere.[7] Nitrate interferes with the determination but chloride and sulphate can be present. This method is very well suited to the determination of trivalent cerium in presence of thorium, of the other rare earth ions and iron (III).

Cerium (III) can be determined by amperometric titration in 3–5 molar K_2CO_3 solution even in the presence of large amounts of cerium (IV) compounds.[8]

Chromium. Chromium (III) compounds are oxidized to chromate with potassium hexacyanoferrate (III) at elevated temperature in concentrated NaOH solution.[9,9a,9b] The presence of small amounts of thallium (I) sulphate, chloride or nitrate promotes smooth titration. It is, however, more advantageous to catalyse the reaction with osmium (VIII) oxide.[2,3] This method has proved well suited in the potentiometric determination of chromium in chromium plating baths.[9c]

In the presence of sodium phosphate, chromium (II) compounds can be oxidized quantitatively with potassium hexacyanoferrate (III) to chromium (III) at a pH below 10·4.[10] In the same medium, molybdenum (III) (at pH 8·8–11) and tungsten (IV) (at pH 6·5–10·6) are oxidized to molybdenum (VI) and tungsten (VI).

Cobalt. The oxidimetric determination of cobalt (II) compounds with hexacyanoferrate (III) was investigated some time ago by Tomíček and Freiberger.[11] The potentiometric titration is carried out in 3 N NH_4OH solution and in a CO_2 atmosphere, also in the presence of large amounts of various metals. This method is above all suitable for the determination of traces of cobalt in nickel salts.

Because of its great practical significance, the oxidation of cobalt (II) with hexacyanoferrate (III) has been studied under various conditions of titration[12–18g] and the first mentioned method has been found to be the most suitable one. It was only the titration in the presence of glycine,[19] which forms with cobalt (II) a complex ion of unlimited stability in air, that brought about an essential improvement of this method. Here, the cobalt (II) is easily oxidized to the respective cobalt (III) complex with hexacyanoferrate (III). It is titrated at a pH of 9·5–12. As little as 5 mg of cobalt in a volume of 50–80 ml can be determined reliably by this method. The titration is carried out at normal temperature in air. This is the great advantage of this method which, moreover, can be applied also in the presence of larger amounts of nickel (II), copper (II), zinc, arsenic (III), arsenic (V), cadmium, lead (II), barium, ammonium, chromium (VI), molybdenum (VI), tungsten (VI), vanadium (V) ions, and in the presence of smaller amounts of iron (II) and manganese (II) ions. The titration is not affected by the presence of 10 per cent alkali chloride, nitrate or sulphate, nor by the presence of larger concentrations of fluoride, diphosphate, tartrate and citrate. This determination proved suitable for the analysis of cobalt in nickel (II) salts and in alloys, steels and ores. Lately, it has been proposed that glutamic acid be used instead of glycine,[20] which is more suitable in the analysis of cobalt in the presence of manganese.

The determination of cobalt can also be carried out amperometrically using a rotating platinum electrode.[20a] This method of indication has been applied to the determination of cobalt in steel, in various alloys and silicates. It offers, however, no significant advantages over the potentiometric determination.

Manganese. The potentiometric determination of manganese (II) is carried out with potassium hexacyanoferrate (III) in 1·5–2 N NaOH solution in the presence of the complex forming substances glycerol, mannite, ethylene glycol and tartaric acid at 10–12°C in a CO_2 atmosphere.[21] The manganese (II) is oxidized to trivalent manganese in tartaric acid solution; in the presence of glycerol or glycol it is oxidized to tetravalent manganese. In a solution containing all three complexing agents, the oxidation of the manganese is expressed by two distinct changes of the potential ($Mn^{2+} \rightarrow Mn^{3+}$ and $Mn^{3+} \rightarrow Mn^{4+}$). The determination can be carried out in the presence of copper (II), zinc, nickel (II), arsenic (III), arsenic (V), antimony (III), iron (III), aluminium, phosphate and molybdate ions. It is suitable for the analysis of ores and alloys.

Divalent manganese can be titrated potentiometrically also in a cyanide solution[22] in the presence of ammonium hydroxide and carbonate following the reaction:

$$[Mn(CN)_6]^{4-} + [Fe(CN)_6]^{3-} \rightleftharpoons [Mn(CN)_6]^{3-} + [Fe(CN)_6]^{4-}$$

In the presence of cobalt, the two ions cannot be determined separately, but only together. Under these circumstances, their determination with the disodium salt of ethylenediamine tetraacetic acid is preferable.[23] The cobalt (III) ion, formed by oxidation with potassium dichromate, forms a complex with EDTA. Through further reaction with ammonium hydroxide and potassium cyanide, the stable hexacyanocobalt (III) is formed. The divalent complex-bound manganese is oxidized quantitatively to trivalent manganese with potassium hexacyanoferrate (III) following the above-mentioned equation. The potentiometric titration can be carried out in the presence of iron (III), copper (II), nickel (II), cadmium (II), zinc, aluminium, titanium (II), molybdenum (VI), tungsten (VI) compounds and of alkaline earth metals. Only vanadate (V) interferes with this determination. This method is especially well suited for the analysis of manganese in ores and alloys.

Molybdenum. The determination of molybdenum (VI) compounds is based on their reduction with zinc amalgam, and on the visual titration of the molybdenum (III) thus formed with potassium hexacyanoferrate (III) in a medium of EDTA, tartrate and 1·5–2 N NaOH. Indigo carmine is used as indicator.[23a]

Thallium. The reaction between thallium (I) ions and hexacyanoferrate (III) proceeds in NaOH solution under the same conditions as described for the determination of arsenic.[5,24,24a,25] The determination can also be carried out coulometrically.[26]

Selenium. Potassium hexacyanoferrate (III) oxidizes selenite (IV) in alkali hydroxide solution quantitatively to selenate (VI).[27,28] Following Solymosi,[27] one titrates directly in 3·5–5 N NaOH or 2–3 N KOH solution at 55–65°C, using osmium (VIII) oxide as catalyst. The equivalence point is detected by the "dead-stop" or potentiometric method. All ions which are oxidized by hexacyanoferrate (III) interfere with this method, furthermore ammonia, tellurite (IV) and tellurate (VI).

Uranium. The determination of uranium (VI) compounds is based on their reduction with zinc amalgam and on the potentiometric titration of the uranium (IV) compounds formed during the reduction with potassium hexacyanoferrate (III), in a medium of sodium hydrogen carbonate, potassium cyanide and ammonium chloride.[29] The determination is not affected by the presence of lead (II), bismuth (III), copper (II), tin (II), tin (IV), arsenic (III), arsenic (V), antimony (III), antimony (V), zinc, nickel (II), cobalt (II), chromium (II), calcium, magnesium, thorium (IV), beryllium (II), titanium (III), iron (III), aluminium, phosphate and tungsten (VI) ions. Compounds of silver, cadmium, manganese, vanadium and cerium interfere. The method is suitable for the analysis of uranium ores. The determination of uranium is also described by Deshmukh and Joshi.[30]

Vanadium. Vanadyl ions are titrated potentiometrically with potassium hexacyanoferrate (III) in 10 N NaOH, 2 N Na_2CO_3, 1 M $Na_2B_4O_7$ and concentrated NH_4OH solution.[5] Tomíček found that the low results of this determination are caused by the partial decomposition of the vanadyl ions in strongly alkaline solution.[31] The unfavourable low results can be reduced or completely removed by applying complex forming reagents. So, e.g. tetravalent vanadium is determined at pH 10·6–13 in the presence of sodium diphosphate.[32] Vanadyl ions can be titrated visually using diphenyl-dianisidine-*o,o'*-dicarbonic acid as indicator.[32a]

The colorimetric titration of vanadium with potassium hexacyanoferrate (III) is of little importance.[33]

The reaction of vanadium (III) ions with hexacyanoferrate (III) proceeds under the same conditions as described for the determination of molybdenum.[23a]

Iron and titanium. Following Wittmann,[34] iron (II) ions are determined potentiometrically with hexacyanoferrate (III) in 12–20 per cent hydrochloric acid, even in the presence of arsenic, antimony, cobalt, nickel, manganese, zinc, copper, lead and cadmium compounds. This method can be used in the determination of pyrite. Kiboku[35] found that iron (II) can be titrated with hexacyanoferrate (III) in a solution of sodium diphosphate at pH 4·3–10. Thus, a mixture of iron (II) and titanium (III) ions can be determined, or the titanium (III) alone, which is titrated at pH 9·8.

Iron (II) ions can be determined in the presence of iron (III) ions using EDTA.[35a]

Peroxide. Hydrogen peroxide is oxidized quantitatively to oxygen with potassium hexacyanoferrate (III) in a 10–40 per cent solution of NaOH.[36] The end-point of the titration is determined potentiometrically. A 0·1 per cent H_2O_2 solution can just be titrated with a 0·01 N standard solution to yield adequate accuracy. Other peroxides are determined similarly.

Sulphides and some other sulphur compounds. The oxidation of sulphides yields different reaction products under different conditions. Thus,

Charlot[37] found that the oxidation of sulphide proceeds quantitatively to sulphur in a buffer solution of ammonium chloride and ammonia at pH 9·4. The end-point of the titration is indicated by the disappearance of the red colour of the iron (II) complex with dimethylglyoxime.

The oxidation of sulphide in sodium hydroxide, sulphite and ammonia solution at pH 10·0–12·0 yields the same products. Here, sodium nitroprusside is used as indicator. Other authors also deal with the same problem.[38,39,40]

Sulphides, sulphites, hydrogen sulphite, disulphite, dithionite, thiosulphate and tetrathionate (after decomposition into thiosulphate and sulphide) are oxidized quantitatively to sulphates in a 4–5 N NaOH solution under the catalytic action of OsO_4.[41–44]

It is titrated at a temperature of 50–60°C either potentiometrically or using the " dead-stop " method. In rapid titrations, atmospheric oxygen does not interfere; only in the determination of dithionite and tetrathionate is a nitrogen atmosphere necessary. With this method, various mixtures can be determined, such as dithionite and sulphite, dithionate and thiosulphate, sulphite and thiosulphate, tetrathionate and thiosulphate.[42,45,46] The determination of sulphides, sulphites and tetrathionates is, compared with other methods (especially iodimetry), more advantageous. The rate of reaction is very much accelerated by the use of catalysts.

Dithionites can likewise be determined in other media.[5,47] Amongst other inorganic sulphur compounds, thiocyanates[48] and CS_2[48a] are determined with potassium hexacyanoferrate (III).

Hydrazine, isonicotinic acid hydrazide and hydroxylamine. The direct potentiometric titration of these three substances with potassium hexacyanoferrate (III) proceeds instantaneously and quantitatively in 10–25 per cent KOH solution, forming nitrogen or dinitrogen monoxide (in the oxidation of hydroxylamine).[49] The titration of hydroxylamine can be applied to the determination of carbonyl groups;[50] the hydroxylamine which is added in excess, is back-titrated in strongly alkaline solution with potassium hexacyanoferrate (III). Hydrazine and hydroxylamine can be determined in a mixture,[51] as well as separately in the presence of zinc (II) sulphate.[51,52] The photometric titration of hydrazine[53] and the indirect determination of hydroxylamine[54] are both of little importance.

Active methylene groups. The potentiometric determination of active methylene groups is on principle an oxidation of diethyl-*p*-phenylene diamine to quinone diimine; this reacts with the active methylene groups to form a leucobase and the respective dye.

An excess of diethyl-*p*-phenylene diamine is added to the sample to be determined and then titrated with 0·1 M potassium hexacyanoferrate (III) in K_2CO_3 solution.[55] With this method can be determined:

Acetoacetanilide, *p*-carboxyacetoacetanilide, *o*-chloroacetoacetanilide, benzoylacetanilide.

Reducing sugars. Considerable attention has been given to the reaction of reducing sugars with potassium hexacyanoferrate(III).[56-61a,61b] This determination is carried out by an empirical method which is based on the oxidation of these substances with hexacyanoferrate (III) in alkaline solution. Following one of these methods,[62,63] one proceeds so that a known amount of the sugar solution is added dropwise from a burette into a boiling 1 per cent alkaline solution of potassium hexacyanoferrate (III). The titration is controlled potentiometrically. It has been recommended that some zinc sulphate[63] be added to the solution to obtain a smooth reaction.

Other methods for determining reducing sugars exploit their reaction with an excess of potassium hexacyanoferrate (III) and, thereupon, back-titration with suitable reducing agents, or titration of the hexacyanoferrate (II) formed during the reaction, with e.g. cerium (IV) sulphate[64] or zinc sulphate.[65] The reaction of potassium hexacyanoferrate (III) with reducing sugars has also been applied in biochemistry.[66]

The titrimetric determination of reducing sugars based on the reduction of Fehling's solution is, however, regarded to be the most reliable method.[61]

Thiourea. The slow rate of the reaction of thiourea with potassium hexacyanoferrate (III) enables only an indirect determination of thiourea. The excess of the oxidizing agent is back-titrated in alkaline solution with sodium arsenite (III),[67] or iodimetrically in the presence of zinc sulphate;[68] in some cases, the potassium hexacyanoferrate (II) formed is determined with cerium (IV) sulphate using ferroin as indicator.[67,68]

Adrenalin, penicillin, hemoglobin, hemin. Adrenalin[69] is determined in 0·2 M sodium hydrogen phosphate solution with " dead-stop " titration. The oxidation proceeds quantitatively at a ratio of 1 mole adrenalin to 4 moles hexacyanoferrate (III). The method has proved suitable in the analysis of " Adrenalium chloratum sol." used in injections for medical treatment.

Hiscox[70] has worked out an indirect method for penicillin, which is based on its oxidation with potassium hexacyanoferrate (III) in alkaline solution and titration of the hexacyanoferrate (II) formed with cerium (IV) sulphate.

Hemoglobin[71] and hemin[72] can be determined likewise.

Cysteine. The determination of cysteine is carried out in the semi-micro-range[73,74] by direct titration with 0·1–0·001 M potassium hexacyanoferrate (III) solution in phosphate buffered solution at pH 7. The reaction is catalysed by copper (II) ions. The end-point of the titration is determined amperometrically.

p-*Nitrosoquinoline-iodomethylate, quinidine-iodoethylate and quinoline-iodomethylate.* p-Nitrosoquinoline-iodomethylate is titrated with potassium hexacyanoferrate (III) in a 2–3 per cent NaOH solution at 50–60°C.[75] Under similar conditions, but indirectly, quinoline-iodomethylate[76] and quinidine-iodoethylate[77] are titrated. These methods should actually be used for the determination of the alkaloids, after their transformation to alkaloid-iodomethylate or ethylate. The preparation of these compounds is, however, lengthy and therefore these methods are not suitable for practical purposes.

Ascorbic acid. Ascorbic acid can be determined with potassium hexacyanoferrate (III) either visually, using potassium iodide and starch as indicator,[78] or applying the " dead-stop " method in sodium hydrogen carbonate solution.[79,79a]

Other substances. Variamine Blue[80] is titrated amperometrically in a buffered solution at pH 9·0. *Acetals, ketals, mercaptans*[80b] and *vinyl ether* are determined potentiometrically,[80a] *hydroquinone*,[81] *morphine*[82] and *uric acid*[82a] visually by indirect titration, *nitrite*,[83] *iodide* in neutral solution,[84] *isacene*,[84a] in alkaline medium, *formaldehyde*(84b) and some other substances[85-88,45] by direct titration.

A review of titrations with potassium hexacyanoferrate (III) is given in the literature.[89-92]

REFERENCES

1. WILLARD, H. H. and MANALO, G. D., *Anal. Chemistry*, **19**, 462 (1947).
1a. KOLTHOFF, I. M. and TOMSICEK, W. J., *J. Phys. Chem.*, **39**, 945 (1935).
1b. WINEFORDNER, J. D. and DAVISON, G. A., *Analyt. Chim. Acta*, **28**, 480 (1963).
2. SOLYMOSI, F., *Magyar Kém. Folyóirat*, **63**, 294 (1957).
3. SOLYMOSI, F., *Naturwiss.*, **44**, 374 (1957).
3a. GLEU, K., *Z. anal. Chem.*, **95**, 305 (1933).
4. LUCENA-CONDE, F. and BELLIDO, I. S., *Talanta*, **1**, 305 (1958).
4a. KOMÁREK, K., *Collection (Czech. Chem. Comm.)*, **9**, 246 (1938).
4b. SOMEYA, K., *Z. anorg. Chem.*, **152**, 386 (1926).
4c. TOMÍČEK, O., *Chemical Indicators*, Butterworths, London 1951.
5. FRESNO, C. DEL, and VALDES, L., *Z. anorg. Chem.*, **183**, 251, 258 (1929).
6. SOLYMOSI, F., *Acta Chim. Acad. Sci. Hung.*, **16**, 267 (1958); *Magyar Kém. Folyóirat*, **62**, 318 (1956).
6a. SANT, S. B., *Z. anal. Chem.*, **168**, 112 (1959).
6b. KIBOKU, M., *Japan Analyst*, **10**, 19 (1961).
7. TOMÍČEK, O., *Rec. Trav. Chim.*, **44**, 410 (1925).
8. LEONARD, G. W., KEILY, H. J. and HUME, D. N., *Analyt. Chim. Acta*, **16**, 185 (1957).
9. HAHN, F. L., *Z. angew. Chem.*, **40**, 349 (1927).
9a. BOLLENBACH, H. and LUCHMAN, E., *Z. anorg. Chem.*, **60**, 446 (1910).
9b. PALMER, H. E., *Z. anorg. Chem.*, **67**, 448 (1910).
9c. KUBIŠTA, Z., *Chem. Průmysl*, **5**, 84 (1955).
10. KIBOKU, M., *Japan Analyst*, **6**, 356 (1957).
11. TOMÍČEK, O. and FREIBERGER, F., *J. Amer. Chem. Soc.*, **57**, 801 (1935).
12. DICKENS, P. and MAASSEN, G., *Arch. Eisenhüttenw.*, **9**, 487 (1936).
13. YARDLEY, J. T., *Analyst*, **75**, 156 (1950).
14. CHIRNSIDE, R. C., CLULEY, J. H. and PROFFIT, P. M. C., *Analyst*, **72**, 351 (1947).
15. BAGSHAVE, B. and HOBSON, J. D., *Analyst*, **73**, 152 (1948).

16. HALL, A. and YOUNG, R. S., *Chem. and Ind.*, **24**, 394 (1946).
17. YOKOSUKA, S., *Japan Analyst*, **6**, 690 (1957).
18. DIEHL, H. and BUTLER, J. P., *Anal. Chemistry*, **27**, 777 (1955).
18a. FUREY, J. J. and CUNNINGHAM, T. R., *Anal. Chemistry*, **20**, 563 (1948).
18b. TOUHEY, W. O. and REDMOND, J. C., *Anal. Chemistry*, **20**, 202 (1948).
18c. ANAND, V. D., *Z. anal. Chem.*, **174**, 192 (1960).
18d. ROBIN, J., DEWASNES, P. and DEVORE, P., *Bull. Soc. Chim. France*, **2**, 223 (1961).
18e. KIRTCHIK, H. and SEVFARINGEN, F. H., *Analyst*, **86**, 188 (1961).
18f. AGASYAN, P. K. and KHAKIMOVA, V. K., *Zavodskaya Lab.*, **28**, 1184 (1962).
18g. KRATOCHVIL, B., *Anal. Chemistry*, **35**, 1314 (1963).
19. KOPANICA, M. and DOLEŽAL, J., *Chem. Listy*, **50**, 1225 (1956); *Collection (Czech. Chem. Comm.)*, **22**, 195 (1957).
20. CHANG-YE-SIA, DOLEŽAL, J. and ZÝKA, J., *Zhur. Anal. Khim.*, **16**, 308 (1961).
20a. BOZSAI, I., *Magyar Kém. Folyóirat*, **65**, 207 (1959); *Magyar Kém. Lapja*, **15**, 423 (1960).
21. TOMÍČEK, O. and KALNÝ, J., *J. Amer. Chem. Soc.*, **57**, 1209 (1935).
22. TOMÍČEK, O., ŠANDL, Z. and SIMON, V., *Collection (Czech. Chem. Comm.)*, **14**, 20 (1949).
23. PŘIBIL, R. and SIMON, V., *Chem. Listy*, **43**, 145 (1945).
23a. HAYASHI, S., *Japan Analyst*, **11**, 442, 951 (1962).
24. MIURA, K., *Sci. Rep. Tohoku Univ.*, **37**, 103 (1953).
24a. MIURA, K., *J. Elektrochem. Assoc. Japan*, **19**, 341 (1951).
25. DESHMUKH, G. S., *Analyt. Chim. Acta*, **12**, 586 (1955).
26. HARTLEY, A. M. and LINGANE, J. J., *Analyt. Chim. Acta*, **13**, 183 (1955).
27. SOLYMOSI, F., *Magyar Kém. Folyóirat*, **63**, 313 (1957); *Acta phys. et chem. Szeged*, **3**, 112 (1957).
28. DESHMUKH, G. S. and BAPAT, M. G., *Z. anal. Chem.*, **156**, 273 (1957).
29. SIMON, V. and PŘÍPLATOVÁ, E., *Chem. Listy*, **50**, 907 (1956).
30. DESHMUKH, G. S. and JOSHI, M. K., *Z. anal. Chem.*, **143**, 334 (1954).
31. TOMÍČEK, O., *Chem. Listy*, **32**, 442 (1938).
32. KIBOKU, M., *Bunseki Kagaku*, **6**, 11 (1957).
32a. FRUMINA, N. S. and MUSTAFIN, I. S., *Zhur. Anal. Khim.*, **15**, 671 (1960).
33. TATWAWADI, S. V., *Z. anal. Chem.*, **168**, 15 (1959).
34. WITTMANN, G., *Z. anal. Chem.*, **141**, 241 (1954).
35. KIBOKU, M., *Bunseki Kagaku*, **5**, 503 (1956).
35a. HALL, L. C. and FLANIGAN, D. A., *Anal. Chemistry*, **33**, 1495 (1961).
36. VULTERIN, J. and ZÝKA, J., *Chem. Listy*, **48**, 619 (1954).
37. CHARLOT, G., *Bull. Soc. Chim. France*, **6**, 1447 (1939).
38. KIBOKU, M., *Japan Analyst*, **6**, 491 (1957).
39. SCAGLIARINI, S., *Atti X. Congress internat. Chim.*, **3**, 466 (1939).
40. IONESCO-MATIU, A. I. and POPESCO, A., *Bull. Acad. med. Roumanie*, **4**, 385 (1939).
41. SOLYMOSI, F. and VARGA, A., *Magyar Kém. Folyóirat*, **64**, 443 (1958).
42. SOLYMOSI, F. and VARGA, A., *Analyt. Chim. Acta*, **17**, 608 (1957).
43. SOLYMOSI, F. and VARGA, A., *Acta Chim. Acad. Sci. Hung.*, **20**, 295 (1959).
44. DESHMUKH, G. S. and BAPAT, M. G., *Z. anal. Chem.*, **156**, 105 (1957).
45. SOLYMOSI, F. and VARGA, A., *Magyar Kém. Folyóirat*, **64**, 245 (1958).
46. SOLYMOSI, F. and VARGA, A., *Acta Chim. Acad. Sci. Hung.*, **20**, 339 (1959).
47. CHARLOT, G., *Bull. Soc. Chim. France*, **6**, 977 (1939).
48. SUSEELA, B., *Z. anal. Chem.*, **145**, 175 (1955).
48a. KIBOKU, M., *Japan Analyst*, **12**, 797 (1963).

49. VULTERIN, J. and ZÝKA, J., *Chem. Listy*, **48**, 839 (1954).
50. BUDĚŠÍNSKÝ, B. and KÖRBL, J., *Mikrochim. Acta*, **922** (1959).
51. SANT, B. R., *Analyt. Chim. Acta*, **20**, 371 (1959).
52. SANT, B. R., *Analyt. Chim. Acta*, **19**, 205 (1958).
53. BAPAT, M. G. and TATWAWADI, S. V., *Naturwiss*, **44**, 557 (1957).
54. SANT, B. R., *Z. anal. Chem.*, **145**, 257 (1955).
55. RYBA, O., *Collection (Czech. Chem. Comm.)*, **24**, 1950 (1959).
56. GENTELE, J. G., *Dinglers polytechn. J.*, **152**, 68, 139 (1959).
57. HÁLA, E., *Chemický Obzor*, **19**, 193 (1944).
58. ADAMS, R. N., REILLEY, Ch. N. and FURMAN, N. H., *Anal. Chemistry*, **24**, 1200 (1952).
59. IONESCO-MATIU, A., *Produits pharmac.*, **12**, 247 (1957).
60. IONESCU, A. and VARGOLICI, V., *Bull. Soc. Chim. Romania*, **2**, 38 (1920).
61. BLOM, J. and ROSTED, C. O., *Acta Chem. Scand.*, **1**, 32 (1947).
61a. BRITTON, H. T. S. and PHILLIPS, L., *Analyst*, **65**, 149 (1940).
61b. ZELAWSKI, W., WASILEWSKA, D. and KACZKOWSKI, J., *Chem. Analit. (Warszawa)* **6**, 882 (1961).
62. POLUBNAYA, E. T., BUKHAROV, P. S., *Zhur. Anal. Khim.*, **3**, 131 (1948).
63. POLUBNAYA, E. T., BUKHAROV, P. S., *Zhur. Anal. Khim.*, **5**, 300 (1950).
64. WHITMOYER, R. B., *Ind. Engng. Chem. Anal. Ed.*, **6**, 268 (1934).
65. PINXTEREN, J. A. C., *Pharm. Weekbl.*, **93**, 753 (1958).
66. HAGEDORN, H. C. and JENSEN, B. N., *Biochem. Z.*, **242**, 43 (1931).
67. BAPAT, M. G. and SHARMA, B., *J. Sci. Res. Banaras Hindu Univ.*, **7**, 262 (1957).
68. JOSHI, M. K., *Naturwiss*, **44**, 537 (1957).
69. DUŠINSKÝ, G., *Chem. Zvesti*, **9**, 149 (1955).
70. HISCOX, D. J., *Anal. Chemistry*, **21**, 658 (1949).
71. CONANT, J. B., *J. biol. Chemistry*, **57**, 401 (1923).
72. CONANT, J. B., ALLES, J. A. and TONGBERG, C. O., *J. biol. Chemistry*, **79**, 189 (1928).
73. WADDIL, H. G. and GORIN, G., *Anal. Chemistry*, **30**, 1069 (1958).
74. GORIN, G., Abstracts of Papers presented before the Division of Organic Chemistry at the 127th Meeting of the Amer. Chem. Soc., 1956, p.10.
75. TOMÍČEK, O. and ŠANDL, Z., *Chem. Listy*, **40**, 219 (1946).
76. LHOTA, Z., Doctoral Thesis, Charles University, Prague, 1953.
77. TOMÍČEK, O. and ŠIMON, J., *Českoslov. Farmac.*, **1**, 25 (1952).
78. SANT, B. R., *Chemist-Analyst*, **47**, 65 (1958).
79. PINXTEREN, VAN, J. A. and VERLOOP, E., *Pharm. Weekbl.*, **93**, 982 (1958).
79a. SHIMOMURA, S., *J. pharmac. Soc. Japan (Yakugaku Zasshi)*, **81**, 1415 (1961).
80. BÁNYAI, E. and ZUMAN, P., *Collection (Czech. Chem. Comm.)*, **24**, 522 (1959).
80a. BUDĚŠÍNSKÝ, B. and KÖRBL, J., *Mikrochim. Acta*, 697 (1960).
80b. KONDÔ, G. and MURAKAMI, J., *Osaka Ind. Res. Inst.*, **9**, 277 (1958).
81. MIKKELSON, V. J., *Trudy Tallinsk. Polytechn. Inst.*, A **63**, 127 (1955).
82. PRUNER, G., *Rend. Inst. super Sanita*, **19**, 492 (1956).
82a. BECCARI, E., *Bull. Soc. Ital. biol. sperim.*, **19**, 372 (1940).
83. SANT, B. R., *Analyt. Chim. Acta*, **19**, 523 (1958).
84. MÜLLER, E., *Z. anorg. Chem.*, **135**, 265 (1924).
84a. DUŠINSKÝ, G. and TYLOVÁ, M., *Českoslov. Farmac.*, **11**, 359 (1962).
84b. SOLYMOSI, F., *Chemist-Analyst*, **51**, 71 (1962).
85. WILLARD, H. H. and MANALO, G. D., *Ind. Engng. Chem. Anal. Ed.*, **19**, 167 (1947).

86. THYAGARAJAN, B. S., *Chem. Reviews*, **58**, 439 (1958).
87. BROWNING, P. E. and PALMER, H. E., *Z. anorg. Chem.*, **29**, 71 (1908).
88. BAPAT, M. G. and TATWAWADI, S. V., *J. Sci. Res. Banaras Hindu Univ.*, **7**, 235 (1957).
89. BERKA, A. and VULTERIN, J., *Chemie*, **10**, 113 (1958).
90. SANT, B. R. and SANT, S. B., *Talanta*, **3**, 261 (1960).
91. SOLYMOSI, F., *Talanta*, **4**, 211 (1960).
92. SANT, B. R. and SANT, S. B., *Talanta*, **4**, 212 (1960).

CHAPTER 5

HYPOHALITES (HYPOCHLORITE, HYPOBROMITE)

JELLINEK and KRESTEFF[1] were the first investigators to recommend hypohalites for inorganic volumetric analysis; they used hypochlorites of alkali metals as titrant. Kolthoff and Stenger[2] later replaced the less stable sodium hypochlorite by calcium hypochlorite and recommended the addition of potassium bromide to the solution to be titrated. They obtained in this way " hypobromite *in statu nascendi* " because the solution contains now the hypobromite ion as the active component.

Tomíček and co-workers[3,4] studied the direct potentiometric titrations with hypochlorite, hypobromite and " hypobromite *in statu nascendi* ".

The oxidation processes due to hypochlorite and hypobromite are described by the following equations:

$$ClO^- + H_2O + 2e^- \rightarrow Cl^- + 2\,OH^-$$
$$BrO^- + H_2O + 2e^- \rightarrow Br^- + 2\,OH^-$$

The reactions do not always proceed stoichiometrically following these equations. Titrations with alkaline hypochlorite or hypobromite solutions (by dissolving bromine in alkali hydroxide) usually give higher errors. Firstly, this is due to the low stability of the two reagents and, secondly, to the bromate, which can be formed additionally in the preparation of the hypobromite solution:

$$3\,NaBrO \rightarrow 2\,NaBr + NaBrO_3$$

this reacts also as an oxidant, but in a different stoichiometric ratio from hypobromite. The same would be the case—to an even higher degree—in the preparation of hypoiodite solutions which, for that reason, have not been investigated in detail.

Titrations with calcium hypochlorite in the presence of bromide have proved best. Because of the presence of bromide in the solution to be titrated, the more stable calcium hypochlorite is transformed into the less stable but more active oxidation agent, namely hypobromite.

Direct and indirect titrations with hypohalogenite are mostly used in oxidations in weakly alkaline solution. In this way, several inorganic substances, particularly ammonium compounds and organic substances (urea) can be determined. More recently, *chloramine-T* which reacts as titrant in a similar way, has been recommended; it is more easily accessible and more stable than the hypochlorites (see Chapter 6, p. 37).

Because so many articles have been published in the literature on oxidations with hypohalites, they cannot all be listed in this book; only the most important ones are mentioned here. For further details—especially on earlier publications—the book of Kolthoff and Belcher[5] should be consulted.

STANDARD SOLUTION

A 0·1 N (0·025 M) calcium hypochlorite solution is prepared from chloride of lime, which contains usually 25 per cent of active chlorine. The weight amount (corresponding to the chlorine content previously determined) is finely ground with water in a mortar, transferred into a suitable flask, well shaken and then set aside. The turbid solution is filtered into a measuring flask, using a Buchner funnel, and diluted to 1 l. The standard solution is stored in the dark (best in a dark bottle).

For the preparation of a 0·1 N (0·05 M) sodium hypobromite solution, 24 g of pure NaOH are dissolved in 1 l. distilled water and cooled in a freezing mixture to − 4°C. Then, 8 g bromine are added carefully. After mixing it is cooled again for 2 hr and after 12 hr the solution is standardized.[6]

The stability of the solution depends on various factors. The action of light, a too high alkalinity and too high temperature and, especially with hypobromite, traces of various metal ions, lower the normality.[5] Alkali hypobromite solutions are stable only for a few days; hypochlorites are more stable, among them, calcium hypochlorite is the most stable solution (the factor of these solutions is lowered after several months by less than 1 per cent).[6]

Standardization

The titre can usually be determined iodimetrically or by potentiometric titration in a sodium hydrogen carbonate solution with arsenite.[5,6,16] Photometric titration with potassium hexacyanoferrate (II) has also been suggested for standardization;[42] here, the colour intensity of the potassium hexacyanoferrate (III) formed is measured. This method is of little practical significance.

INDICATOR

In the indirect determination, various reversible or even irreversible redox indicators can be used for the back-titration of the unconsumed excess of reagent. In one of the recent publications, the end-point of the titration of hypobromite with hydrogen peroxide in alkaline solution is indicated by the change of the green colour of divalent to the black colour of trivalent nickel hydroxide.[17,50]

The end-point determination in direct titrations is more important; indigo, methyl red, methyl orange, potassium iodide–starch solution,[1,7] quinoline yellow,[8,9] tartrazine,[10] brasilin,[36,41,46] bromothymol blue[53] and luminescence indicators, e.g. luminol[15] are used.

Amongst electrometric methods, mainly potentiometry[3,4] or amperometry[11–14,43,54] are applied for the indication of the end-point.

REVIEW OF DETERMINATIONS

Arsenic, antimony and tin. Of the numerous determinations of arsenic (III), antimony (III) and tin (II), the titration using luminol as indicator has to be mentioned in the first instance; this is carried out in alkaline solution (in the determination of antimony in the presence of tartrate) with hypochlorite at 80°C or with hypobromite at room temperature.[15] In hydrochloric acid medium, methyl orange is used as indicator;[15] as little as 0·1 mg arsenic can be determined in this way. Trivalent arsenic and antimony compounds can also be titrated potentiometrically or amperometrically.[3,4,37,52,54]

The titration of arsenite can be used for the indirect determination of permanganate or dichromate; both anions are reduced by an excess of arsenite, and the excess is titrated with hypochlorite.[7]

The oxidation of tin (II) with hypochlorite or hypobromite proceeds in an inert atmosphere exactly as with other oxidizing agents.[1,3–5]

Ammonium salts. The reaction of ammonium salts, as e.g. ammonia with hypohalites has been often investigated to establish suitable conditions for quantitative oxidation following the equation:[5]

$$2\ NH_3 + 3\ ClO^- \rightarrow N_2 + 3\ H_2O + 3\ Cl^-$$

in some cases nitrogen oxides can be formed besides nitrogen. Kolthoff and Laur[18] and later Kolthoff and Stenger[2] have studied these questions very thoroughly. They developed a method which gives precise results. The ammonium salt is oxidized in a hydrogen carbonate solution with a small excess of calcium hypochlorite in the presence of an excess of bromide; the unconsumed reagent is titrated with arsenite solution precisely five minutes later. The method is also suitable for the determination of nitrogen in organic substances after they have been transformed into ammonium salt; it has also been used in the analysis of mineral material.[19] Recently, Köszegi and Salgo[20] revised this method. They carried out the oxidation with bromine, which is formed in the reaction of bromate and bromide in acid medium; after the oxidation the solution is made alkaline and hypobromite is formed. The excess of hypobromite is determined iodimetrically.

Ammonium salts can also be titrated directly in a hydrogen carbonate solution potentiometrically[3,4,6] or amperometrically with a rotating platinum electrode.[14] Ions which precipitate as MNH_4PO_4 (M = Mg, Zn,Mn,Cd) can be determined by both methods. The separated precipitate is dissolved in EDTA and the ammonium is titrated with calcium hypochlorite in the presence of bromide. The method is used fairly widely and has proved suitable in the determination of uranium in the presence of beryllium and titanium.[51]

Ammonium salts in fertilizers are determined by "dead-stop" titration.[44,52,53]

Iodide, bromide, cyanide and thiocyanate. The oxidation of iodide to iodate, of cyanide to cyanate, of thiocyanate to cyanate and sulphate proceed quantitatively with hypohalite in weakly alkaline solution:

$$I^- + 3\,BrO^- \to IO_3^- + 3\,Br^-$$

$$CN^- + BrO^- \to CNO^- + Br^-$$

$$CNS^- + 4\,BrO^- + H_2O \to CNO^- + 4\,Br^- + 2\,H^+ + SO_4^{2-}$$

The equivalence point is determined either visually or, better, potentiometrically[3,4,6]

Iodides are titrated with hypochlorite in sodium hydrogen carbonate solution; iodide can also be determined in the presence of chloride and bromide in this way.[21,40]

In the same medium, cyanide and thiocyanate determinations also proceed well; it is better to titrate the thiocyanate with hypochlorite than with hypobromite, because this reaction yields more precise results.[3,4,6,54] In the cyanide determination, it is advisable first to add the greater part of the reagent and then to finish the titration dropwise. Cyanide can also be determined in complex compounds with mercury, using hypobromite.[22]

The determination of bromide is based on its transformation to hypobromite by an excess of hypochlorite according to the reaction:[23]

$$ClO^- + Br^- \to BrO^- + Cl^-$$

The velocity of this reaction depends very much on the medium and has its maximum at a pH 9–10. In acid solution, bromate can be partially formed. According to Tomíček,[6] iodides and bromides can be titrated in the presence of chloride with hypochlorite at 50–60°C. The method is well suited for the determination of small amounts of bromide in the presence of chloride.

Sulphide, sulphite and thiosulphate. The oxidation of sulphides to sulphate with hypohalites usually does not follow exactly the equation:

$$S^{2-} + 4\,ClO^- \to SO_4^{2-} + 4\,Cl^-$$

but it is mostly accompanied by the formation of colloidal sulphur. Several authors studied the influence of the pH on the stoichiometry of this reaction;[5] from their publications it would seem that the best results are obtained by indirect titration in alkaline solution. In this way, sulphur in steel can be determined if it is first separated as H_2S from the steel; the excess of hypobromite is titrated with hydrogen peroxide in the presence of nickel (II) hydroxide as indicator.[17,50] A similar method is recommended for the determination of sulphur in organic compounds; the excess of the reagent is determined iodimetrically.[50a] The direct titration of sulphite solutions with hypohalite does not give satisfactory results.

Sulphite can be determined precisely with an excess of hypobromite in alkaline solution:[24]

$$SO_3^{2-} + BrO^- \rightarrow SO_4^{2-} + Br^-$$

Erdey and Buzás[15] describe the possibility of a direct sulphite titration using luminol as indicator, whereas, according to Tomíček and co-workers, good results can be only obtained if the titrant is standardized precisely with hypochlorite.[3,4]

The titration of thiosulphate with hypobromite in the presence of bromide yields precise results. The reaction follows the equation

$$S_2O_3^{2-} + 4\,BrO^- + 2\,OH^- \rightarrow 2\,SO_4^{2-} + 4\,Br^- + H_2O$$

and can be carried out either directly or indirectly,[2-5,15,25,58] visually or potentiometrically.

Thallium. Thallium (I) salts can be titrated potentiometrically with hypochlorite; the solution should be 1·5–2 N NaOH at the end of the titration. The titration is carried out in an inert atmosphere.[3,4,27] The determination is precise; because thallium (I) carbonate is used as a standard substance in analytical chemistry, this method can be used for standardizing hypohalite solutions.[3,4,6] The titration of thallium (I) with hypobromite *in statu nascendi* is less suitable, because the accuracy of the results is impaired by thallium bromide, formed with the excess of bromide present in the solution.

Cerium (III) salts[3,4,27] can be titrated directly potentiometrically and *selenium (IV)* or *tellurium (IV) salts* amperometrically.[4,6,13] In this way, *silver, lead* and *mercury* can be determined indirectly by precipitating them as *selenites* (IV), followed by an amperometric titration.[45]

In the determination of some other inorganic compounds with hypohalites, the excess of reagent added is back-titrated nearly without exception. In this way, *peroxides,*[26] *nitrites,*[5] *phosphites*[5,21] and *chromium (III) salts,*[15] even in the presence of an excess of chromate (VI), are oxidized quantitatively.

Singh and co-worker[38] determined *iron (II), mercury (I)* and some hydrazine derivatives[38,39] by oxidizing them first with iodine chloride and then titrating them with sodium hypochlorite. Hydrazine and its salts can also be titrated directly potentiometrically[3,4,6,55,56] or using luminol as indicator.[15] The titration follows exactly the reaction:

$$N_2H_4 + 2\,ClO^- \rightarrow N_2 + 2\,H_2O + 2\,Cl^-$$

whereas this titration is not suited for the quantitative determination of *hydroxylamine*, because nitrogen oxides are formed besides nitrogen.[28]

Amongst the oxidations of organic compounds with hypobromite, main consideration has been given to the determination of *urea*. This titration is very sensitive to conditions; as Tomíček and Filipovič state, the oxidation does not correspond exactly to the reaction

$$(NH_2)_2 CO + 3 BrO^- \rightarrow CO_2 + N_2 + 2 H_2O + 3 Br^-$$

a potentiometric investigation of the reaction revealed, that hydrazine is formed intermediately; the results have usually a constant error. For precise determinations only the indirect titration is therefore suitable in which an excess of reagent is added. A very detailed procedure is given in the publications of Wölfel,[29] Bitskei[46] and Grover.[47]

Of other organic compounds, only *hydroquinone* can be directly titrated potentiometrically with hypobromite *in statu nascendi*;[6] the same reagent can also be used for the determination of 3-*thioketo*-5-*keto*-6-*benzyl*-1,2,4-*triazine*[30] and *thioacetamide*.[57] Some *aromatic sulphonic acids*[31] are determined with hypochlorite.

The other volumetric determinations using hypohalites are without exception lengthy oxidations, e.g. the determination of *amino acids*,[32,33] *glycerol*,[34] *benzene-sulphinic acid*, *isopropanol* and *acetone*[48] or *sugar*.[49]

A recent communication deals with the determination of *formaldehyde* and *benzaldehyde* using an excess of NaClO in alkaline medium (OsO_4 as catalyst) and back-titration with arsenite.[59]

REFERENCES

1. JELLINEK, K. and KRESTEFF, W., *Z. anorg. Chem.*, **137**, 333 (1924).
2. KOLTHOFF, I. M. and STENGER, V. A., *Ind. Engng. Chem. Anal. Ed.*, **7**, 79 (1935).
3. TOMÍČEK, O. and JAŠEK, M., *Collection (Czech. Chem. Comm.)*, **10**, 353 (1938).
4. TOMÍČEK, O. and FILIPOVIČ, P., *Collection (Czech. Chem. Comm.)*, **10**, 340, 415 (1938).
5. KOLTHOFF, I. M., STENGER, V. A., BELCHER, R. and MATSUYAMA, G., *Volumetric Analysis*, III. Titration Methods: Oxidation–Reduction Reactions, pp. 573–595, Interscience, New York 1957.
6. TOMÍČEK, O., *Potenciometrické titrace*, JČMF, 1941, Prague.
7. JELLINEK, K. and KÜHN, W., *Z. anorg. Chem.*, **138**, 99 (1924).
8. SINN, V., *Chim. analytique*, **29**, 58 (1947).
9. BELCHER, R., *Analyt. Chim. Acta*, **5**, 27 (1951).
10. BELCHER, R., *Analyt. Chim. Acta*, **4**, 468 (1950).
11. KOLTHOFF, I. M., STRICKS, W. and MORREN, L., *Analyst*, **78**, 405 (1953).
12. LAITINEN, H. A. and WOERNER, D. E., *Anal. Chemistry*, **27**, 214 (1955).
13. DESHMUKH, G. S., BAPAT, M. G., BALKRISHNAN, E. and ESHWAR, M. C., *Naturwiss.*, **45**, 129 (1958).
14. SIMON, V., SEKERKA, I. and DOLEŽAL, J., *Chem. Listy*, **46**, 617 (1952).
15. ERDEY, L. and BUZÁS, I., *Acta Chim. Acad. Sci. Hung.*, **6**, 93, 115 (1955).

HYPOHALITES (HYPOCHLORITE, HYPOBROMITE) 35

16. GOLDSTONE, N. I. and JACOBS, M. B., *Ind. Eng. Chem., Anal. Ed.*, **16**, 206 (1944).
17. ERDEY, L. and BUZÁS, I., *Acta Chim. Acad. Sci. Hung.*, **6**, 77 (1955).
18. KOLTHOFF, I. M. and LAUR, A., *Z. anal. Chem.*, **73**, 177 (1928).
19. BELCHER, R. and BHATTY, M. K., *Mikrochim. Acta*, 1183 (1956).
20. KÖSZEGI, D. and SALGÓ, E., *Z. anal. Chem.*, **143**, 423 (1954).
21. SCHWICKER, A., *Z. anal. Chem.*, **110**, 161, 173 (1937).
22. GOLSE, J., *Bull. Soc. Chim. France*, **45**, 177 (1929).
23. FARKAS, L. and LEWIN, M., *Anal. Chemistry*, **19**, 662, 665 (1947).
24. BONNER, W. D. and YOST, D. M., *Ind. Engng. Chem.*, **18**, 55 (1926).
25. BITSKEI, J. and FORHENCZ, M., *Mag. Kém. Lapja*, **2**, 117 (1947).
26. BITSKEI, J., *Acta Chim. Acad. Sci. Hung.*, **8**, 203 (1955), **10**, 327 (1957).
27. TOMÍČEK, O. and JAŠEK, M., *J. Amer. Chem. Soc.*, **57**, 2409 (1935).
28. RILEY, R. F., RICHTER, E., ROTHEMAN, M., TODD, N., MYERS, L. S. and NUSBAUM, R., *J. Amer. Chem. Soc.*, **76**, 3301 (1954).
29. WÖLFEL, K., *Z. anal. Chem.*, **90**, 170 (1932).
30. CATTELAIN, J., *Ann. Chim. anal.*, **24**, 150 (1942).
31. SCHÄFER, H. and WILIE, E., *Z. anal. Chem.*, **130**, 396 (1950).
32. GORDON, H. T., *Anal. Chemistry*, **23**, 1853 (1951).
33. FRIEDMAN, A. H. and MORGULIS, S., *J. Amer. Chem. Soc.*, **58**, 909 (1936).
34. CUTHILL, R. and ATKINS, C., *J. Soc. Chem. Ind.*, **57**, 89 (1938).
35. ACKERMANN, L., *Ind. Engng. Chem. Anal. Ed.*, **18**, 243 (1946).
36. BITSKEI, J., *Acta Chim. Acad. Sci. Hung.*, **10**, 313 (1957).
37. KHATUM, S. and KHUNDHAR, M. H., *J. Indian Chem. Soc.*, **34**, 114 (1957).
38. SINGH, B. and SINGH, S., *Analyt. Chim. Acta*, **19**, 10 (1958).
39. SINGH, B., SAHOTA, S. S. and SINGH, S., *Z. anal. Chem.*, **160**, 429 (1958).
40. BITSKEI, J., *Z. anal. Chem.*, **150**, 267 (1956).
41. BITSKEI, J., *Z. anal. Chem.*, **151**, 422 (1956).
42. DESHMUKH, G. S. and TATWAWADI, S. V., *Z. anal. Chem.*, **168**, 411 (1959).
43. KULKARNI, V. P. and NABARS, G. M., *J. sci. ind. Res. (New Delhi), Sect. B* **15**, 708 (1956).
44. ILJIN, W. S., *Agron. trop.*, **4**, 191 (1958); *Chem. Abstr.*, **53**, 17754 (1959).
45. DESHMUKH, G. S., BAPAT, M. G., BALKRISHNAN, E. and ESHWAR, M. C., *Z. anal. Chem.*, **170**, 381 (1959).
46. BITSKEI, J., *Mag. Kém. Folyóirat*, **62**, 71 (1956).
47. GROVER, K. C., *Agra Univ. J. Res.*, **4**, 627 (1955); *Chem. Abstr.*, **52**, 10805 (1958).
48. GROVER, K. C. and MEHROTRA, R. C., *Z. anal. Chem.*, **160**, 274 (1958).
49. YOSHIMURA, CH. and KIBOKU, M., *J. chem. Soc. Japan, pure Chem. Sect.*, **77**, 1546 (1956).
50. ERDEY, L. and INCZÉDY, I., *Z. anal. Chem.*, **166**, 410 (1959).
50a. HASHMI, M. H., MANZOOR, ELAHI and EHSAN, ALI, *Analyst*, **88**, 68 (1963).
51. SEKERKA, I. and VORLÍČEK, J., *Chem. Listy*, **47**, 512 (1952).
52. ŘEZÁČ, Z. and FIGAROVÁ, M., *Z. anal. Chem.*, **176**, 115 (1960).
53. CERANA, A., *Rev. Fac. Ing. Quim. Univ. Nacl. Litoral. Santa Fé* (Argentine), **29**, 105 (1960); *Chem. Abstr.*, **56**, 5384 (1962).
54. DESHMUKH, G. S. and ESHWAR, M. C., *J. Sci. Res. (India)*, **29**, 388 (1960).
55. TATSUZAVA, M., *Eishei Shikenjo Hôkoku*, **77**, 131 (1959); *Chem. Abstr.*, **55**, 9781 (1961).
56. SINGH, B. and SAHOTA, S. S., *J. Indian Chem. Soc.*, **38**, 569 (1961).
57. CLAEYS, A., SION, H., CAMPE, A. and THUN, H., *Bull. Soc. chim. Belgique*, **70**, 576 (1961).

58. CLAEYS, A. and SION, H., *Bull. Soc. chim. Belgique*, **70**, 154 (1961); *Z. anal. Chem.*, **187**, 149 (1962).
59. NORKUS, P. K., *Zhur. Anal. Khim.*, **18**, 650 (1963).

CHAPTER 6

CHLORAMINE-T

THE sodium salt of p-toluene-sulphochloramide $CH_3C_6H_4SO_2$ $NNaCl \cdot 3H_2O$, known as chloramine-T, acts as an oxidant in alkaline as well as in acid solution;[1] the reaction mechanism is not yet fully elucidated. In aqueous solution, chloramine-T is considered to hydrolyse according to the following equation:

$$CH_3C_6H_4SO_2NClNa + H_2O \rightleftharpoons CH_3C_6H_4SO_2NH_2 + NaClO \quad (1)$$

The hydrolysis is actually more complicated; the following partial reactions can take place:[2]

$$2\,CH_3C_6H_4SO_2NClNa + 2\,H_2O \rightleftharpoons 2\,CH_3C_6H_4SO_2NClH + 2\,NaOH \quad (2)$$

$$2\,CH_3C_6H_4SO_2NClH \rightleftharpoons CH_3C_6H_4SO_2NCl_2 + CH_3C_6H_4SO_2NH_2 \quad (3)$$

$$CH_3C_6H_4SO_2NCl_2 + NaOH \rightleftharpoons CH_3C_6H_4SO_2NClH + NaClO \quad (4)$$

$$CH_3C_6H_4SO_2NClH + NaOH \rightleftharpoons CH_3C_6H_4SO_2NH_2 + NaClO \quad (5)$$

When a dilute chloramine-T solution is acidified, a white precipitate forms which disappears on the addition of an oxidizable compound. The precipitate has been identified as a mixture, consisting of p-toluene-sulphoamide and p-toluene-sulphodichloramide, according to equation (3). Because 1 mole of p-toluene-sulphodichloramide contains twice as much active chlorine as the chloramine, the formation of the precipitate is not connected with a loss of active chlorine. The formation of much precipitate should nevertheless be avoided in volumetric determinations. The p-toluene-sulphodichloramide separates first from the acidified solution in the form of small droplets; in this form, the compound is very reactive. The crystals which are gradually formed, however, react only very slowly with reducing substances, so that the determination can yield results which are too low. Therefore, one titrates with chloramine-T in acid solution only if the reaction proceeds rapidly.

The redox potential of chloramine-T in 1 N H_2SO_4 is 1·52 V; in neutral solution 0·90 V.[3] The titration with chloramine-T is in some cases preferred to the more costly iodimetric determination.

STANDARD SOLUTION

The standard solution is prepared by dissolving the pure substance. If the chloramine is impure, it has to be recrystallized from hot water and dried over concentrated sulphuric acid. The aqueous solution reacts alkaline (pH of a 1 per cent solution = 10·05) and shows a very faint opalescence.

Aqueous chloramine-T solutions are very stable. If stored in dark bottles, they do not change their true titre for several months; boiling for several hours likewise does not change their true titre. The normality has to be controlled only after two to three months.[7,8] Lately, Bishop and Jennings[52] have occupied themselves with the standardization and the stability of chloramine solutions.

Standardization

The solution is standardized iodimetrically or by titrating with it an arsenite solution.
1. Add 25 ml of 0·5 N KI and 25 ml of HCl (1:4) to 30 ml of 0·1 N chloramine-T solution and titrate with 0·1 N $Na_2S_2O_3$.[4]
2. To 25 ml of 0·1 N arsenite solution (in the presence of $NaHCO_3$) add 0·5 g of KI and 1 ml of starch solution and titrate with the chloramine solution to a blue colour.[5]

INDICATOR

In the visual titration, the blue colour can be used for indicating the equivalence point, which is formed by the iodine liberated at the equivalence point from the excess of iodide present in the solution, with starch. The substance to be determined must be a stronger reducing agent than the iodide present, otherwise iodine would separate on adding the titrant. This end-point determination has proved suitable in the titration of As^{3+}, Sb^{3+}, Sn^{2+}, Fe^{2+}, N_2H_4, CS_2, etc.

Another method of visual end-point determination is based on the destruction of a dyestuff indicator by the first drop in excess of the titrant. For this purpose, methyl red,[9,10] *p*-ethoxychrysoidine and brilliant carmoisine[11] have been proposed as indicators for the titration of As^{3+}, Sb^{3+} and N_2H_4. When standard solutions of lower molarity are used, it is advisable to determine the amount of reagent necessary for the destruction of the indicator in a blank test; this is then deducted from the result of the actual titration.

Indigo carmine can be used as indicator for the direct titrimetric determination of Fe^{2+} (colour change from blue to green).[12]

In solutions acidified with hydrochloric acid, iodide can be titrated so that the equivalence point is recognized by the colour change in the chloroform layer from violet (owing to I_2 formed) to pale yellow (ICl). This method has been applied to the indirect determination of a number of substances, which are oxidized first with iodine monochloride forming iodide; the iodide thus formed is then determined in the way mentioned above.[13,14]

In the indirect determination with chloramine-T, the excess of reagent is mostly back-titrated iodimetrically.

All the systems mentioned above, as well as a number of other substances, for the determination of which the visual method is not suitable, can be titrated potentiometrically. In some cases, the amperometric end-point has proved satisfactory.[14a,b]

REVIEW OF DETERMINATIONS

Arsenic and antimony. The oxidation of trivalent arsenic with chloramine-T proceeds according to the equation

$$AsO_3^{3-} + CH_3C_6H_4SO_2NClNa + H_2O \rightarrow AsO_4^{3-} + \\ + CH_3C_6H_4SO_2NH_2 + NaCl.$$

In a sodium hydrogen carbonate solution, arsenite can be titrated directly with chloramine-T after adding potassium iodide. Starch solution serves as indicator.[5] In the presence of iodide, As^{3+} can also be determined potentiometrically;[9] if the concentration of the iodide near the end of the titration is 0·005 M, it is possible to titrate at pH 4–9.[15a] In acid solution, the arsenic (III) is only very slowly oxidized by chloramine in the absence of halides. It has therefore been recommended that the potentiometric determination be titrated in 0·35–1·75 N HCl at 55 to 65°C.[9,15]

Recently, the potentiometric[15a] and the visual[15b] titration of arsenic (III) compounds with chloramine-T have been again thoroughly studied. It has been found that the reaction proceeds satisfactory if the chloride ion concentration in the solution is more than 0·5 M and the hydrogen ion concentration 0·5–5·0 M. In the presence of bromide ions (concentration near the end-point 0·1 M) the reaction is quantitative in solutions with a hydrogen ion concentration of 10^{-5} to 5 M.

For visual titrations, the following conclusions can be drawn:

In 1–2 N HCl solutions, amaranth, bordeaux and o-dianisidine can be used for visual indication; p-ethoxy-chrysoidine as well as methyl orange or methyl red can be used, if a correction of 0·02 ml of 0·1 N standard solution is applied. If the same solution is moreover 0·1 M, with regard to bromide ions, rosaniline, amaranth and α-naphthol-flavone can serve as indicators. The titration in 1 M H_2SO_4 solution, which is 0·01–0·1 M in respect of bromide ions, proceeds very well using rosaniline as indicator. In the same medium, amaranth, bordeaux, methyl red, methyl orange, the reversible indicators quinoline yellow and p-ethoxy-chrysoidine can likewise be used. In an acetate buffered solution, 0·1 M with regard to bromide

D

ions, quinoline yellow and tartrazine proved best. The most accurate results are obtained in sulphuric acid solution in the presence of bromide ions. Titrating dilute solutions with 0·01 N standard solution, it is advisable to buffer the solution with $NaHCO_3$, and to use starch as indicator in the presence of iodide.

The potentiometric determination of Sb^{3+} with chloramine-T[9,15] as well as its visual titration using starch solution in the presence of tartaric acid,[16] are both carried out as the arsenic determination.

Very recently, Bishop and Jennings,[61] studied the reaction of trivalent antimony with chloramine-T. According to the authors, precise results are achieved by potentiometric titration in 1–4 M HCl solution or in 1–3 M HCl using amaranth as indicator. The best results are, however, achieved in solutions of pH 6·5–7·5 which are $5 \times 10^{4-}$ to 5×10^{-3} M with regard to iodide ions. The titration proceeds also satisfactory in 3 M HCl solution after addition of 2–5 ml of 0·1 M iodine monochloride solution, irrespective of whether the end-point is determined potentiometrically or using amaranth as indicator.

Both As^{3+} as well as Sb^{3+} can be titrated amperometrically,[14a,b,c] or spectrophotometrically (in the presence of bromide).[16a]

Tin. The titrations of tin (II) salts with chloramine-T are carried out in HCl solution in an inert atmosphere. The acid concentration must not be more than 3–5 per cent using a visual end-point with iodide starch solution.[17,18] The titration, based on the destruction of the indicator, is not precise.[9] The potentiometric titration is best done in 1 N hydrochloric acid at 55–65°C. The end-point can also be determined amperometrically.[14a]

Iron. The titration of iron (II) salts with chloramine-T is carried out potentiometrically in 2–3 N HCl solution. Higher concentrations of H_2SO_4 interfere with this determination.

Iron (II) can be titrated directly in acid solution at 40°C using indigo carmine or iodide–starch as indicator, if the iron (III) ions formed are bound by a suitable complex forming reagent such as fluoride. Because the redox potential of the system Fe^{2+}/Fe^{3+} is thus diminished, the iron (II) ions can already be oxidized by atmospheric oxygen under these conditions. From a practical point of view it is therefore recommendable to apply the indirect determination:

A known amount of chloramine-T is added to the iron (II) solution. The unconsumed chloramine-T is reduced by an excess of arsenite, which is then back-titrated with chloramine-T against *p*-ethoxy-chrysoidine or brilliant carmoisine. The oxidation of the iron (II) ions is carried out in 10 per cent HCl solution. In the back-titration, the HCl concentration should be about 5 per cent. It is not necessary to remove the Fe^{3+} ions formed by means of a complexing reagent. Only in the titration of small iron (II) concentrations with 0·01 N chloramine-T solution, the addition of fluoride facilitates the recognition of the equivalence point.

Hexacyanoferrate (II). The determination of hexacyanoferrate (II) with chloramine-T is carried out in 1 N HCl solution by direct potentiometric titration. Before the equivalence point, one has to wait for 5 min in order to allow the potential to adjust correctly. In acid solution, the potential adjusts itself rapidly, the result shows, however, a positive error.[9] The titration end-point can also be determined amperometrically.[14a]

Hexacyanoferrate (II) can be determined visually by indirect titration in hydrochloric acid solution. Chloramine-T is added in excess to the weakly acid hexacyanoferrate (II) solution and heated to 40°C (or set aside for 1 hr at room temperature). Then, an excess of standard arsenite solution is added, the unconsumed part is back-titrated with chloramine-T after the addition of iodide–starch solution.[19]

According to other publications,[7] the determination is carried out in more strongly hydrochloric solutions; here, the oxidation is complete in 2–3 min. The excess of arsenite solution is titrated with chloramine-T using brilliant carmoisine as indicator.

Titanium and vanadium. They can be likewise titrated in a mixture with chloramine-T.[18a]

Hydrazine. Hydrazine is oxidized with chloramine-T following the equation:

$$N_2H_4 + 2\ CH_3C_6H_4SO_2NClNa \rightarrow N_2 + 2\ CH_3C_6H_4NH_2 + 2\ NaCl.$$

Except by direct potentiometric[20] and, eventually, amperometric titration[14a] in hydrochloric acid solution, the determination of hydrazine can also be carried out by direct titration in sodium hydrogen carbonate solution in the presence of iodide–starch as indicator. The indirect titration has been recommended in neutral or sodium acetate buffered solution; the excess of chloramine-T is back-titrated with arsenite or thiosulphate.[19]

Hydrazine can be titrated directly with chloramine-T in acid, as well as in neutral solution, in the presence of potassium bromide.[6] Beyond that it is possible to titrate in acid solution against methyl red where the methyl red colour is destroyed at the equivalence point.

A mixture of hydrazine and methyl hydrazine can be determined using chloramine-T and hypochlorite in the presence of bromide.[20a]

Carbon disulphide. Dissolve the sample in 5 per cent alcoholic KOH solution, just acidify with acetic acid and neutralize with sodium hydrogen carbonate. Titrate with chloramine-T in the presence of iodide starch solution as indicator. The results of this method are higher[19] than those of the iodimetric determination.

Thiosulphate. The potentiometric titration of thiosulphate with chloramine-T (oxidation to sulphate) proceeds best at 50–60°C in solutions containing boric acid.[6]

Sulphite. The potentiometric determinations of sulphite with chloramine-T yield results too low by several per cent. Titrating a known amount of chloramine-T with the sulphite solution gives better results.[6]

Cyanide. The determination of cyanide (oxidation to $[OCN]^-$) is carried out potentiometrically in a potassium hydrogen carbonate solution in the presence of potassium iodide (at most 0·3 g KI and 100 ml). The titration should be done as rapidly as possible because the cyanide can be oxidized already by atmospheric oxygen under these conditions. It is necessary to add more iodide if using starch solution as indicator. This, however, brings about lower results.

Thiocyanate. The direct titration of thiocyanate (oxidation to $[OCN]^-$ and SO_4^{2-}) can be carried out potentiometrically in the presence of potassium bromide at 70°C in solutions containing boric acid.[6]

The indirect determination is done in sodium hydrogen carbonate solution; an excess of chloramine-T is added and, after 4 hr, the excess is determined iodimetrically; or, an excess of arsenite is added and the unconsumed amount is back-titrated with chloramine-T in the presence of potassium iodide.[19]

Hydrogen sulphide, sodium sulphide and sulphites. These substances are determined under the same conditions as described in the methods for determining thiocyanate. The actual oxidation is carried out in alkaline solution.[21]

Thiosulphate, sulphite, sulphide,[21a] *dixanthate.*[21b] These substances can be oxidized by an excess of chloramine-T in acid solution,[21b] the excess of reagent is determined iodimetrically.

Iodide,[9,20] *hydroquinone, quinhydrone, sodium hydrogen sulphite.*[20] These compounds are titrated directly in hydrochloric acid solution potentiometrically.

Iodide, sulphite, thiosulphate, ascorbic acid, hydroxy-quinoline. These substances can be titrated amperometrically.[14a]

The oxidation of iodide with chloramine-T has been recently studied extensively.[21c] It has been found that the oxidation proceeds very well in 1 M sulphuric or acetic acid solution forming iodine and in 3·0–4·0 N hydrochloric acid solution forming iodine monochloride. The end-point is detected either potentiometrically or visually with the aid of a carbon tetrachloride layer.

Thallium. Thallium (I) is directly titrated potentiometrically with standard chloramine-T solution in hydrochloric acid solution and in the presence of potassium bromide.[22]

Hypophosphorous acid. Hypophosphorous acid is determined indirectly. The sulphuric acid solution is treated with an excess of chloramine-T and set aside for 24 hr; the unconsumed part is back-titrated iodimetrically.[19]

Iodide, hydrazine, trivalent arsenic and antimony, divalent tin and iron, monovalent mercury, thiocyanate.[13] After having been oxidized with iodine monochloride, these compounds can be determined in hydrochloric acid solution with chloramine-T. During the oxidation with iodine monochloride, the equivalent amount of iodide is formed, which is then titrated with chloramine-T. The change of colour of the chloroform layer from violet (I_2 is formed intermediately) to pale yellow (formation of ICl) gives the end-point of the titration.

Hydrogen peroxide, lead (IV) oxide, selenium (IV) oxide, sodium formate, permanganate, periodate, potassium dichromate (VI). These substances are determined indirectly so that they are first quantitatively reduced by one of the above mentioned reducing agents; the excess of the reducing agent is back-titrated with chloramine-T.[14]

Bromate (V), chlorate (V), hexacyanoferrate (III), hydrogen peroxide. The substances mentioned are titrated indirectly with chloramine-T using a potentiometric end-point.[22a]

Nitrite. Nitrite is oxidized to nitrate with chloramine-T. The titration of nitrite with chloramine-T does not give good results. Therefore, one titrates a known amount of chloramine-T in acetic acid solution with the nitrite solution to be determined.[20]

In another method, nitrite is oxidized with an excess of chloramine-T in *N,N*-dimethylformamide solution; the excess of reagent is titrated iodimetrically[22b] or amperometrically.[22c]

Ascorbic acid. Ascorbic acid can be titrated visually with chloramine-T, using starch solution as indicator in the presence of iodide in hydrochloric acid solution[23] or using Variamine Blue in the presence of bromide, best at a pH of 3.[24]

The amperometric titration is carried out at pH 4–4·5 and + 0·3 V (rot. Pt–S.C.E.).[24a]

Aldehyde. Some aldehydes are oxidized quantitatively with chloramine-T in alkaline solution:

$$RCHO + CH_3C_6H_4SO_2NClNa + NaOH \rightarrow RCOONa +$$
$$+ CH_3C_6H_4SO_2NH_2 + NaCl.$$

This reaction has been used for the direct potentiometric titration of *acetaldehyde, benzaldehyde,* p-*hydroxy-benzaldehyde, furfurole, salicylaldehyde* and *vanillin,* using a platinum indicator electrode and a calomel reference electrode.[25] Because the change of potential at the equivalence point is only very small, it is recommended that the aldehyde be determined indirectly:

An excess of chloramine-T is added and the unconsumed amount is determined iodimetrically.[26,27]

Uric acid. Uric acid can be titrated with chloramine-T in sodium hydrogen carbonate solution in the presence of potassium iodide, using α-naphthoflavone, to a blue colour.[28]

p-*Nitrophenylhydrazine, 2,4-dinitrophenylhydrazine, semicarbazide, thiosemicarbazide, isonicotinylhydrazide and thiourea.* These substances are determined directly in hydrochloric acid solution by potentiometric titration with chloramine-T in the presence of potassium bromide.[26] The titrations proceed sufficiently rapidly so that they can be carried out at room temperature. Only the determination of thiosemicarbazide is done at 60°C.

Urea, thiourea, oxalic acid, ethylene glycol, glycerol, mannitol, glucose, phenylhydrazine, some aldehydes, xanthates, lactic acid. These compounds are titrated directly with chloramine-T, either potentiometrically[25] or visually, using methyl red or indigo carmine as indicators (which are destroyed at the equivalence point); these titrations are mostly carried out at elevated temperatures (eventually in a boiling water bath).[29]

Derivatives of thiobarbituric acid,[26] *organic sulphide compounds,*[30] *some sugars,*[31] *phenols and salicylic acid.*[32] These substances are determined indirectly. The excess of chloramine-T is back-titrated iodimetrically.

In the determination of *phenol* and *salicylic acid*, the excess of chloramine-T reacts in the presence of potassium bromide so that bromine, used for the bromination, is liberated. Thereby, the compounds mentioned above are transformed into the respective tribromo-derivatives.

Some recently published work deals with catalytic action of osmium (VIII) oxide on the oxidation with chloramine-T in alkaline solution,[53] furthermore with the determination of *thallium*[54] (using a visual endpoint), of *organic derivatives of hydrazine,*[55] of *carbon disulphide,*[57] of *polythionates,*[56] of *nitro-furfurane*[58] and with the indirect determination of *cadmium.*[59]

A method has been proposed for the determination of *zinc, magnesium, iron, aluminium, vanadium* and *titanium*; these metal ions are precipitated with 8-hydroxy-quinoline; after dissolving the precipitates, the hydroxyquinoline liberated is oxidized with an excess of chloramine-T and the excess of reagent back-titrated iodimetrically.[60]

Very recently there have been published critical studies on the titration of trivalent *arsenic* and *antimony, hydrazine, thiocyanate* and *monovalent thallium* in the presence of iodine monochloride,[62] on the titration of *hydrazine*[63] and on the determination of *thallium, thiocyanate, sulphite, hexacyanoferrate* (II), *oxine* and some other substances,[64] without exception using an excess of reagent.

CHLORAMINE-B

Chloramine-B, $C_6H_5SO_2NNaCl \cdot 3\ H_2O$, is also suitable for titrimetric analysis. This compound was first proposed by Afanasjev[33] in 1950 for volumetric purposes instead of chloramine-T. Since then, numerous publications deal with the use of chloramine-B in volumetric analysis.[34-50] It would seem that chloramine-B is as useful a titrimetric reagent as chloramine-T. Chloramine-B has been used also for the indirect *determination of some metals* which precipitate with anthranilic acid or oxine.[35] After formation of the precipitate the anthranilic acid or the oxine is determined:

To an excess of potassium bromide, an excess of chloramine-B is added (thus an equivalent amount of bromine is liberated); the excess is backtitrated iodimetrically.

Chloramine-T has been used for the indirect determination of some metals in a similar way, as described in subsequent publications.[51]

Other chloramines, e.g. dichloramine-T ($CH_3C_6H_4SO_2NCl_2$) or dichloramine-B ($C_6H_5SO_2NCl_2$) cannot be used in analytical chemistry, despite their analogous properties and despite their higher content of active chlorine, because they are nearly insoluble in water as well as in dilute acids.

REFERENCES

1. BERKA, A., *Chemie*, **10**, 121 (1958).
2. DIETZEL, R. and TÄUFEL, K., *Apotheker-Ztg.*, **44**, 989, 1007 (1929).
3. AFANASJEV, B. N., *Zhur. Fiz. Khim.*, **22**, 499 (1948).
4. BÖTTGER, K. and BÖTTGER, W., *Z. anal. Chem.*, **70**, 225 (1927).
5. NOLL, A., *Chemiker-Ztg.*, **48**, 845 (1924); **64**, 308 (1940).
6. SAMEK, B., *Časopis Českého Lékarnictva*, **21**, 77 (1941).
7. POETHKE, W. and WOLF, F., *Z. anorg. allg. Chem.*, **268**, 244 (1952).
8. SCHIEMANN, G. and NOVAK, P., *Angew. Chem.*, **40**, 1032 (1927).
9. TOMÍČEK, O. and SUCHARDA, B., *Collection (Czech. Chem. Comm.)*, **4**, 285 (1932).
10. BRUCHHAUSEN, V. F. and HANZLÍK, E., *Apotheker-Ztg.*, **40**, 1115 (1925).
11. POETHKE, W., *Pharm. Zentralhalle*, **86**, 2 (1947).
12. AFANASJEV, B. N. and URALSKAYA, A. V., *Zavodskaya Lab.*, **15**, 407 (1949).
13. SINGH, B. and SOOD, K. CH., *Analyt. Chim. Acta*, **13**, 301 (1955).
14. SINGH, B. and SOOD, K. CH., *Analyt. Chim. Acta*, **13**, 305 (1955).
14a. KHADEEV, V. A., ZHDANOV, A. K. and RECHKINA, L. G., *Uzbek. Khim. Zhur.*, **6**, 28 (1960); *Ref. Zhur. Khim.*, **15 D 16**, 117 (1961).
14b. DOLCETTA, M., *Leybold Ber.*, **3**, 50 (1955).
14c. DESHMUKH, G. S., ESHWAR, M. C. and RAO, M. S. V., *Chemist-Analyst*, **51**, 70, (1962).
15. MCMILLAN, A. and EASTON, W., *J. Soc. Chem. Ind.*, **46**, 472 (1927).
15a. BISHOP, E. and JENNINGS, V. J., *Talanta*, **8**, 22 (1961).
15b. BISHOP, E. and JENNINGS, V. J., *Talanta*, **8**, 34 (1961).
16. RUPP, E., SIEBLER, G. and BRACHMANN, W., *Pharm. Zentralhalle*, **66**, 33 (1925).
16a. OGAWA, K. and MUSHA, S., *Bull. Univ. Osaka Prefect., Ser. A.*, **8**, 57 (1960); *Chem. Abstr.*, **55**, 7156 (1961).
17. RUPP, E., *Z. anal. Chem.*, **73**, 51 (1928).
18. RUPP, E. and LEVY, F., *Z. anal. Chem.*, **77**, 1 (1929).

18a. GOTO, H. and KALZITA, J., *J. Chem. Soc. Japan*, **63**, 229 (1942); DUVAL, C., *Traité de Microanalyse Minerale* II, p. 144, Presses Scientifiques Internationales, Paris 1955.
19. KOMAROVSKI, A. S., FILONOVA, V. F. and KORENMAN, I. M., *Zhur. Priklad. Khim.*, **6**, 742 (1933).
20. SINGH, B. and REHMANN, A., *J. Indian Chem. Soc.*, **17**, 169 (1940).
20a. CLARK, J. D. and SMITH, J. R., *Anal. Chemistry*, **33**, 1186 (1961).
21. LEHMKE, O., Doctoral Thesis, Wroclaw 1926; See ref. [23].
21a. MURTHY, A. R. V., *Current Sci.*, **22**, 342 (1953); Ref. *Zhur. Khim.*, **4**, 5822 (1955).
21b. RAO, S. R., *Talanta*, **8**, 746 (1961).
21c. BISHOP, E. and JENNINGS, V. J., *Talanta*, **8**, 697 (1961).
22. FRESNO, C. and AQUADO, A., *Ann. Soc. españ. fis. quim.*, **34**, 818 (1936).
22a. AZIM, S. M. A. and REHMAN, R. A., *Pakistan J. sci. Res.*, **12**, 74 (1960); *Chem. Abstr.*, **55**, 14159 (1960).
22b. KILLHEFFER, J. V. and JUNGERMANN, E., *Anal. Chemistry*, **31**, 581 (1959).
22c. DESHMUKH, G. S. and MURTY, S. V. S. S., *Indian J. Chem.*, **1** 316 (1963).
23. MERCK, E., *Ausgewählte oxydimetrische Titrationen unter Ersatz oder Einsparung von Jod*, Darmstadt, p. 23.
24. ERDEY, L. and KAPLÁR, L., *Z. anal. Chem.*, **162**, 180 (1958).
24a. DESHMUKH, G. S. and ESHWAR, M. C., *J. Sci. Ind. Res. (India)*, **19**, 502 (1960); *J. Electroanal. Chem.*, **3**, A 309 (1962).
25. SINGH, B., SINGH, A. and SINGH, M., *Res. Bull. of the East Panjab Univ.*, N 30, 55 (1953).
26. BERKA, A. and ZÝKA, J., *Českoslov. Farm.*, **5**, 335 (1956).
27. CARLI, B. and AIROLDI, R., *Ann. Chim. applicata*, **27**, 56 (1937).
28. VŠETEČKA, L., *Časopis Českoslov. Lékarnictva*, **15**, 51 (1935).
29. AFANASJEV, B. N., *Zavodskaya Lab.*, **15**, 1271 (1949).
30. SEASE, J. W., LEE, T., HOLZMAN, G., SWIFT, E. H. and NIEMANN, G., *Anal. Chemistry*, **20**, 431 (1948).
31. HINTON, H. D. and MACARA, T., *Analyst*, **49**, 2 (1924).
32. MÜLLER, R., Doctoral Thesis, Wroclaw 1926; See ref. [23].
33. AFANASJEV, B. N., *Zavodskaya Lab.*, **16**, 1011 (1950).
34. AFANASJEV, B. N., *Uspekhi Khim.*, **21**, 69 (1952).
35. SINGH, B., SINGH, A. and KAPUR, S. R., *Res. Bull. of the East Panjab University*, N 41, 205 (1953).
36. SINGH, B. and SINGH, G., *Analyt. Chim. Acta*, **10**, 81 (1954).
37. SINGH, B. and SINGH, A., *Analyt. Chim. Acta*, **11**, 569 (1954).
38. SINGH, A., *Res. Bull. of the East Panjab University*, N 43, 17 (1954).
39. SINGH, A., *Res. Bull. of the East Panjab University*, N 47, 59 (1954).
40. SINGH, A., *J. Indian Chem. Soc.*, **31**, 327 (1954).
41. SINGH, A., *J. Indian Chem. Soc.*, **31**, 605 (1954).
42. SINGH, A., *J. Indian Chem. Soc.*, **31**, 609 (1954).
43. SINGH, A., *J. Indian Chem. Soc.*, **31**, 647 (1954).
44. SINGH, A., *J. Indian Chem. Soc.*, **31**, 648 (1954).
45. SINGH, A., *J. Indian Chem. Soc.*, **32**, 473 (1955).
46. SINGH, A., *J. Indian Chem. Soc.*, **32**, 544 (1955).
47. PAUL, R. CH. and SINGH, A., *J. Indian Chem. Soc.*, **32**, 599 (1955).
48. PAUL, R. CH. and SINGH, A., *J. Indian Chem. Soc.*, **32**, 694 (1955).
49. SINGH, B. and SOOD, K. CH., *Analyt. Chim. Acta*, **11**, 313 (1954).
50. SINGH, B. and SOOD, K. CH., *Analyt. Chim. Acta*, **11**, 317 (1954).

51. SINGH, B. and SONI, S. K., *Res. Bull. of the East Panjab University*, **N 115**, 341 (1957).
52. BISHOP, E. and JENNINGS, V. J., *Talanta*, **1**, 197 (1958).
53. SOLYMOSI, F. and CSIK, I., *Chemist-Analyst*, **49**, 12 (1960).
54. DRAGULESCU, C. and DRAGOI, I., *Bull. stiint, Techn. Inst. Polithn. Timisoara*, **1**, 271 (1956); *Chem. Zbl.*, 10164 (1958).
55. SINGH, B., SAHOTA, S. S. and SINGH, R. P., *J. Indian Chem. Soc.*, **37**, 392 (1960); *Chem. Abstr.*, **55**, 214 (1961).
56. SHARADA, K. and MURTHY, A. R. V., *Z. anal. Chem.*, **177**, 401 (1960).
57. RAO, V. R. S. and MURTHY, A. R. V., *Talanta*, **4**, 206 (1960).
58. SPACU, P. and TEODORESCU, G., *Z. anal. Chem.*, **174**, 321 (1960).
59. KANKANYAN, A. G. and MELOYAN, P. G., *Izvest. Akad. Nauk. Armen. S.S.R.*, *Khim. Nauk*, **12**, 261 (1959); *Ref. Zhur. Khim.*, **20**, 131 (1960).
60. SPACU, P., OVANESIAN, A. and GAVANESCU, D., *Bul. Inst. Politekn. Bucuresti*, **19**, 183 (1957).
61. BISHOP, E. and JENNINGS, V. J., *Talanta*, **9**, 593 (1962).
62. BISHOP, E. and JENNINGS, V. J., *Talanta*, **9**, 581 (1962).
63. BISHOP, E. and JENNINGS, V. J., *Talanta*, **9**, 603 (1962).
64. BISHOP, E. and JENNINGS, V. J., *Talanta*, **9**, 679 (1962).

CHAPTER 7

BROMINE

THE various direct and indirect determinations with standard bromate solution in acid medium and in the presence of bromide should be regarded as bromometric titrations, because the bromine, liberated in the reaction between bromide and bromate is the active substance. It is, however, usual to call only such determinations bromometric titrations in which elementary bromine in a suitable solution is used as standard solution. When the redox potentials are compared, it is seen that bromine is a weaker oxidant than bromate. The redox potential, corresponding to the reaction

$$Br_2 + 2e^- \rightleftharpoons 2\,Br^-$$

is stated to be + 1·07 V. Because elementary bromine is very reactive, it is more powerful than bromate (V) under identical conditions. Owing to their relatively low stability, bromine solutions have only been used on a limited scale in titrimetric analysis; one makes use of them only in titrations in weakly acid or alkaline solutions.

Most bromometric titrations are carried out by adding an excess of the reagent and back-titrating iodimetrically or with arsenite.

STANDARD SOLUTIONS

Manchot and Oberhauser[1] studied extensively the stability of bromine solutions in various media and their use in titrimetric analysis. They found that aqueous bromine solutions are unstable and therefore not suitable for titrimetric purposes. More stable are solutions of bromine in 20–22 per cent hydrochloric acid, in 1 N KBr solution or in glacial acetic acid. Kaufmann[23] recommends a bromine solution containing bromide and methanol.[2–5] Bromine in glacial acetic acid is used in most cases, because glacial acetic acid is a suitable solvent for many organic substances. For the preparation of a 0·1 N standard solution, dissolve 7·99 g bromine in 1 l. glacial acetic acid.

Standardization

The solution is standardized either iodimetrically or with arsenite. According to Levy[10] a standard arsenite solution is titrated with the bromine solution to be standardized, using indigo carmine as indicator. It is advisable to check the titre daily even if the solution is stored in dark bottles or in a place protected against light.

INDICATOR

In direct titrations, the same method for determining the end-point can be applied as in titrations with bromate, i.e. the destruction of the colour of the indicator used (e.g. methyl orange, indigo carmine, crystal violet or malachite green in aqueous solution) by the first excess of bromine. The potentiometric determination of the equivalence point is however more precise.

REVIEW OF DETERMINATIONS

Amongst the bromometric determinations of inorganic compounds, the titration of *arsenic (III)*[6] has gained some significance in the first instance. It is an advantage of this titration that it can be carried out in acid solution, contrary to the iodimetric titration. The reverse titration may be used for determining the unconsumed amount of bromine and so used for indirect titrations, in which bromine is added in excess. Titrations of arsenic (III) with bromine against methyl orange as indicator have been used also for determining arsenic in organic compounds, especially poison gases.[7] Arsenic (III) compounds can be oxidized quantitatively in glacial acetic acid using also a glacial acetic acid solution of the reagent.[8] The equivalence point is determined potentiometrically.

In the determination of *ammonium salt* an excess of bromine is added (according to Manchot and Oberhauser[9] an aqueous bromine solution containing bromide). The reaction is carried out in a sodium hydrogen carbonate solution. After a few minutes, the solution is acidified and the unconsumed bromine is determined iodimetrically or by titration with arsenite.[10] This method can also be used for microanalytical determination of ammonium salts and has even been applied (in a modified form) to determinations of nitrogen in organic compounds.[11]

There are some more publications about the use of direct or indirect titrations with bromine solution, e.g. the determination of *antimony, tin, iron (II), peroxides, sulphides, sulphites, hypophosphites* and so on, which are however mostly of theoretical interest only. In a monograph edited by Kolthoff and Belcher[12] references are to be found to original publications. The standard bromine solution in glacial acetic acid has proved well in the determination of *antimony (III), mercury (I), iron (II)* and *thallium (I)*.[8] The errors are always higher than ± 1 per cent. Attempts to use bromine solutions for the titrimetric determination of selenium (IV) and iodides were less satisfactory.

As regards the oxidation of organic substances with bromine solutions much attention has been paid to the determination of *formic acid*. The oxidation was originally carried out in hydrogen carbonate solution with an excess of bromine[13,14] and was then applied in a modified form[15,18] adding pyridine to accelerate the reaction. *Glycerol* can be oxidized in neutral solution with an excess of bromine solution.[16] *Hydroquinone* and *ascorbic acid* can be titrated with bromine in glacial acetic acid.[17] Aqueous bromine solutions have been used for the potentiometric investigations of the oxidation of N,N-*dimethyl benzidine*[17a] and N,N,N,N-*tetramethyl naphthidine*.[17b]

Most organic compounds are determined, however, with the aid of substitution or addition based on the action of an excess of bromine and back-titration, or on the direct potentiometric titration with bromine. These determinations are mostly dependent on several factors. The conditions, sometimes complicated, have to be followed precisely. It would, however, exceed the scope of this book, to mention these conditions in detail. The reader is referred to the monograph of Dusinsky and Gruntová.[19] With regard to some recent publications in this field we refer to the original literature.[20–27a] Very recently, titration with hydrazine sulphate has been suggested for back-titrating the excess of reagent.[28]

REFERENCES

1. MANCHOT, W. and OBERHAUSER, F., *Z. anorg. Chem.*, **130**, 161 (1923); **138**, 357 (1924); **139**, 40 (1924).
2. KAUFMANN, H. P., *Z. Unters. Lebensmittel*, **51**, 3 (1926).
3. CRITCHFIELD, F. E., *Anal. Chemistry*, **31**, 1406 (1939).
4. ORLOWSKI, M. and SIMON, J., *Acta biochim. pol.*, **1**, 231 (1954).
5. KAUFMANN, H. P. and ARENDS, W., *Z. anal. Chem.*, **150**, 60 (1956).
6. MANCHOT, W. and OBERHAUSER, F., *Z. anal. Chem.*, **130**, 161 (1923); *Z. anorg. Chem.*, **138**, 357 (1924).
7. NORTHROP, J. P., *Proc. Soc. exp. Biol. Med.*, **67**, 15 (1948).
8. TOMÍČEK, O. and HEYROVSKÝ, A., *Collection (Czech. Chem. Comm.)*, **15**, 997 (1950).
9. MANCHOT, W. and OBERHAUSER, F., *Ber. dt. chem. Ges.*, **57B**, 29 (1924).
10. LEVY, B., *Z. anal. Chem.*, **84**, 98 (1931).
11. TSCHEPELEWETZKY, M. and POSDNIAKOWA, A., *Z. anal. Chem.*, **84**, 106 (1931).
12. KOLTHOFF, I. M., STENGER, V. A., BELCHER, R. and MATSUYAMA, G., *Volumetric Analysis*, III, Titration Methods: Oxidation–Reduction Reactions, Interscience, New York 1957.
13. OBERHAUSER, F. and HEUSINGER, W., *Z. anorg. Chem.*, **160**, 366 (1927).
14. MEULEN, VAN DER, J. H., *Chem. Weekbl.*, **27**, 550 (1930).
15. LONGSTAFF, J. V. and SINGER, K, *Analyst*, **78**, 491 (1953).
16. JUHLIN, O., *Z. anal. Chem.*, **113**, 339 (1938).
17. TOMÍČEK, O. and VALCHA, J., *Collection (Czech. Chem. Comm.)*, **16/17**, 113 (1951/52).
17a. MATRKA, M., *Chem. Průmysl*, **12**, 138 (1962).
17b. MATRKA, M., SÁGNER, Z. and NAVRÁTIL, F., *Chem. Průmysl*, **12**, 178 (1962).
18. BINOUN, L. and PERLMULTES-HEYMANN, B., *Bull. Res. Council, Israel, Sect. A*, **5A**, 52 (1957).

19. DUŠÍNSKÝ, G. and GRUNTOVÁ, Z., *Potentiometrické titrcáie vo farmácii a príbuzných odboroch*, Nakl. SAV, Bratislava (1956).
20. SINGH, B. and SINGH, M., *Res. Bull. Panjab Univ.*, **73**, 73 (1955); *Chem. Abstr.*, **50**, 7667 (1956).
21. KAUFMANN, H. P. and ARENDS, W., *Arch. Pharm.*, **287**, 590 (1954).
22. ORLOWSKI, M. and SIMON, J., *Acta biochim. pol.*, **1**, 231 (1954).
23. SODOMKA, J., *Chem. Průmysl*, **9**, 363 (1959).
24. SCHULEK, E. and BURGER, K., *Magyar Kém. Folyóirat*, **61**, 359 (1955).
25. BUDĚŠÍNSKÝ, B. and VANÍČKOVÁ, E., *Českoslov. Farm.*, **6**, 305 (1957).
26. LUCKA, B. and TABORSKA, H., *Przem. chem.*, **11**, 706 (1955).
27. INGBERMANN, A. K., *Anal. Chemistry*, **30**, 1003 (1958).
27a. HUBER, C. O. and GILBERT, J. M., *Anal. Chemistry*, **34**, 247 (1962).
28. BERKA, A. and ZÝKA, J., *Českoslov. Farm.*, **8**, 17 (1959).

CHAPTER 8

N-BROMOSUCCINIMIDE

N-BROMOSUCCINIMIDE contains unstably bound bromine and is used for brominations and dehydrogenations in organic chemistry.[1] The solution of the reagent has been used for oxidimetric titrations and for titrations based on addition reactions.

$$\begin{array}{c} CH_2-C \\ | \\ CH_2-C \end{array} \!\!\! \begin{array}{c} O \\ \diagdown \\ NBr \\ \diagup \\ O \end{array}$$

In oxidimetric titrations it can be assumed that the actual oxidizing agent is either the hypobromous acid, formed by hydrolysis of N-bromosuccinimide

$$\begin{array}{c} CH_2-C \\ | \\ CH_2-C \end{array} \!\!\! \begin{array}{c} O \\ \diagdown \\ NBr \\ \diagup \\ O \end{array} + H_2O \rightarrow HBrO + \begin{array}{c} CH_2-C \\ | \\ CH_2-C \end{array} \!\!\! \begin{array}{c} O \\ \diagdown \\ NH \\ \diagup \\ O \end{array}$$

or the monovalent positive bromine of the N-bromosuccinimide (the bond between bromine and nitrogen is polarized by the two neighbouring carbonyl groups). In both cases, bromide is formed during the titration and the stoichiometric relations are therefore the same:

$$BrO^- + H_2O + 2e^- \rightarrow Br^- + 2\,OH^-$$

$$\begin{array}{c} CH_2-C \\ | \\ CH_2-C \end{array} \!\!\! \begin{array}{c} O \\ \diagdown \\ NBr \\ \diagup \\ O \end{array} + 2H^+ + 2e^- \rightarrow \begin{array}{c} CH_2-C \\ | \\ CH_2-C \end{array} \!\!\! \begin{array}{c} O \\ \diagdown \\ NH \\ \diagup \\ O \end{array} + HBr$$

The second interpretation is however more likely, especially for explaining the reaction in acid solution. Against the formation of hypobromous acid stands the fact that N-bromosuccinimide can be recrystallized from hot water without hydrolytic decomposition.[2]

STANDARD SOLUTION

For the preparation of the standard solution, dissolve the compound in distilled water, eventually filter off any insoluble residue and bring to the respective volume. The standard solutions are not very stable and after a few hours bromine separates. 0·01 M aqueous solutions maintain their titre for about three days if stored at 4°C.

Standardization

The standardization is carried out by potentiometric titration of a standard arsenite solution.[3]

INDICATOR

The equivalence point is nearly always determined potentiometrically, but in some exceptional cases visually by destruction of the indicator, methyl red. Ascorbic acid is determined in the presence of iodide; the first drop in excess of reagent liberates elementary iodine which forms the well-known blue colour with starch.[7]

REVIEW OF DETERMINATIONS

Table 1 gives a review of the direct oxidimetric determinations.[3]

TABLE 1. SUBSTANCES DETERMINED WITH
N-BROMOSUCCINIMIDE

Substance	Titration medium
As^{3+}	0·05 N NaOH–18% HCl*
Sb^{3+}	12% HCl*
Sn^{2+}	2% HCl
Ti^{3+}	15% HCl
Tl^+	10% H_2SO_4
I^-	13–18% HCl
I_2	15% HCl
SCN^-	$NaHCO_3$
S^{2-}	pH 5–7
Hydroquinone	pH 7, 3% HCl
Quinhydrone	pH 7, 3% HCl
Hydrazine	pH 7, 5% HCl*
Isonicotinyl hydrazide	pH 7, 5% HCl*
Ascorbic acid	0·5% CH_3COOH*
Thiourea	$NaHCO_3$

* Can also be titrated visually.

In the oxidations, the cations listed are transformed into the respective higher valency states:

Iodine and *iodide* to iodine monochloride, *thiocyanate* to cyanate and sulphate, *sulphide* to elementary sulphur, *thiourea* to urea and sulphate. The other substances are oxidized as usual.

The titration of *arsenic (III) salts* has been applied to the determination of Fowler's solution and to the micro-determination in blood and urine;[7a] the titration of *iodides* has been used to control their purity (the determination can be carried out in the presence of numerous anions) and to the determination of iodide in ointments and to the determination of iodine and iodide in Lugol's solution.[5] The determination of *ascorbic acid* is suitable for the determination of the proportion of it in injections, tablets, in blood, urine, fruits,[4] milk[7b] and in plants.[6]

The following compounds have been determined by direct titration with *N*-bromosuccinimide based on addition reactions:

Allyl alcohol, allyl amine, crotonic acid, diallyl, allylisobutyl, allylisopropyl-barbituric acid and *tetrahydro-benzoic acid amide*.[7] To determine the *iodine numbers* of fats and oils, the reagent is added in excess and the unconsumed amount is titrated iodimetrically.[8]

According to recent publications, *N*-bromosuccinimide has been used for the direct visual titration of *arsenite*,[9] *hydroquinone*[10] and *ascorbic acid* in coloured solutions.[11] The titration is carried out in the presence of iodide; the elementary iodine, liberated at the equivalence point, is recognized by starch solution or by extraction with diethylether.[11]

Another recently published communication deals with the micro titration of *hydrazine* and some of its derivatives.[12]

REFERENCES

1. HUDLICKÝ, M., *Preparativní reakce v organické chemii* II, *Halogenace a dehalogenace*, p. 369, NČSAV, Prague 1955.
2. BARAKAT, M. Z. and MOUSSA, G. M., *J. Pharm. Pharmacol.*, 4, 115 (1952).
3. BERKA, A. and ZÝKA, J., *Chem. Listy*, 51, 1823 (1957); *Collection (Czech. Chem. Comm.)*, 23, 402 (1958).
4. BARAKAT, M. Z., WAHAB, M. F. and SADR, M. M., *Anal. Chemistry*, 27, 536 (1955).
5. BERKA, A. and ZÝKA, J., *Českoslov. Farm.*, 6, 110 (1957).
6. SIRANSI, KHAJASI, SAITO, ASO, *Japan. J. Med. Progr.*, 43, 621 (1956); *Ref. Zhur. Khim.* No. 14, 252 (1957).
7. BERKA, A. and ZÝKA, J., *Českoslov. Farm.*, 6, 212 (1957).
7a. BARAKAT, M. Z. and ABDALLA, A., *Egyptian Veterinary Medical Journal*, 7, 159 (1961).
7b. BARAKAT, M. Z. and ABDEL-WAHAB, M. F., *Egyptian Veterinary Medical Journal*, 7, 275 (1961).
8. JOVTSCHEFF, A., *Die Nahrung*, 3, 153 (1959).
9. BARAKAT, M. Z. and ABDALLA, A., *Analyst*, 85, 288 (1960).
10. BARAKAT, M. Z., SHEHAB, S. K. and ABDALLA, A., *J. Amer. Pharm. Assoc. sci. Ed.*, 49, 360 (1960); *Chem. Abstr.*, 54, 17800 (1960).
11. EVERED, D. F., *Analyst*, 85, 515 (1960).
12. BARAKAT, M. Z. and SHAKER, M., *Analyst*, 88, 59 (1963).

CHAPTER 9

IODINE MONOCHLORIDE

AN IODINE monochloride solution is used in analytical chemistry (in titrations with oxidimetric reagents) often as catalyst, and eventually as indicator.

As a titrimetric reagent, it was used first in the indirect determination of iron[1] and arsine.[2] Gengrinovič and co-workers worked out volumetric determinations for many organic substances with iodine monochloride in aqueous solutions, mainly based on substitution and addition reactions (for more details see refs. [3–12]). Within the last years, Číhalík and co-workers pointed out the possibility of using iodine monochloride as an oxidimetric reagent.[13-23]

The direct titration with iodine monochloride can be done in acid as well as in weakly alkaline solutions. Iodine monochloride is not suitable for indirect determinations in more alkaline solutions, because it is transformed to hypoiodite, which is then subject to further valency changes.

In acid solution, the reagent can be reduced as far as iodide, when the real redox potential of the titrated system is lower than $+$ 0·4 V. The iodide thus formed can be oxidized by iodine monochloride to iodine. Two potential jumps can therefore be noticed in potentiometric titrations. The calculated standard redox potential for the system I^-/I^+ is $+$ 0·795 V.[23]

In the titration of other systems, iodine monochloride is reduced to elementary iodine. According to Latimer[24] the standard potential is $+$ 1·06 V for the reaction

$$ICl_2^- + e^- \rightarrow 1/2\ I_2 + 2\ Cl^-$$

Figure 1 shows the relation between the redox potential of iodine monochloride and the hydrochloric acid concentration.[25]

The application of iodine monochloride in titrimetric analysis is similar to that of iodine. The iodine monochloride solution in contrast with an iodine solution is, however, more stable, less costly and more widely applicable.

STANDARD SOLUTION

The standard iodine monochloride solution is best prepared by oxidation of iodide with iodate in hydrochloric acid solution:

$$2\,I^- + IO_3^- + 6\,H^+ \rightarrow 3\,I^+ + 3\,H_2O$$

To prepare a 0·1 M solution, dissolve 11·068 g KI in 50 ml of water, add a solution of 7·134 g KIO_3 in 250 ml water and, immediately, 200 ml concentrated hydrochloric acid (the hydrochloric acid must be free from iron (III) and chlorine); then make the solution up to 1000 ml. The potential of this solution is 750–800 mV, measured against a saturated calomel electrode. If the potential should be too low, add 0·1 M KIO_3 solution dropwise, until a noticeable potential jump occurs; should the potential be too high, add some drops of 0·1 M KI solution, until a distinct

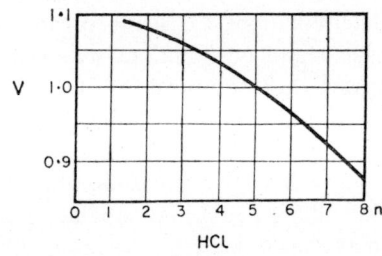

Fig. 1

regression in potential takes place; then bring the potential to the value mentioned above by adding a drop of iodate. If the preparation of the solution is controlled visually, chloroform has to be added to the solution. If a violet colour forms on shaking, add iodate dropwise until the chloroform layer decolorizes. In the opposite case, add iodide until a violet colour appears and then just decolorize by adding iodate.

A 0·5–0·0005 M standard solution is practically stable if it is sufficiently acid (at least 100 ml of concentrated HCl in 1000 ml).

Standardization

The solution is standardized with known solutions of potassium iodide, arsenic (III) oxide, potassium hexacyanoferrate (II) or hydrazine sulphate (see below).

INDICATOR

All oxidimetric determinations with iodine monochloride can be carried out potentiometrically using a platinum indicator electrode and calomel reference electrode (in the reduction of the iodine monochloride to iodide, an additional potential change due to oxidation of the latter to iodine can be detected). The titration end-point can be determined visually where the reagent is reduced to iodide. The first drop of reagent in excess liberates elementary iodine from the iodide present, which forms the well-known blue colour with starch solution or colours a chloroform layer violet. In the determination of some substances, an intermediate separation of iodine occurs before the equivalence point, so that the end-point is screened. In such cases, mercury (II) chloride is added, which complexes the iodide formed during the titration. The oxidation power of the standard reagent is thus enhanced, the reaction proceeds more rapidly at the equivalence point and the visual end-point indication is sharp. In some titrations, Variamine Blue is used as redox indicator.

A very recent publication points to the possibility of indicating the end-point amperometrically.[25a,b]

REVIEW OF DETERMINATIONS

Arsenic and antimony. Arsenite is oxidized with an excess of iodine monochloride to arsenate (V). The iodine thus formed is titrated with iodate in the presence of cyanide until the blue colour (due to added starch) disappears.[2] The direct titration of arsenite[14] in alkaline solution ($NaHCO_3$ $Na_2B_4O_7$, Na_2HPO_4 or Na_3PO_4) with iodine monochloride can be carried out potentiometrically as well as visually. The following reactions take place:

$$I^+ + 2 HCO_3^- \rightarrow IO^- + 2 CO_2 + H_2O$$
$$AsO_3^{3-} + IO^- \rightarrow AsO_4^{3-} + I^-$$

In the visual titration, the equivalence point is indicated by the well-known blue colour of starch solution with the iodine liberated from iodide by the excess of reagent.

In the potentiometric titration, two potential changes can be noticed in neutral and acid solution; the first one corresponds to the oxidation of As^{3+} to As^{5+}, the second one to the oxidation of the I^- formed to I_2. The best results are achieved if the pH is about 7 at the beginning of the titration. A potentiometric micro-titration can be carried out in a sodium hydrogen carbonate solution. As little as 1 μg As can be determined in 5 ml.[22]

The determination of antimony is carried out under the same conditions as that of arsenic, but in the presence of tartaric acid.

The end-point of the titration of arsenic and antimony can also be easily recognized by the violet colour of the liberated iodine extracted into a chloroform layer, if mercury (II) chloride is added to the 1–2·5 N hydrochloric acid solution for making this visual indication more distinct.[26] The same end-point indication can be applied to titrations in solutions buffered with sodium acetate (pH 6·5–7·5).[27]

Both systems can be titrated amperometrically at + 0·2 to − 0·2 V (versus a standard calomel reference electrode) using a rotating platinum indicator electrode.[25a]

Tin. In the potentiometric titration of tin (II) salts in weakly acid solution with iodine monochloride, two changes of potential are achieved ($Sn^{2+} \rightarrow Sn^{4+}$ and $2 I^- \rightarrow I_2$).

In alkaline solution, in which the intermediately precipitated tin (II) hydroxide is just soluble, a potential change can be seen corresponding to the oxidation to stannate (IV) (the iodine monochloride is reduced to iodide). The titration is carried out in an inert atmosphere. For the amperometric titration a rotating platinum indicator electrode was used.[25b]

Iron. The indirect determination of iron is based on the oxidation of iron (II) with an excess of iodine monochloride. The iodine formed during the reaction is titrated in strongly hydrochloric acid solution with standard iodate solution.

The direct potentiometric titration of iron (II) ions[21] is carried out at pH 0·5–1 in the presence of EDTA. The reaction proceeds in two steps:

First Fe^{2+} is oxidized to iron (III) EDTA and the iodine monochloride is reduced to iodide

$$2\,Fe^{2+} + 2\,H_2Y^{2-} + I^+ \rightarrow 2\,FeY^- + 4\,H^+ + I^- \qquad (1)$$

Thereupon the iodide formed is oxidized to iodine. The complete procedure can be expressed by the following equation:

$$2\,Fe^{2+} + 2\,H_2Y^{2-} + 2\,I^+ \rightarrow 2\,FeY^- + 4\,H^+ + I_2.$$

From the graph, two changes of the potential sequence result; one of them corresponds to reaction (1) and is much more distinct than the second, which corresponds to the oxidation of iodide to iodine. A small excess of EDTA is enough.

This reaction can also be used for the determination of EDTA; an excess of iron (II) is added to the EDTA and the amount of complex bound divalent iron (equivalent to the EDTA present) is titrated with iodine monochloride.

Titanium. The potentiometric titration of titanium (III)[21] proceeds in 1–3 M HCl or in 3 M H_2SO_4 in an inert atmosphere according to the equation:

$$2\,Ti^{3+} + I^+ \rightarrow 2\,Ti^{4+} + I^-$$

The end of the further titration (oxidation of the iodide) for which no longer an inert atmosphere is necessary, gives a less distinct potential change.

If to the titanium (III) solution, ammonium fluoride is added, which forms a stable complex with titanium (IV) ions, the first potential difference ($Ti^{3+} \rightarrow Ti^{4+}$) amounts to 1000 mV for 0·02 ml of standard solution. The determination can be carried out over a wide range of acid concentration.

If the solution does not contain complex forming substances, the determination of titanium (III) ions can also be carried out in the presence of iron (II) ions. In solutions containing EDTA and fluoride (pH 2−3), Ti^{3+} and Fe^{2+} can vice versa be determined side by side. The potentiometric curve shows three potential changes; the first one corresponds to the oxidation $Ti^{3+} \rightarrow Ti^{4+}$ (for 0·02 ml standard solution 180 mV), the second one to the oxidation $Fe^{2+} \rightarrow Fe^{3+}$ (340 mV) and the third one to the oxidation $2 I^- \rightarrow I_2$ (40 mV).

Chromium. The potentiometric titration of chromium (II) salts with iodine monochloride[23] proceeds in two steps. The first reaction follows the equation

$$2 Cr^{2+} + I^+ \rightarrow 2 Cr^{3+} + I^-$$

The iodide is then oxidized to iodine. The titration is carried out in an inert atmosphere, best in acid solution, containing 2–5 ml concentrated H_2SO_4 in 100 ml.

Iodide. Iodide is titrated potentiometrically with iodine monochloride in neutral or weakly acid solution (0·05 N HCl).[13] It is necessary to dilute the solution to be titrated sufficiently with water, because standard iodine monochloride solutions contain considerable amounts of hydrochloric acid. The titration of iodide is not affected by chlorides up to the ratio 1:100 and by bromides up to the ratio 1:10. 0·1 μg iodide can be determined in 5 ml of solution with an error of ± 0·75 per cent.[22]

Cyanide. Cyanides can be titrated with iodine monochloride either potentiometrically or visually to the blue colour with starch solution, best in a sodium hydrogen carbonate medium.[17] The reaction proceeds according to the equation:

$$CN^- + I^+ \rightleftharpoons ICN$$

The potentiometric titration is not even affected by large amounts of chloride or bromide, whereas iodide interferes already in tenfold excess. In the presence of thiocyanate, both ions are titrated simultaneously.

Thiocyanate. Thiocyanate can be determined by direct potentiometric titration with iodine monochloride in weakly acid or weakly alkaline solutions.[17] In weakly acid and neutral medium, the reaction proceeds following the equation

$$SCN^- + 7 I^+ + 4 H_2O \rightarrow SO_4^{2-} + ICN + 3 I_2 + 8 H^+$$

in weakly alkaline solution ($NaHCO_3$) according to

$$SCN^- + 4 I^+ + 4 OH^- \rightarrow SO_4^{2-} + ICN + 3 I^- + 4 H^+$$

In sodium hydrogen carbonate solution, the potential change at the equivalence point is more distinct. Contrary to bromides and iodides, chlorides do not interfere with this determination.

Hexacyanoferrate (II). The potentiometric titration of hexacyanoferrate (II) proceeds in weakly acid and weakly alkaline solution following the equation:

$$2[Fe(CN)_6]^{4-} + 2 I^+ \rightarrow 2[Fe(CN)_6]^{3-} + I_2$$

A 0·001–0·005 N hexacyanoferrate (II) solution is best suited for the titration. At higher concentrations, elementary iodine can separate which has an unfavourable influence on the titration.[13] As little as 2 μg hexacyanoferrate (II) can be determined in a volume of 3–5 ml, preferably in sodium hydrogen carbonate solution.[22] In solutions buffered with sodium acetate to pH 6·5–7·5, hexacyanoferrate (II) can be titrated visually with iodine monochloride at 40°C (reduction to iodide), to the coloration of the chloroform layer by the iodine formed.[27]

Thiosulphate. Thiosulphate is determined indirectly. About 1 g of KI is added to a known amount of iodine monochloride. The iodine formed is titrated with the thiosulphate solution either visually using starch as indicator or potentiometrically.[13]

Sulphite. Sulphite is oxidized with iodine monochloride in neutral or weakly acid solution according to the following equation:

$$SO_3^{2-} + 2I^+ + H_2O \rightarrow SO_4^{2-} + I_2 + 2H^+$$

A known volume of a 0·1 N ICl solution is diluted with water and the sulphite solution to be determined is added, using a pipette (the tip of the pipette just touching the surface of the liquid). The excess of iodine monochloride is back-titrated with arsenite potentiometrically.

In buffered solutions, sulphite can be titrated directly;[27] amperometric indication of the end-point is recommended in recent communications.[25a,b]

Ascorbic acid. The determination of ascorbic acid, proceeding according to the equation

$$\underset{\underset{O\ \ OH\ OH}{\shortparallel\ \ |\ \ \ |}}{C-C=C-CH-CH-CH_2} + I^+ \rightarrow \underset{\underset{O\ \ O\ \ O}{\shortparallel\ \ \shortparallel\ \ \shortparallel}}{C-C-C-CH-CH-CH_2}$$
$$+ I^- + 2H^+$$

(with OH OH on the right side CH groups)

can be done in aqueous solution by a direct potentiometric titration with iodine monochloride, without specially adjusting the medium. In the course of the further titration, the iodide formed is oxidized quantitatively to iodine.

Under the same conditions the titration can also be done with the "dead-stop" method at a polarization potential of 20–50 mV, with practically the same accuracy. The equivalence point is indicated by a remaining deflection of the pointer of the galvanometer.[23]

Ascorbic acid can be titrated amperometrically in acid solution with a rotating platinum electrode at a potential of $+0.2$ to -0.2 V. The acidity should not exceed 1·5 M.[25a]

According to the above mentioned equation, ascorbic acid can also be titrated visually, using starch or Variamine Blue as indicator.[23] In the presence of an excess of $HgCl_2$ it can also be titrated to the coloration of a chloroform layer by liberated iodine[26] in 1·0–2·5 N HCl solution.

Hydrazine. Hydrazine is oxidized by iodine monochloride to elementary nitrogen. It can be titrated in acid as well as in alkaline, best in sodium acetate buffered solution, potentiometrically or also visually to the colour of the chloroform layer,[27] or using starch solution[18] (eventually in the presence of $HgCl_2$[26] in 1·0–2·5 N NCl solution).

Phenylhydrazine. The direct potentiometric titration of phenylhydrazine with iodine monochloride[18] proceeds in acid solution according to:

$$C_6H_5NHNH_2 + 3\,I^+ \to N_2 + C_6H_5I + 3\,H^+ + I_2$$

and in alkaline solution (best in a sodium acetate buffered solution) according to:

$$C_6H_5NHNH_2 + I^+ \to N_2 + C_6H_6 + I^- + 2\,H^+$$

Phenylhydrazine can also be titrated in the presence of mercury (II) chloride[26] in 1·0–2·5 N HCl solution visually to the colour of a chloroform layer.

Hydroxylamine. Hydroxylamine is oxidized by iodine monochloride according to the equation:

$$NH_2OH + 4\,I^+ + H_2O \to NO_2^- + 2\,I_2 + 5\,H^+$$

The sample solution is added to an excess of iodine monochloride solution; the unconsumed excess is determined by titrating it with standard hydrazine sulphate solution.[18]

Hydroquinone. The potentiometric titration of hydroquinone with iodine monochloride proceeds in weakly acid medium according to equation (1), and in weakly alkaline solution according to equation (2):

$$\text{HO-C}_6\text{H}_4\text{-OH} + 2\,I^+ \to \text{O=C}_6\text{H}_4\text{=O} + I_2 + 2\,H^+ \qquad (1)$$

$$\text{HO-C}_6\text{H}_4\text{-OH} + I^+ \to \text{O=C}_6\text{H}_4\text{=O} + I^- + 2\,H^+ \qquad (2)$$

The titration end-point can be recognized visually with starch[15] or, in sodium acetate buffered solutions, by the change of colour of a chloroform layer from yellow to orange.[27] Hydroquinone can be titrated potentiometrically in glacial acetic acid in the presence of alkali acetate.[28] The standard iodine monochloride solution is, however, unstable in glacial acetic acid. This titration is therefore of little significance.

Metol. The potentiometric titration of metol (*N*-methyl-*p*-aminophenol) with iodine monochloride proceeds in weakly acid and neutral solutions according to the equation:[15]

$$\underset{\underset{NHCH_3}{|}}{\underset{|}{\overset{OH}{\bigcirc}}} + 2\,I^+ \rightarrow \underset{\underset{NCH_3}{\|}}{\overset{O}{\underset{\|}{\bigcirc}}} + I_2 + 2\,H^+$$

Phenylenediamine. A study of the reaction between iodine monochloride and *o*-, *m*- and *p*-phenylenediamine, has shown that the phenylenediamines are oxidized if they are present in basic form, whereas the hydrochlorides (with the exception of that of *p*-phenylenediamine) are subject to iodine substitution. Only the hydrochloride is suitable for the practical determination of *o*-phenylenediamine; of the *m*- and *p*-derivatives, the hydrochloride as well as the respective free base can be determined.[15]

Thiourea. Thiourea can be titrated directly with iodine monochloride at pH 4–7 either potentiometrically or visually, using starch solution. The colour is due to the iodine formed by oxidation of iodide by the first excess drop of reagent. Thiourea is oxidized to the corresponding disulphite.[16] A similar method is described for alkyl- and aryl-derivatives of thiourea.[28a]

Thiosemicarbazide. The potentiometric determination of thiosemicarbazide with iodine monochloride proceeds as follows:

$$\underset{\underset{NHNH_2}{\diagdown}}{\overset{\overset{NH}{\diagup}}{C}}\!\!-\!\!SH + 10\,I^+ + 4\,H_2O \rightarrow N_2 + HCN + H_2SO_4 + 5\,I_2 + 10\,H^+$$

At the beginning, the solution has to be weakly acid or weakly alkaline ($NaHCO_3$).

Thiosemicarbazide can be determined visually with iodine monochloride in 1·0–2·5 N HCl in the presence of $HgCl_2$. The end-point is indicated by the colour of the chloroform layer.[26]

Mercaptobenzothiazol. Iodine monochloride oxidizes mercaptobenzothiazol[20] according to the equation:

$$2\,C_7H_4SNSH + I^+ \rightarrow C_7H_4SNS\!\!-\!\!SNSH_4C_7 + I^- + 2\,H^+.$$

Two changes of potential appear in the potentiometric titration; the first one corresponds to the above mentioned reaction and the second one to the oxidation of the iodide formed to elementary iodine. Starch is used as indicator in the visual titration. The determination proceeds best in aqueous neutral or weakly acid solutions (in which no mercaptobenzothiazol is precipitated).

Bismuthon. Bismuthon (potassium salt of 5-mercapto-3-phenyl-2-thio-1, 3,4-thiodiazolone-(2)) is oxidized by iodine monochloride to the respective disulphide, with simultaneous reduction of the iodine monochloride to iodide.[19] In the potentiometric titration, the pH should not be lower than 4–5 at the beginning.

In the *indirect determination of heavy metals,* the cation is precipitated by an excess of bismuthon solution and the unconsumed amount is titrated with iodine monochloride. It is not necessary to remove the precipitate, because the bismuthon, bound in the precipitate, is not oxidized during the titration. Silver and copper have been thus determined.

Dimercaptopropanol. The potentiometric titration of dimercaptopropanol proceeds in a solution saturated with sodium acetate according to the equation:

$$2\ C_3H_8OS_2 + 2\ I^+ \rightarrow C_6H_{12}O_2S_4 + 2\ I^- + 4\ H^+$$

Methionine. In the direct potentiometric titration, methionine is oxidized in hydrochloric acid solution to the respective sulphoxide.[29]

Some more substances containing thio- or mercapto-groups can be determined applying an excess of reagent, followed by iodimetric back-titration.[29a,b] The determination of methylene blue is based on the same principle.[29c]

Aminoguanodinium chloride, 4-phenylsemicarbazone chloride, semicarbazide chloride. These substances can be titrated with iodine monochloride in 1–2·5 N hydrochloric acid solution in the presence of mercury (II) chloride.[26] The end-point is recognized from the colour of the chloroform layer with elementary iodine. Semicarbazide is titrated at 30°C.

In a recent communication, the oxidation of *arsenic (III), antimony (III), iron (II), mercury (I)* salts and of *sulphite* (in glacial acetic acid medium) is described.[29d]

Iodine monochloride is used to a great extent in the analysis of organic substances, on the basis of *substitution* and *addition* reactions. These titrations are done either directly or indirectly by determining the excess reagent. Recently there have been thus determined amongst others: *Antipyrine,*[3,28] *analgine,*[36] *amidopyrine,*[28] some *phenols,*[3,28,30] *oxine,*[4,20,25a] *aniline, sphaerophysine,*[12] *antifebrine, novocaine, anaesthesine,* various *sulphonamides,*[5–7] *fat,*[8,33] *phenolphthaleine,*[9] *oleic acid,*[28] *anthranilic acid,*[20] *methyl and ethylsalicylate,*[11] *rivanol,*[34] *salicylaldoxime,*[23] *enoles,*[31] *olefines,*[32] *derivatives of barbituric acid,*[35] *tibon* (*thiosemicarbazone of* p-acetamidobenzaldehyde),[35a] *rutine,*[35b] etc.

Besides iodine monochloride, iodine monobromide and bromine chloride are used in similar manner. The action of ICl and ICl$_3$ has been compared.[36a,b,c,d] In some of the newest publications, the addition and substitution reactions of the interhalogen compounds are discussed from the theoretical viewpoint.[37-44]

REFERENCES

1. HEISING, G. B., J. Amer. chem. Soc., 50, 1687 (1928).
2. KUBINA, H., Z. anal. Chem., 76, 39 (1929).
3. GENGRINOVICH, A. I., Farm. Zhur., 13, 2, 27 (1940); Nr. 3, 15; Nr. 4, 23; Nr. 14, 2, 19 (1941).
4. GENGRINOVICH, A. I., Farmatsyja, 9, 6, 5 (1946); 10, 2, 23 (1947).
5. GENGRINOVICH, A. I. and KADYROV, J. K., Aptechnoe Delo, 1, 1, 46 (1952).
6. GENGRINOVICH, A. I. and IBADOV, A. J., Aptechnoe Delo, 1, 3, 18 (1952).
7. GENGRINOVICH, A. I. and BARON, M. S., Aptechnoe Delo, 1, 4, 27 (1952).
8. GENGRINOVICH, A. I. and JUDEVICH, E. A., Aptechnoe Delo, 1, 5, 17 (1952).
9. GENGRINOVICH, A. I. and MANSURKHANOVA, I., Aptechnoe Delo, 3, 6, 9 (1954).
10. GENGRINOVICH, A. I., KAGAN, F. E. and FIALKOV, J., Trudy Komisii Anal. Khim. Akad. Nauk S.S.S.R., 5, 8, 237 (1954).
11. GENGRINOVICH, A. I. and KADYROV, J. K., Aptechnoe Delo, 6, 2, 68 (1957).
12. GENGRINOVICH, A. I. and IBADOV, A. J., Aptechnoe Delo, 7, 67 (1958).
13. ČÍHALÍK, J. and VAVREJNOVÁ, D., Chem. Listy, 49, 693 (1955); Collection (Czech. Chem. Comm.), 20, 1059 (1955).
14. ČÍHALÍK, J., Chem. Listy, 49, 1167 (1955); Collection (Czech. Chem. Comm.), 21, 181 (1956).
15. ČÍHALÍK, J. and VAVREJNOVÁ, D., Chem. Listy, 49, 1176 (1955); Collection (Czech. Chem. Comm.), 21, 192 (1956).
16. ČÍHALÍK, J. and RŮŽIČKA, J., Chem. Listy, 49, 1731 (1955); Collection (Czech. Chem. Comm.), 21, 262 (1956).
17. ČÍHALÍK, J. and TEREBOVÁ, K., Chem. Listy, 50, 1761 (1956); Collection (Czech. Chem. Comm.), 22, 748 (1957).
18. ČÍHALÍK, J. and TEREBOVÁ, K., Chem. Listy, 50, 1768 (1956); Collection (Czech. Chem. Comm.), 22, 756 (1957).
19. ČÍHALÍK, J. and RŮŽIČKA, J., Chem. Listy, 51, 264 (1957); Collection (Czech. Chem. Comm.), 22, 764 (1957).
20. ČÍHALÍK, J. and TEREBOVÁ, K., Chem. Listy, 51, 272 (1957); Collection (Czech. Chem. Comm.), 23, 110 (1958).
21. ČÍHALÍK, J., Chem. Listy, 52, 1075 (1958); Collection (Czech Chem. Comm.), 24, 708 (1959).
22. ČÍHALÍK, J. and VORÁČEK, J., Chem. Listy, 52, 1269 (1958); Collection (Czech. Chem. Comm.), 24, 1643 (1959).
23. ČÍHALÍK, J., Doctoral Thesis, University of Chem. Technology, Prague 1957.
24. LATIMER, W. M., Oxidation Potentials, Prentice-Hall, New York 1938.
25. WILLARD, H. H. and FURMAN, N. H., Elementary Quantitative Analysis, Van Nostrand, New York 1940.
25a. ZHDANOV, A. K., KHADEEV, V. A., KUBRAKOVA, A. I. and BONDARENKO, N. V., Uzbek. Khim. Zhur. Nr. 2, 44 (1961); Chem. Abstr., 56, 914 (1962).
25b. GENGRINOVICH, A. I., KORNEVA, L. E., MURTAZAEV, A. M., Trudy Taschkk. Pharm. Inst. 2, 355 (1960); Zhur. Khim. 7 D 29 (1961).
26. SINGH, B., KASHYAP, G. P. and SAHOTA, S. S., Z. anal. Chem., 162, 357 (1958).
27. SINGH, B. and KASHYAP, G. P., Z. anal. Chem., 163, 338 (1958).
28. TOMÍČEK, O. and VALCHA, J., Chem. Listy, 44, 283 (1950).

28a. SINGH, B., VERMA, B. CH. and SARAN, M. S., *J. Indian Chem. Soc.*, **39**, 211 (1952).
29. JANČÍK, F., BUBEN, F. and KÖRBL, J., *Českoslov. Farm.*, **5**, 515 (1956).
29a. RAPAPORT, L. I., *Aptechnoe Delo*, **8**, 2, 63 (1959).
29b. RAPAPORT, L. I., RAZNATOVSKA, V. F., *Farm. Zhur. (Kiev)*, **15**, 22 (1960); cit.: *Chem. Abstr.*, **55**, 9793 (1961).
29c. SUPRUN, P. P., *Med. Prom. S.S.S.R.*, **12**, 8, 38 (1958); *Chem. Abstr.*, **55** 12923 (1959).
29d. PICCARDI, G., ALLINI, P., *Anal. Chim. Acta*, **29**, 107 (1963).
30. KAGAN, F. E. and SHAKH, C. I., *Ukrain. Khim. Zhur.*, **23**, 537 (1957).
31. GERO, A., *Anal. Chemistry*, **26**, 609 (1954); *J. organ. Chem.*, **19**, 469 (1954).
32. LEE, T. S., KOLTHOFF, I. M. and JOHNSON, E. H., *Anal. Chemistry*, **22**, 995 (1950).
33. SUPRUN, P. P., *Aptechnoe Delo*, **7**, 3, 48 (1958).
34. SUPRUN, P. P., *Med. Prom. S.S.S.R.*, **11**, 2, 49 (1957).
35. RAPAPORT, L. I., *Zhur. Anal. Khim.*, **12**, 415 (1957).
35a. RAPAPORT, L. I., *Farm. Zhur. (Kiev)*, **14**, 2, 25 (1959); *Chem. Zbl.*, **131**, 1600 (1960).
35b. SUPRUN, P. P., *Med. Prom. S.S.S.R.*, **15**, 3, 46 (1961); *Chem. Abstr.*, **55**, 23484 (1961).
36. RAPAPORT, L. I. and SHVARZBURD, M. M., *Aptechnoe Delo*, **3**, 47 (1954).
36a. SINGH, B. and KASHYAP, G. P., *Indian J. appl. Chem.*, **22**, 15 (1959); *Chem. Abstr.*, **7**, 855 (1960).
36b. KAGAN, F. E., *Farm. Zhur. (Kiev)*, **15**, 13 (1960); *Chem. Abstr.*, **55**, 213 (1961).
36c. SHAKH, C. I. and KAGAN, F. E., *Farm. Zhur. (Kiev)*, **15**, 18 (1960); *Chem. Abstr.*, **55**, 21484 (1961).
36d. SCHULEK, E. and LADÁNYI, L., *Talanta*, **9**, 727 (1962).
37. SCHULEK, E. and PUNGOR, E., *Analyt. Chim. Acta*, **7**, 402 (1952).
38. SCHULEK, E. and BURGER, K., *Magyar Kém. Folyóirat*, **64**, 241 (1958).
39. SCHULEK, E. and BURGER, K., *Talanta*, **1**, 147, 219, 224, 344 (1958).
40. SCHULEK, E. and BURGER, K., *Talanta*, **2**, 280, 322 (1959).
41. PUNGOR, E., BURGER, K. and SCHULEK, E., *Magyar Kém. Folyóirat*, **65**, 301 (1959).
42. SCHULEK, E. and BURGER, K., *Acta chim. Acad. Sci. Hung.*, **19**, 453 (1959).
43. SCHULEK, E., PUNGOR, E. and BURGER, K., *Chem. Zvesti*, **13**, 669 (1959).
44. PUNGOR, E., *Chem. Zvesti*, **13**, 680 (1959).

CHAPTER 10

PERIODIC ACID AND ITS SALTS

PERIODIC acid and its salts, known as strong oxidants, are often used in preparative organic chemistry. Malaprade[1] used periodate for the first time as a titrimetric reagent in the indirect determination of some organic compounds. For the direct titration of inorganic substances, Syrokomskii and co-workers[2-5] used periodate for the first time.

The true redox potential of the system

$$IO_4^- + 2e^- + 2H^+ \rightarrow IO_3^- + H_2O$$

in 1 N H_2SO_4 is + 1·375 V. Figure 2 shows the dependence of the potential changes on the pH.[2]

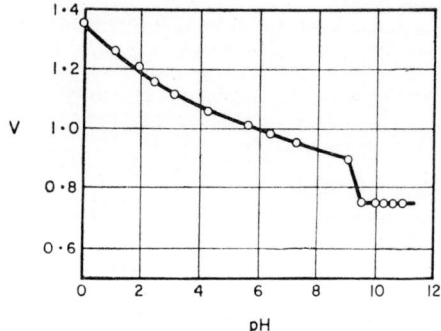

FIG. 2

In titrations with potassium periodate, the periodate is not solely reduced to iodate. It can also be reduced to the I^+ ion, to elementary iodine or to the I^- ion, following the equations:

$$IO_4^- + 6e^- + 8H^+ \rightarrow I^+ + 4H_2O$$
$$2IO_4^- + 16H^+ + 14e^- \rightarrow I_2 + 8H_2O$$
$$IO_4^- + 8H^+ + 8e^- \rightarrow I^- + 4H_2O$$

The standard potential of the system IO_4^-/I^- is $+ 1.4$ V.[6]

The following review of determinations is arranged according to the respective reduction step of the periodate. At the end of the chapter examples are given for the determination of organic substances (in connection with the respective reaction mechanisms) in which the carbon–carbon bond is decomposed. (Malaprade reaction, etc.)

STANDARD SOLUTION

Aqueous solutions (suitably acidified) of ortho-periodic acid, of the tertiary sodium salt of ortho-periodic acid and of the potassium or sodium salt of meta-periodic acid are generally used as standard solutions. The solutions are prepared by dissolving the commercially available substances. If the solution is not clear enough, it is set aside for 24 hr and then filtered. The solutions thus obtained are stable at normal temperature; at boiling temperature they are less stable.

Standardization

The standardization is done by various modifications of the iodimetric method.[7-15]

In buffered alkaline solutions periodate is reduced with iodide to iodate:

$$IO_4^- + 2I^- + H_2O \rightarrow IO_3^- + I_2 + 2OH^-$$

The iodine formed is thereupon titrated with arsenite.[7,13]

Procedure. Add 20 ml of saturated borax solution containing 0·5 g boric acid to a periodate solution. Then add 2 g of potassium iodide. Titrate the solution with a 0·1 N arsenite solution.

In buffered alkaline solutions periodate can also be quantitatively reduced to iodate with an excess of standard arsenite solution in the presence of potassium iodide as a catalyst. The unconsumed arsenite is then determined with a known iodine solution. In this determination consequently two standard solutions are used.[10,14,15]

Procedure. Add 5–10 ml of a saturated sodium hydrogen carbonate solution to 5–25 ml of a periodate solution (containing about 0·5 millimoles of periodate) which has been neutralized against phenolphthalein; then add 15 ml of 0·1 N arsenite solution and 1 ml of 20 per cent potassium iodide solution, set aside for 15 min at room temperature and titrate the excess of arsenite with a 0·1 N iodine solution.

In acid solution, iodide reduces periodate to iodine

$$IO_4 + 7I^- + 8H^+ \rightarrow 4I_2 + 4H_2O$$

The iodine thus formed is titrated with thiosulphate or arsenite.

INDICATOR

Because periodate as well as iodate ions are colourless, redox-indicators can be used for the visual detection of the end-point in direct titrations in which periodate is reduced to iodate.[16] Diphenylamine has proved best; it was applied in the determination of divalent iron, of tervalent arsenic and of tervalent antimony (under the catalytic action of iron (II) salts).[3] In the indirect determination of larger amounts of tellurium, phenylanthranilic acid is more suitable as an indicator than diphenylamine.[4]

In determinations in which periodate is reduced to the I^+ ion, some substances can be titrated to the colour change of a chloroform or carbon tetrachloride layer.[17,18] The colour, owing to the intermediately formed iodine, disappears at the equivalence point, because the iodine is quantitatively oxidized to the I^+ ion. The decolorization of the blue iodine–starch solution can also be used for the detection of the end-point.[19]

Vice versa, in cases where periodate is reduced to iodide, the blue colour of the starch solution with iodine, formed from the iodide by oxidation with the first excess drop of reagent, serves to detect the equivalence point.[20]

The use of p-ethoxy chrysoidine, safranine-T and neutral red in the determination of vanadium (II) ions has no advantage over the above described indication method.

In many cases, the end-point is determined potentiometrically.

REVIEW OF DETERMINATIONS

Titrations in which periodate is reduced to iodate

Arsenic and antimony.[3]   Arsenic and antimony (III) salts can be titrated in weakly acid solutions using diphenylamine as indicator; 2 drops of a 0·02 N solution of Mohr's salt are added as catalyst.

Iron.   The titration of iron (II) salts is carried out visually against diphenylamine in 1 N sulphuric acid solution in the presence of phosphoric acid. It is titrated to the permanent blue colour. This method has been used for the determination of iron in ores.[3] In other publications[21] the titration end-point is detected by spot reactions with ferricyanide.

The reaction between periodate and divalent iron can also be used for the determination of *zinc* and *cadmium* which are both precipitated as sparingly soluble periodates. The excess of periodate is titrated with a solution of Mohr's salt.[5] It is also possible to dissolve the precipitate and to determine the dissolved periodate with a solution of Mohr's salt.

Tellurium.   The oxidation of tellurium (IV) compounds with periodate proceeds at room temperature and is almost independent of the acidity of the solution.[4] The titration is controlled potentiometrically using a platinum electrode as indicator electrode and a mercury (I) sulphate electrode as reference electrode.

In the visual determination of the equivalence point tellurous acid is oxidized by an excess of periodate. The excess is then back-titrated with standard iron (II) solution in the presence of phosphoric acid in 1 N H_2SO_4 against diphenylamine, or in 5–6 N H_2SO_4 against ferroin or phenylanthranilic acid as indicator. With higher amounts of tellurium the titration against phenylanthranilic acid is more accurate than that against diphenylamine. The determination is not affected by the presence of selenium. This method has been applied to the determination of tellurium in technical tellurium.

Cerium. The determination of tervalent cerium by direct potentiometric titration proceeds best in a 2 M carbonate buffered solution at pH 9·8.[39]

Titrations, in which periodate is reduced to positive monovalent iodine

Periodate is reduced by some substances to positive monovalent iodine. It is always titrated in hydrochloric acid solution, when *iodine monochloride* is formed (A). In some determinations an excess of cyanide or bromide is added to the solution to be titrated, in which case the I $^+$ ion is bound as *iodine cyanide* (B) or as *iodine monobromide* (C). In a number of cases, the velocity as well as the quantitative course of the reaction depend on the formation of these compounds.

(A) *Hydrazine sulphate, arsenic (III), antimony (III), tin (II), iron (II) and mercury (I) salts, iodide and thiocyanate* are titrated directly in 4–7 N hydrochloric acid solution; in the determination of arsenic and antimony the solution should be at the utmost 5 N with regard to hydrochloric acid. The cations listed above are transformed to the respective highest valencies. *Iodide* is oxidized to I $^+$, *thiocyanate* to sulphate and cyanate and *hydrazine* to nitrogen. The end-point of the titration is detected by the colour change of the chloroform layer from violet (owing to intermediately liberated iodine) to pale yellow (iodine monochloride)[17] or potentiometrically.[23] Several determinations of organic hydrazine derivatives are based on the same principle.[23a,b]

The titrations mentioned have also been used for some indirect determinations.[22] On the addition of iodide to *manganese (IV) oxide* and *lead (IV) oxide* an equivalent amount of iodine is liberated which is then titrated with periodate.

Formates react with an excess of mercury (II) chloride forming an equivalent amount of mercury (I) chloride which is determined volumetrically with periodate.

Sulphides are precipitated with arsenite as arsenic (III) sulphide. The unconsumed arsenite is titrated with periodate.

Copper, mercury, zinc and cobalt are precipitated either as simple or as complex thiocyanates. The excess of the precipitation reagent is determined with periodate or the precipitate is filtered off, again dissolved and the thiocyanate contained herein is titrated with periodate.[21]

The *hydrazide of isonicotinic acid, phenylhydrazine, semicarbazide, thiosemicarbazide* (after hydrolysis), *ascorbic acid and hydroquinone* are determined by direct potentiometric titration[24] in 9 N hydrochloric acid solution; only the hydroquinone is titrated in 6–7 N hydrochloric acid solution.

Thallium. The direct potentiometric micro-titration of Tl^+ ions with periodate is carried out in concentrated or in 6–9 N hydrochloric acid solution.[25,25a] As little as 25 μg of thallium can still be determined in 25 ml.

(B) *Hydrazine, hydroquinone, potassium antimony tartrate, iodide, arsenic (III), antimony (III), tin (II) and iron (II) salts, ferrocyanide, lead (II), silver (I) and mercury (I) iodide, hydrogen peroxide, thiosulphate, hydrogen sulphite and sulphite.* In the determination of these substances with periodate an excess of cyanide can be added to bind the I^+ ion formed. All the substances mentioned (with the exception of hydrazine, tin (II) and iodide) have to be oxidized before the actual titration. Ferrocyanide and hydrogen peroxide are at first oxidized with hypoiodous acid in alkaline medium and then acidified with hydrochloric acid. The other compounds are oxidized with iodine monochloride in hydrochloric acid solution. An equivalent amount of iodine is liberated in both cases; after adding cyanide, this is titrated with periodate in 1 N hydrochloric acid solution until the blue colour disappears, using starch solution as indicator.

(C) Periodate is reduced to iodine monobromide in the presence of an excess of bromide. In hydrochloric acid solution the end-point of the titration is determined potentiometrically[26] or by the decolorization of the carbon tetrachloride layer,[18] which is coloured red by elementary iodine before the equivalence point. This elementary iodine is formed either by the intermediate oxidation with iodine monochloride or by the reduction of the periodate during the titration. In this way, the following substances can be determined:

Hydrazine, phenylhydrazine, hydroquinone, iodide, arsenic (III), antimony (III), tin (II), iron (II) and mercury (I) salts, thiocyanate, sulphite, thiosulphate and tetrathionate. Iron (II), mercury (I) salts and sulphite are oxidized at first with iodine monochloride.

Titrations in which periodate is reduced to iodine

Periodate can be reduced quantitatively to iodine, if the acidity of the solution is adequately adjusted. The equivalence point is determined potentiometrically. On continuing the titration a second potential jump can be seen, corresponding to the oxidation of the iodine formed to iodine monochloride.

Ascorbic acid and cystein chloride can be thus titrated with periodate in 9 N HCl solution and *hydroquinone* in 1 N HCl solution.[24]

In the titration of *penicillin* the iodine liberated is bound by acetone, forming iodo-acetone and iodide ions. The end-point of the titration is

PERIODIC ACID AND ITS SALTS

determined potentiometrically or using Variamine Blue as redox indicator.[24a]

Titrations in which periodate is reduced to iodide

Vanadium. The reaction between periodate and divalent vanadium[20] has been thoroughly studied. It was found that the divalent vanadium is oxidized quantitatively in 1 N sulphuric acid medium to tervalent vanadium and the periodate is reduced to iodide. A distinct potential jump is recognized at the end of the titration if this reaction is followed up potentiometrically. The first excessive drop of periodate reacts with the iodide formed, liberating iodine. The end-point of the titration can then be recognized with starch solution; *p*-ethoxy-chrysoidine, safranine-T or neutral red can be likewise used as indicators.

After complete oxidation of V^{2+} to V^{3+}, iodide is oxidized to iodine and, simultaneously, V^{3+} to V^{4+}. When all the V^{3+} is oxidized to V^{4+}, the oxidation of the iodine formed to iodate follows. If mercury (II) chloride is added to the solution of the vanadium (II) salt before the titration, this binds the iodide formed as HgI_2; consequently the second potential jump which corresponds only to the oxidation of V^{3+} to V^{4+} can be observed in the potentiometric titration. The reaction proceeds more slowly than the oxidation of V^{2+} to V^{3+}, but it can be enhanced catalytically by iodide or manganese (II) salt. For quantitative determinations the first reaction (V^{2+} to V^{3+}) is best suited.

Thiourea and thiosine amine. Both substances are titrated potentiometrically with periodate in a 1 N HCl solution. The end-point of the titration is recognized by the sudden increase of the potential which lasts for about 30 sec.[25] The *allyl isothiocyanate* has been determined likewise in oil of mustard, in mustard seed and mustard plaster[27] (after transformation of the isothiocyanate to thiosine amine by ammonia).

Oxidation of organic substances in which the carbon–carbon bond is split (Malaprade reaction, etc.)

One of the most important reactions of periodate is the oxidation of organic substances which have neighbouring carbons with free hydroxyl groups; a splitting of the carbon–carbon bond takes place. Compounds with a related structure (such as hydroxy-aldehydes, amino-alcohols, etc.) react similarly, but perhaps not so readily. Only lead (IV) acetate equals periodate so far as the range of action and definite specificity are concerned.

The splitting of α-dihydroxy compounds, which Malaprade[28] noted for the first time, proceeds according to the equation:

$$\begin{array}{c} H \\ | \\ R-C-OH \\ | \\ R'-C-OH \\ | \\ H \end{array} + H_5IO_6 \rightarrow R-\overset{H}{\underset{|}{C}}=O + R'-\overset{H}{\underset{|}{C}}=O + HIO_3 + 3 H_2O$$

F

The oxidation mechanism has not yet been completely clarified. For example, Criegee attributed the selective oxidizing actions of periodic acid to its ability to form intermediately cyclic esters with α-diols.[29]

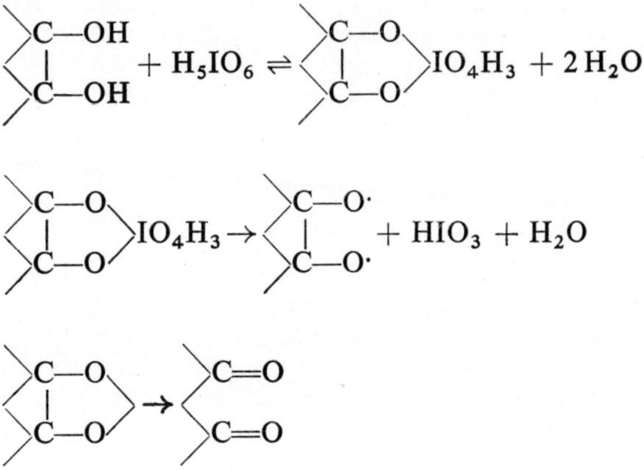

$$\begin{matrix} >\!C\!-\!OH \\ | \\ >\!C\!-\!OH \end{matrix} + H_5IO_6 \rightleftarrows \begin{matrix} >\!C\!-\!O \\ | \!\!\!\!\!\!\!\!\!\!\!\!\!\!>\!IO_4H_3 \\ >\!C\!-\!O \end{matrix} + 2H_2O$$

$$\begin{matrix} >\!C\!-\!O \\ | \!\!\!\!\!\!\!\!\!\!\!\!\!\!>\!IO_4H_3 \\ >\!C\!-\!O \end{matrix} \rightarrow \begin{matrix} >\!C\!-\!O\cdot \\ | \\ >\!C\!-\!O\cdot \end{matrix} + HIO_3 + H_2O$$

$$\begin{matrix} >\!C\!-\!O \\ | \!\!\!\!\!\!\!\!\!\!\!\!\!\!> \\ >\!C\!-\!O \end{matrix} \rightarrow \begin{matrix} >\!C\!=\!O \\ \\ >\!C\!=\!O \end{matrix}$$

Waters, on the other hand, proposes the following scheme:[30]

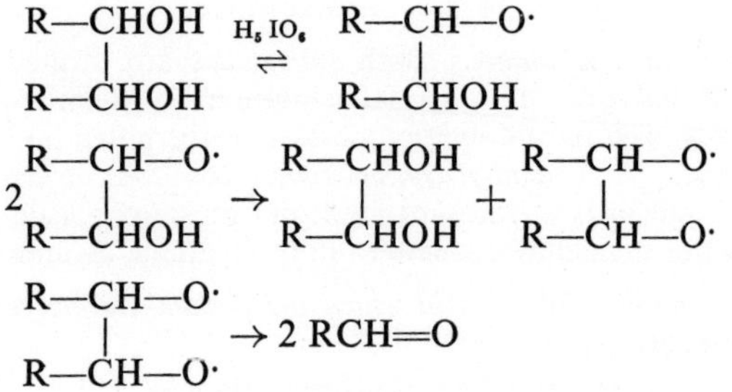

$$\begin{matrix} R\!-\!CHOH \\ | \\ R\!-\!CHOH \end{matrix} \underset{\rightleftarrows}{H_5IO_6} \begin{matrix} R\!-\!CH\!-\!O\cdot \\ | \\ R\!-\!CHOH \end{matrix}$$

$$2\begin{matrix} R\!-\!CH\!-\!O\cdot \\ | \\ R\!-\!CHOH \end{matrix} \rightarrow \begin{matrix} R\!-\!CHOH \\ | \\ R\!-\!CHOH \end{matrix} + \begin{matrix} R\!-\!CH\!-\!O\cdot \\ | \\ R\!-\!CH\!-\!O\cdot \end{matrix}$$

$$\begin{matrix} R\!-\!CH\!-\!O\cdot \\ | \\ R\!-\!CH\!-\!O\cdot \end{matrix} \rightarrow 2\ RCH\!=\!O$$

Polyhydroxy compounds are oxidized with periodic acid according to the equation:

$$CH_2OH\!-\!(CHOH)_n\!-\!CH_2OH + (n+1)\,H_5IO_6 \rightarrow 2\,HCHO + \\ + n\,HCOOH + (n+1)\,HIO_3 + (2n+3)\,H_2O$$

The oxidation of *α-amino alcohols* having primary or secondary amino groups can be formulated as follows:[31]

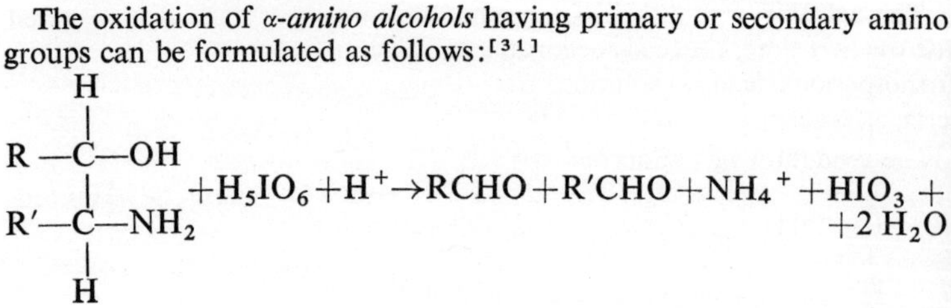

$$\begin{matrix} H \\ | \\ R\!-\!C\!-\!OH \\ | \\ R'\!-\!C\!-\!NH_2 \\ | \\ H \end{matrix} + H_5IO_6 + H^+ \rightarrow RCHO + R'CHO + NH_4^+ + HIO_3 + 2H_2O$$

α-*Ketols*, α-*diketones*, α-*ketoaldehydes*, α-*hydroxy aldehydes and glyoxal*[32] (the carbonyl group is oxidized to the carboxyl group) are easily oxidized with periodic acid (respectively periodate). The oxidation of α-*hydroxy acids*, α-*keto acids and* α-*amino acids*[33] proceed more slowly, as does the oxidation of compounds with *active hydrogen atoms*,[34] of α-*diamines* and of *acyl derivatives of* α-*amino alcohols*. The oxidations are accelerated by raising the temperature. *Simple alcohols, ketones* and *aldehydes* are oxidized only very slowly even at 100°C.

Periodic acid and periodates oxidize many substances at high temperatures; hence these oxidations lose their analytical significance at high temperatures. On the contrary, some oxidations are carried out at temperatures below room temperature in order to improve the specificity of the oxidizing action of periodic acid and periodates. The velocity of the oxidation of various types of organic substances also depends very much on the pH. Whereas the oxidations of polyhydroxy compounds proceed best at pH 4, α-amino acids are oxidized at pH from 7–9. With some substances the oxidation can also be influenced by the action of light.[35] Oxidations with periodic acid or periodate are usually carried out in aqueous solution; but organic solvents (such as methanol, ethanol, tert. butanol, dioxan, acetic acid, etc.) can also be used.

An excess of the reagent is generally applied in the determinations mentioned above. After a certain time either the unconsumed amount of the titrant or the reaction product is determined.

In an analysis of a sample composed of several substances, the total oxidizability by periodate is determined following the first procedure, whereas the various single substances are determined side by side following the second procedure. A necessary condition is, however, that the various substances yield different reaction products during the analysis. One has therefore to choose always the most suitable conditions of reaction for the particular compound to be determined. These conditions have to favour the actual oxidizing action of periodate (the reaction has to proceed sufficiently quickly and selectively) as well as the quantitative determination of the reaction product. Organic acids, carbonyl compounds and ammonia are the reaction products most often determined.

The oxidation of organic substances is usually carried out with sodium or potassium metaperiodate in neutral solution in cases where the final determination is based on the titration of acids formed.

The oxidation of *polyalcohols, sugars, polyhydroxy mono-* and *polyhydroxy dicarboxylic acids* is carried out in dilute sulphuric acid solution. Ortho-periodic acid (in solutions free of metal ions) or periodate solutions serve as reagent.

The oxidation of amino acids is carried out in sodium hydrogen carbonate buffered solutions. $Na_3H_2IO_6$ is applied as reagent suitable for that case. The same reagent is used in the oxidation of *amino alcohols* and *amino acids* in NaOH solution, where, after the oxidation of these substances, the ammonia content is determined.

Use is also made of periodic acid to verify the structure of sugars, steroids and other compounds.

The determinations of *glycerol, ethylene glycol* and *1,2-propylene glycol* are mentioned as examples.[37] All three compounds can be determined side by side, based on the different reaction products formed in the oxidation with periodate. The three substances react with periodate according to the equations:

$$CH_2OH—CHOH—CH_2OH + 2\,KIO_4 \rightarrow 2\,HCHO + HCOOH +$$
$$+ 2\,KIO_3 + H_2O$$

$$CH_2OH—CH_2OH + KIO_4 \rightarrow 2\,HCHO + KIO_3 + H_2O$$

$$CH_3—CHOH—CH_2OH + KIO_4 \rightarrow CH_3CHO + HCHO + KIO_3 +$$
$$+ H_2O$$

Glycerol is determined by titrating the formic acid produced with standard hydroxide solution. 1,2-Propylene glycol contains the group CH_3CHOH bound to the hydroxyl (bearing) carbon atom, thus yielding acetaldehyde on oxidation with periodate, unlike the two other substances. Acetaldehyde (and so 1,2-propylene glycol) is determined by the sulphite method after having been separated from formaldehyde. Ethylene glycol is determined by the difference from the total periodate consumed or from the total aldehyde formed.

The concentration of the three substances mentioned can also be found by determining the formic acid produced, the total aldehyde, and by determining the formaldehyde by the cyanide method in the presence of acetaldehyde.

A survey of the use of periodic acid and its salts in analytical chemistry is given in the publications of Smith,[36] Kolthoff and Belcher,[37] Fleury and Courtois.[38] The reader will find there also numerous references to original publications and reports, the number of which amounts already today to the hundreds.

REFERENCES

1. MALAPRADE, L., *Bull. Soc. chim. France* (4), **43**, 683 (1928).
2. SYROKOMSKII, V. S. and MELAMED, S. I., *Zavodskaya Lab.*, **16**, 131 (1950).
3. SYROKOMSKII, V. S. and MELAMED, S. I., *Zavodskaya Lab.*, **16**, 273 (1950).
4. SYROKOMSKII, V. S. and KNYAZEVA, R. N., *Zavodskaya Lab.*, **16**, 1041 (1950).
5. SYROKOMSKII, V. S. and MELAMED, S. I., *Zavodskaya Lab.*, **16**, 398 (1950).
6. JANDER, G. and co-workers, *Neuere massanalytische Methoden*, Ferdinand Enke Verlag, Stuttgart 1956.
7. MÜLLER, E. and WEGELIN, E., *Z. anal. Chem.*, **52**, 755 (1913).
8. SCHWAIBOLD, J., *Z. anal. Chem.*, **78**, 161 (1929).
9. SCHWICKER, A., *Z. anal. Chem.*, **110**, 182 (1937).
10. FLEURY, P. and LANGE, J., *J. Pharmac. Chim.* (8) **17**, 107, 196 (1937).
11. RAPPAPORT, F., REIFER, I. and WEINMANN, H., *Mikrochim. Acta*, **1**, 290 (1937).

12. WILLARD, H. H. and MERRITT, L. L., *J. Ind. Eng. Chem. Anal. Ed.*, **14**, 489 (1942).
13. WILLARD, H. H. and GREATHOUSE, L. H., *J. Amer. Chem. Soc.*, **60**, 2869 (1938).
14. MÜLLER, E. and FRIEDBERGER, F., *Ber. dt. chem. Ges.*, **35**, 2652 (1902).
15. MÜLLER, E. and JACOB, W., *Z. anorg. Chem.*, **82**, 308 (1913).
16. BERKA, A. and VULTERIN, J., *Chemie*, **10**, 40 (1958).
17. SINGH, B. and SINGH, A., *J. Indian Chem. Soc.*, **29**, 34 (1952).
18. SINGH, B. and SINGH, A., *Analyt. Chim. Acta*, **9**, 22 (1953).
19. SINGH, B. and SINGH, A., *J. Indian Chem. Soc.*, **30**, 143 (1953).
20. MÁZOR, I. and ERDEY, L., *Acta chim. Acad. Sci. Hung.*, **2**, 331 (1952).
21. SINGH, B. and SINGH, A., *J. Indian Chem. Soc.*, **29**, 517 (1952).
22. SINGH, B. and SINGH, A., *Res. Bull. East Panjab Univ.*, **17**, 51 (1951).
23. SINGH, A., SINGH, B. and SINGH, R., *J. Indian Chem. Soc.*, **30**, 147 (1953).
23a. SINGH, B., SAHOTA, S. S. and GUPTA, M. P., *J. Indian Chem. Soc.*, **38**, 189 (1961).
23b. SINGH, B., SAHOTA, S. S., *J. Sci. Ind. Res. (New Delhi), Sect. B* **17**, 386 (1958).
24. BERKA, A. and ZÝKA, J., *Chem. Listy*, **50**, 314 (1956).
24a. MÁZOR, L. and PÁPAY, M. K., *Acta chim. Acad. Sci. Hung.*, **26**, 473 (1961).
25. SIMON, V., *Chem. Listy*, **49**, 1727 (1955).
25a. ALIMARIN, I. P., TSENG, J. A. and PUZDRENKOVA, I. V., *Chem. analit. (Warszawa)*, **3**, 245 (1958); *Z. anal. Chem.*, **172**, 285 (1960).
26. SINGH, B. and SINGH, A., *J. Indian Chem. Soc.*, **30**, 786 (1953).
27. BERKA, A. and ZÝKA, J., *Českoslov. Farm.*, **4**, 222 (1955).
28. MALAPRADE, L., *Compt. Rend.*, **186**, 382 (1928); *Bull. Soc. Chim. France*, **43** (4), 683 (1928).
29. CRIEGEE, R., *Angew. Chem.*, **50**, 153 (1937).
30. WATERS, W. A., *Trans. Faraday Soc.*, **42**, 184 (1946).
31. NICOLET, B. N. and SHINN, L. A., *J. Amer. Chem. Soc.*, **61**, 1615 (1939); FLEURY, P., COURTOIS, J. and GRANDSCHAMP, M., *Bull. Soc. Chim. France*, **88** (1949).
32. FLEURY, P. and LANGE, J., *Compt. Rend.*, **195**, 1395 (1932); *J. Pharmac. Chim.* (8) **17**, 409 (1933); MALAPRADE, L., *Bull. Soc. Chim. France* (5) **1**, 833 (1934); CLUTTERBUCK, P. W. and REUTER, F., *J. Chem. Soc.*, 1467 (1935).
33. FLEURY, P. and BOISSON, S., *Compt. Rend.*, **204**, 1264 (1937); FLEURY, P. and BOISSON, S., *J. Pharmac. Chim.*, **30**, 145, 307 (1939); *Compt. Rend.*, **208**, 1509 (1939).
34. HEUBNER, C. F., AMES, S. R. and BUBL, E. C., *J. Amer. Chem. Soc.*, **68**, 1621 (1946); FLEURY, P. and COURTOIS, J., *Compt. Rend.*, **223**, 633 (1946); *Bull. Soc. Chim. France*, 358 (1947); 190 (1948).
35. HEAD, F. S. H., *Nature (London)*, **165**, 236 (1950).
36. SMITH, G. F., *Analytical Application of Periodic Acid and Their Salts*, Smith Chemical Company, Columbus (Ohio) 1950.
37. KOLTHOFF, I. M., BELCHER, R., STENGER, V. A. and MATSUYAMA, G., *Volumetric Analysis*, III, Titration Methods, Interscience, New York, 1957.
38. FLEURY, P. and COURTOIS, J., *Inst. Internat. Chim. Solvay*, 8ème Conseil de Chimie, p. 279, R. Stoops, Bruxelles 1950.
39. DOLEŽAL, J. RÖSSLER, S. and ZÝKA, J., *Collection (Czech. Chem. Comm.)*, **27**, 1031 (1962).

CHAPTER 11

LEAD (IV) ACETATE

LEAD (IV) ACETATE is commonly used as a reagent in preparative organic chemistry. It was synthesized for the first time by Jaquelain.[1] Its oxidizing action was first studied by Dimroth and coworkers.[2,3] They found that lead (IV) acetate exhibits a different and, particularly, more selective action than the lead (IV) oxide used until that time. Criegee[4-10] hinted at the characteristic action of lead (IV) acetate towards α-glycols and substances having a similar structure where the carbon–carbon bond is ruptured. Lead (IV) acetate can oxidize the same types of organic compounds as periodic acid and its salts, yielding also nearly always the same reaction products. A lucid description of the action of lead (IV) acetate on various organic compounds is given by Fleury and Courtois.[11]

Lead (IV) acetate and periodic acid react differently with α-monohydroxy acids and some *cis–trans* isomers of α-glycols. Whereas periodic acid oxidizes α-monohydroxy acids at room temperature but slowly, their oxidation with lead (IV) acetate mostly proceeds quickly and quantitatively. Periodic acid oxides *cis-* as well as *trans*-isomers of α-glycols about equally fast; lead (IV) acetate, on the contrary, oxidizes the *cis*-isomers much faster than the *trans*-isomers. Lead (IV) acetate is therefore frequently used to differentiate between the two isomers.

Because the oxidations with lead (IV) acetate proceed extremely quickly, it can be used for the indirect determination of α-glycols and substances of similar character. Either the reaction products or the excess of reagent is determined, as in the oxidation with periodic acid. For analytical purposes, lead (IV) acetate is less suited than periodic acid and its salts, because unlike the periodates it also oxidizes slowly the reaction products in many cases. The reactions with an excess of reagent have to be done in non-aqueous medium (to avoid hydrolysis of the lead (IV) acetate). Acetic acid produced in the reduction of the reagent sometimes renders the determination difficult, which is based on determining acid oxidation products.

$$(CH_3COO)_4Pb + 2H^+ + 2e^- \rightarrow (CH_3COO)_2Pb + 2CH_3COOH$$

Tomíček and Valcha[12] studied the direct titration with lead (IV) acetate in glacial acetic acid and studied also the changes of the redox potential Pb^{4+}/Pb^{2+} according to the concentration of sodium acetate and perchloric acid. The redox potential is more than 1·44 V in the presence of perchloric acid and at the utmost 1·32 V in the presence of acetate.

The presence of water accelerates most of the oxidations carried out with lead (IV) acetate; this reaction has been applied in the direct titration of suitably acidified aqueous solutions of organic[13,16] and inorganic substances.[14,19−20] It is, however, a condition necessary for these titrations, that the oxidation of the substances to be determined proceeds faster than the hydrolysis of the lead (IV) salt.

Recently it has been found[18−22] that a number of substances which correspond in their character to the vicinal glycols, or have a similar structure can be determined with an excess of reagent in 50–80 per cent acetic acid in the presence of alkali acetate; the unconsumed amount of reagent is back-titrated potentiometrically with a standard hydroquinone solution after a suitable time. It is an advantage of this method that the oxidation of the substances to be determined, proceeds (compared with other reagents) quickly and quantitatively to carbon dioxide, or to the respective aldehyde even at room temperature. Very small amounts of compounds to be investigated can therefore be determined in a fast and simple way.[22a]

Because of the good results obtained by the use of lead (IV) acetate in aqueous solutions, the formal redox potentials of the reagent have been thoroughly studied in dilute mineral acid as well as in dilute acetic acid (eventually in the presence of alkali acetate).[23] The redox potential in hydrochloric acid (in 0·03 N HCl about 1·38 V) decreases with increasing acid concentration. On the contrary in perchloric acid solution the redox potential depends only very little on the acid concentration; nevertheless it reaches 1·65 to 1·70 V. In acetic acid solution the formal redox potential depends on the acetic acid concentration, as well as on the concentration of lead (IV) and lead (II) ions and changes with time. This can be attributed to complex formation of the lead (IV) acetate with water. In solutions containing less than 30 per cent of acetic acid, the redox potential is about 1·45 V. In 0·4 M sodium acetate solutions, the redox potential is about 100 mV lower (without consideration of the acetic acid concentration).

STANDARD SOLUTION

Lead (IV) acetate is prepared by reaction of red lead (Pb_3O_4) with glacial acetic acid at elevated temperature.[3] Standard lead (IV) acetate solutions can be made either by directly dissolving the right amount of Pb_3O_4 in the respective volume of glacial acetic acid, or by first isolating the lead (IV) acetate crystals; these are then recrystallized and dissolved in the respective amount of glacial acetic acid. Solutions thus prepared no longer contain divalent lead. Also the other method of preparation is rapid and does not require complicated operations.[14]

The normality of these lead (IV) acetate solutions in glacial acetic acid is stable for several months if stored in dark bottles. The stability of the solutions is reduced noticeably by acetic acid anhydride and alkali acetate.

Standardization

The normality of the standard solution is determined iodimetrically: Add an aqueous solution of potassium iodide and sodium acetate to the lead (IV) acetate solution and titrate the liberated iodine with thiosulphate. The standardization can also be carried out by titrating arsenic (III) salt in the presence of iodide in a potassium acetate buffered medium. The endpoint is indicated visually by the blue colour formed by the reaction of starch solution with iodine, liberated from iodide by the first drop in excess of the lead (IV) acetate solution.[23a]

The potentiometric standardization can be carried out by titrating a known amount of hydrazine sulphate[13] or a standard hydroquinone solution,[12] the titre of which has been determined with dichromate.

INDICATOR

In direct titrations with lead (IV) acetate the potentiometric determination of the equivalence point has proved suitable in aqueous as well as in non-aqueous media. In the titration of benzyl mercaptan in glacial acetic acid, quinalizarin is used as redox indicator,[12] in the titration of ascorbic acid, N,N'-bis [4-(4-methoxyphenyl amine)-phenyl] thiourea.[12a]

In indirect titrations, the excess of reagent is determined iodimetrically or by titration with hydroquinone, using a potentiometric end-point[12] or ferroin as indicator.[15]

REVIEW OF DETERMINATIONS

Substances, determined by direct potentiometric titration in aqueous solution,[13,14,16,19-21,29] are listed in Table 2.

The titrations mentioned in Table 2 proceed very fast and with an adequate accuracy. The equivalence point is marked by a great potential change. *Mandelic acid, tartaric acid* and *mannitol*[16] are exceptions; the prescribed conditions have to be followed precisely and they only have to be titrated slowly.

In the titration of *trivalent molybdenum compounds*,[20] the oxidation goes in 10 N hydrochloric acid to molybdenum (V) or in 5–7 N hydrochloric acid to molybdenum (VI). In the micro-titration of thallium (I) salts,[19] as little as 10 μg thallium can be determined very accurately; hence this method provides a useful means of standardizing lead (IV) acetate solutions to be used in the micro-range. *Ruthenium (IV) salts*[21] are quantitatively oxidized to RuO_4; as little as 2 mg ruthenium in 50 ml can be thus determined.

TABLE 2. REVIEW OF DETERMINATIONS WITH LEAD (IV) ACETATE

Substance	Titration medium
As^{3+}	0·05–8 N H_2SO_4; 0·005–8 N HCl
Sb^{3+}	0·25–4 N H_2SO_4; 0·35–6 N HCl
Sn^{2+}	0·25–4 N H_2SO_4; 0·25–2 N HCl
Fe^{2+}	1–4 N HCl
Ti^{3+}	0·5–4 N H_2SO_4; 0·05–4 N HCl
Cr^{2+}	0·08–3 N HCl
Tl^+	8–10 N HCl; 4·1–4·7 N HCl
Mo^{3+}	5–7 N HCl; 10 N HCl
Ru^{4+}	0·03–0·1 N $HClO_4$
NO_2^-	1 M NaCl
Hydrazine sulphate	without addition of acid
Phenyl hydrazine chloride	without addition of acid
p-Nitrophenylhydrazine	20% H_2SO_4
Isonicotinic acid hydrazide	10% H_2SO_4
Semicarbazide chloride	without addition of acid
Ascorbic acid	pH : 4·8
Mandelic acid	pH : 4·8
Tartaric acid	pH : 4·8
Mannitol	2–10% H_2SO_4

For the micro-titration of *tri- and tetravalent uranium compounds* a method is used in which the sample to be determined is oxidized at a definite hydrochloric acid concentration with an excess of trivalent iron; the iron (II) ions thus produced are then titrated potentiometrically with lead (IV) acetate. As little as 20 μg uranium can so be determined.[24]

Hydroquinone, tetrachlorohydroquinone, catechol and *ascorbic acid*[12] can be determined by direct potentiometric titration in glacial acetic acid. *Benzylmercaptan* is titrated in glacial acetic acid using quinalizarin as redox indicator; the colour changes from orange to blue. *Ascorbic acid* is titrated in glacial acetic acid against N,N'-bis [4-(4-methoxy phenylamine)-phenyl] thiourea from yellow to violet.

Whereas the oxidation of substances containing SH-groups proceeds in glacial acetic acid to the respective disulphides, *thioglycolic acid* is oxidized in dilute acetic acid solution by an excess of the reagent to the respective sulphinic acid or, in the presence of hydrochloric acid, to sulphochloride. The oxidation of *cystein, thiourea* and *thiosemicarbazide* has also been thoroughly studied.[25]

In the literature, there is mentioned also the possibility of determining hydrogen peroxide in the presence of other peroxides.[28]

1,4-*Anhydro erythritol, methyl-α-D-mannofuranoside*, 1,4-*anhydro-mannitol, methyl*-2,6-*anhydro-α-D-altropyranoside* can be titrated potentiometrically in glacial acetic acid or in acetic acid solution with lead (IV) acetate using platinum or lead indicator electrodes.[17]

As already mentioned in the introduction to this chapter, an excess of lead (IV) acetate is less suited for quantitative titrimetric analysis. Most of these determinations are based on the fact that the excess of reagent is determined after a precisely defined time of reaction.

The determination of the *glycol bond* may be mentioned as an example. The oxidation proceeds according to the equation:

$$\begin{array}{c} R_1 \\ | \\ R_2\text{—C—OH} \\ | \\ R_3\text{—C—OH} \\ | \\ R_4 \end{array} + (CH_3COO)_4Pb \rightarrow \begin{array}{c} R_1 \\ | \\ R_2\text{—C=O} \\ | \\ R_3\text{—C=O} \\ | \\ R_4 \end{array} + (CH_3COO)_2Pb + 2\,CH_3COOH$$

Procedure. Add an excess of 0·1 N lead (IV) acetate solution in glacial acetic acid to a precisely weighed amount of glycol. Set aside for 20 hr at room temperature. Then add an excess of potassium iodide and potassium acetate and titrate the liberated iodine with thiosulphate.[4]

The determination of a number of organic compounds can be carried out faster in a medium of dilute acetic acid solution in the presence of alkali acetate. In this case, the oxidation of *vicinal* glycols by an excess of reagent is complete after 30 min. The excess of reagent is then titrated iodimetrically or with standard hydroquinone solution, or the divalent lead, produced by the reaction, is determined complexometrically.[30] *Formic acid,*[26] *tartaric acid,*[18] *ethylene glycol, glycerol, mannitol,*

LEAD (IV) ACETATE

calcium glucoside, mandelic acid and *malic acid* have been determined in this way and the oxidation of *citric acid*[22] has been investigated. This method has been used for the determination of *glycerol* in pharmaceutical products[27] and for the determination of *tartrate* in tartrate complexes.[18,31]

REFERENCES

1. JACQUELAIN, C. R., *Trav. Chim.*, **7**, 1 (1851).
2. DIMROTH, O. and FRISTER, F., *Ber. dt. chem. Ges.*, **55**, 1231 (1922).
3. DIMROTH, O. and SCHWEITZER, R., *Ber. dt. chem. Ges.*, **56**, 1376 (1923).
4. CRIEGEE, R., *Ber. dt. chem. Ges.*, **64**, 260 (1931).
5. CRIEGEE, R., KRAFT, L. and RANK, B., *Liebigs Ann. Chem.*, **507**, 159 (1933).
6. CRIEGEE, R., *Chem. Zbl.* II, 2515 (1934).
7. CRIEGEE, R., *Angew. Chem.*, **50**, 153 (1937).
8. CRIEGEE, R. and BÜCHNER, E., *Ber. dt. chem. Ges.*, **73**, 563 (1940).
9. CRIEGEE, R., BÜCHNER, E. and WERNER, W., *Ber. dt. chem. Ges.*, **73**, 571 (1940).
10. CRIEGEE, R., *Ber. dt. chem. Ges.*, **68**, 665 (1935).
11. FLEURY, P. and COURTOIS, J., *Inst. Internat. Chim. Solvay*, 8ème Conseil de Chimie, p. 279, R. Stoops, Bruxelles 1950.
12. TOMÍČEK, O. and VALCHA, J., *Chem. Listy*, **44**, 283 (1950); *Collection (Czech. Chem. Comm.)*, **16**, 113 (1951).
12a. ERDEY, L., MEISEL, T. and RÁDY, G., *Acta chim. Acad. Sci. Hung.*, **26**, 71 (1961).
13. BERKA, A. and ZÝKA, J., *Chem. Listy*, **52**, 926 (1958); *Collection (Czech. Chem. Comm.)*, **24**, 105 (1959).
14. BERKA, A., DVOŘÁK, V., NĚMEC, I. and ZÝKA, J., *Analyt. Chim. Acta*, **23**, 380 (1960).
15. BERKA, A., *Českoslov. Farm.*, **8**, 561 (1959).
16. BERKA, A. and ZÝKA, J., *Chem. Listy*, **52**, 930 (1958); *Collection (Czech. Chem. Comm.)*, **23**, 2005 (1958).
17. REEVES, R. E., *Anal. Chemistry*, **21**, 751 (1949).
18. BERKA, A., *Analyt. Chim. Acta*, **24**, 171 (1961).
19. BERKA, A., DOLEŽAL, J., NĚMEC, I. and ZÝKA, J., *Analyt. Chim. Acta*, **25**, 533 (1961).
20. BERKA, A., DOLEŽAL, J., NĚMEC, I. and ZÝKA, J., *J. electroanalyt. Chem.*, **3**, 278 (1962).
21. NĚMEC, I., BERKA, A. and ZÝKA, J., *Mikrochem. J.*, **6**, 525 (1962).
22. BERKA, A., DVOŘÁK, V. and ZÝKA, J., *Mikrochim. Acta*, 541 (1962).
22a. ZÝKA, J. and BERKA, A., Use of Lead Tetraacetate as Volumetric Oxidimetric Reagent for Microanalytical Purposes, *Mikrochem. J.*, Symposium Series, Vol. 2, Proceedings 1961—International Symposium on Microchemical Techniques, University Park 1961, Interscience, New York, 1962, p. 789.
23. BERKA, A., DVOŘÁK, V., NĚMEC, I. and ZÝKA, J., *J. electroanalyt. Chem.*, **4**, 150 (1962).
23a. BERKA, A., JANATA, J. and ZÝKA, J., *Collection (Czech. Chem. Comm.)*, **29**, 2242 (1964).
24. BERKA, A., DOLEŽAL, J., NĚMEC, I. and ZÝKA, J., *Analyt. Chim. Acta*, **26**, 148 (1962).
25. SUCHOMELOVÁ, L. and ZÝKA, J., *J. electroanalyt. Chem.*, **5**, 57 (1963).

26. PERLIN, A. S., *Anal. Chemistry*, **26**, 1053 (1954).
27. BERKA, A., FARA, M. and ZÝKA, J., *Československ. Farm.*, **12**, 366 (1963).
28. FRISONA, G. and KING, S., Determination of Hydroperoxide in the Presence of Peroxides with Lead Tetraacetate, Lecture at the Delaware Valley Meeting, 25th and 26th January 1962; *Anal. Chemistry*, **34**, 40A (1962).
29. MORALES, A. and ZÝKA, J., *Collection (Czech. Chem. Comm.)*, **27**, 1029 (1962).
30. BERKA, A., *Z. anal. Chem.*, **195**, 263 (1963).
31. LOUB, J. and FREI, V., *Chem. Zvesti*, **16**, 802 (1962).

CHAPTER 12

COMPOUNDS OF PENTAVALENT VANADIUM

Compounds of pentavalent vanadium are distinguished by their great oxidizing power. This property of vanadate (V) has been examined in recent years for its applicability in titrimetric analysis. Lang and Gottlieb as well as Syrokomskii and co-workers suggested the use of vanadate (V) solution as a titrant. Many inorganic and organic substances can be oxidized with this standard solution. The normal redox potential of the system

$$VO_3^- + 4 H^+ + e^- \rightleftharpoons VO^{2+} + 2 H_2O$$

is + 1·00 V. The redox potential increases with increasing acid concentration.[1-3] It has been found to be, e.g. in 1 N H_2SO_4 1·02 V[1], in 2 N H_2SO_4 1·07 V, in 4 N H_2SO_4 1·14 V and in 8 N H_2SO_4 1·30 V.

Most determinations are carried out in sulphuric, hydrochloric or acetic acid solution at room or at elevated temperature. Some reactions have to be catalytically accelerated because they proceed only very slowly even at elevated temperatures. In most of the reactions the pentavalent vanadium is reduced to the tetravalent state according to the above mentioned equation.

STANDARD SOLUTION

Sodium or ammonium meta-vanadate (V) or also meta-vanadic acid (V) are commonly used to prepare the standard solution. Solutions of the two first mentioned compounds are made by dissolving the analytically pure substances in the necessary volume of distilled water. Sodium vanadate is formed from ammonium vanadate by adding an excess of sodium carbonate to the solution and boiling until ammonia is completely expelled. The meta-vanadic acid solution can be prepared by dissolving ammonium meta-vanadate in sulphuric acid (1 + 1) and then diluting with water.

All these standard solutions are stable for several months. The stability is still greater in the presence of about 0·1 per cent sodium carbonate. The titre of the solution does not even change after one hour's boiling in 10 N H_2SO_4.[4]

Standardization

The recrystallized and dried ammonium meta-vanadate can be used as primary standard,[5] but the vanadate or vanadic acid solutions are usually standardized before use, mostly by a visual or potentiometric titration with iron (II) sulphate or Mohr's salt.

The solution can also be standardized against permanganate or iodine. In the manganometric standardization, the pentavalent vanadium is reduced to vanadyl ion in 5 per cent sulphuric acid with sulphur dioxide; after boiling free from the excess of sulphur dioxide, it is titrated with standard permanganate solution at elevated temperature. The iodimetric determination is based on the reduction of vanadate (V) ion with zinc, cadmium or their amalgams to hypovanadous ion (vanadate (II) ion), which then is oxidized with iodine solution to vanadous ion (vanadate (III) ion) using safranine as indicator.

Sodium oxalate is less suitable as primary standard.[6]

INDICATOR

N-Phenylanthranilic acid is commonly used as indicator in visual titrations with vanadate (V). Diphenylamine, diphenylbenzidine and diphenylamine-p-sulphonic acid, as well as compounds of phthalocyanine tetrasulphonic acid with copper and iodine monochloride, are also often used. The end-point indication with iodine monochloride in the presence of chloroform or carbon tetrachloride proceeds in two steps. First, the iodine monochloride is reduced to iodine, which colours the organic layer. On further addition of the titrant, the colour intensity is lowered because the second reaction starts, in which the iodine produced is again oxidized to iodine monochloride. At the equivalence point, the colour disappears completely.

Most of the compounds can also be titrated potentiometrically.

REVIEW OF DETERMINATIONS

Arsenic. Trivalent arsenic can be easily oxidized to the pentavalent state with vanadate (V) in at least 7 N hydrochloric acid medium.[7] Iodine monochloride in a chloroform layer serves as indicator.

Trivalent arsenic can be similarly titrated in phosphoric acid solution; potentiometric detection of the end-point is to be preferred.[8]

In the indirect determination, the sample is oxidized with an excess of ferric ammonium sulphate,[9] under the catalytic action of osmium (VIII) oxide or iodine monochloride, the iron (II) ions thus produced are then titrated with vanadate (V).

Arsenite can be oxidized by an excess of sodium vanadate (V).[10,11] The excess is determined with Mohr's salt, using diphenylbenzidine or N-phenylanthranilic acid as indicator.

Mercury. Mercury (I) ions are determined with sodium vanadate (V) in at least 7 N hydrochloric acid solution.[7] The reaction proceeds as follows:

$$Hg_2^{2+} + 2\,VO_3^- + 8\,H^+ \rightarrow 2\,Hg^{2+} + 2\,VO^{2+} + 4\,H_2O$$

Iodine monochloride in the presence of chloroform serves as indicator. In the indirect determination of monovalent mercury according to Suranova and Olenovich,[12] the excess of potassium vanadate (V) is back-titrated with Mohr's salt using N-phenylanthranilic acid as indicator. Divalent mercury is likewise determined, after having been reduced to the monovalent state. Of only little significance is the indirect determination of mercury (II) salts,[13] based on their precipitation with ammonium thiocyanate and zinc sulphate; the complex salt $Zn[Hg(SCN)_4]$ is formed and the thiocyanate ions are titrated.

Thallium. The oxidation of thallium (I) ions to thallium (III) with sodium vanadate (V) proceeds quantitatively at room temperature in hydrochloric acid solution.[14] The solution should be at least 7 N HCl near the end of the titration. Iodine monochloride in the presence of an organic solvent is used as indicator.

Vanadium. Trivalent vanadium can be determined by direct titration with vanadate (V) in an inert atmosphere, best at 45–50°C in 1–3 N hydrochloric acid solution and in the presence of 5 ml of concentrated phosphoric acid per 50 ml of sample solution. The end-point of the titration is determined using diphenylbenzidine, barium diphenylamine-p-sulphonate or N-phenylanthranilic acid.[14a]

Titanium. Titanium (III) ions can be directly titrated with vanadate (V) in hydrochloric and oxalic acid solution.[14b] The determination is carried out at room temperature against neutral red, phenosafranine or safranine-T as redox indicators; it can also be done potentiometrically.[48]

Molybdenum. Lang and Gottlieb[15] used meta-vanadic acid to oxidize pentavalent molybdenum. Molybdenum (VI) is reduced to molybdenum (V) with tin (II) chloride. The excess of tin (II) chloride is oxidized with bromine and the bromine is removed with arsenite. The pentavalent molybdenum is then titrated with vanadic acid; a 1 per cent solution of diphenylamine-p-sulphonic acid, which has previously been oxidized with potassium dichromate, is used as indicator.

Special precautions have to be taken in this determination, because molybdate (VI) cannot only be reduced to molybdenum (V) by a larger excess of tin (II) chloride, but also to lower valency states which are then re-oxidized to molybdate (VI) by bromine solution; this fact renders the determination impossible. The sample solution is therefore neutralized against methyl orange and a weakly acid and not too concentrated tin (II) chloride solution is used for the reduction.

The determination is not affected by higher concentrations of lead, chromium, arsenic, tin and by small amounts of copper, vanadium, tungsten, sulphate, phosphate and fluoride.

Other determinations[16] are based on the reduction of molybdate (VI) with mercury, and oxidation of the molybdenum (V) produced with standard ammonium vanadate solution in 2–3 N hydrochloric acid medium using N-phenylanthranilic acid as indicator. Because vanadyl ions are not oxidized by vanadate (V) solutions, molybdenum can also be determined in the presence of considerable amounts of vanadyl ions.

Hexavalent molybdenum can also be reduced by metallic bismuth to the pentavalent state and that in 1–1.5 N hydrochloric acid solution.[17] Zinc, bismuth or cadmium amalgams can likewise be used.[18,19] The titration is done with ammonium vanadate (V), N-phenylanthranilic acid serving as indicator. This method is suitable for molybdenum determinations in alloys and steels.

According to another method,[20] molybdate (VI) can be reduced with an excess of hydrazine sulphate in 1–2 N hydrochloric acid medium. The molybdate (V) thus produced is titrated with sodium vanadate (V) in 8 N sulphuric acid solution against N-phenylanthranilic acid or diphenylbenzidine as indicator. The indirect determination of molybdenum after its precipitation with 5,7-dibromoxine is only of little significance.[21]

Tungsten. Tungsten compounds can be determined with ammonium vanadate (V).[16,22] The hexavalent tungsten has to be reduced first by suitable means. Stepin and Silaeva[22] performed the reduction with metallic cadmium in sulphuric and phosphoric acid solution. The reduced tungsten is then titrated with ammonium vanadate (V) against N-phenylanthranilic acid as indicator. The method is suited particularly to the determination of tungsten in steels and Scheelite concentrates.

Uranium. Uranium (IV) produced by reduction of uranium (VI) can be titrated with ammonium vanadate (V) in 5–13.5 N sulphuric acid, using N-phenylanthranilic acid as indicator.[23,24] Rao and co-workers[25] found that this oxidation is catalytically influenced by oxalid acid. Thereby it is possible to titrate in 4 N sulphuric acid.

Diphenylbenzidine can also be used as indicator instead of N-phenylanthranilic acid.[25,26,26a] The titration can be done with this indicator either at elevated temperature (50°C) under the catalytic action of oxalic acid or at room temperature, under the catalytic action of oxalic and phosphoric acid.

Small amounts of tetravalent uranium are titrated potentiometrically[26b] or amperometrically.[26c]

Rao and Sastri[27] used compounds of phthalocyanine tetrasulphonic acid with copper as indicator in determining tetravalent uranium.

Uranium can be determined with ammonium vanadate (V) in a solution containing fluoride also in the presence of iron, vanadium, molybdenum and titanium.[28] This method is suitable for the determination of uranium in rocks.[29]

The more sensitive amperometric method has been proposed instead of the visual determination.[30,31] In this case, one titrates in hydrochloric or perchloric acid solution; as little as 1 μg uranium can be still determined with a sufficient accuracy. The interference of iron is removed by adding o-phenanthroline; lead, bismuth, chromium and nickel are masked with EDTA.

Compounds of hexavalent uranium have to be reduced before the actual determination in a reductor, eventually with reducing agents such as iron (II) sulphate.

The photochemical reduction of hexavalent uranium in the presence of ethanol[32] or ether[33] and also the indirect determination of tetravalent uranium, in which the excess of ammonium vanadate (V) is back-titrated with iron (II) ammonium sulphate[10] or potassium permanganate,[34] are both of little practical significance.

Iron. Iron (II) salts are quantitatively oxidized to iron (III) salts with vanadate (V) in 4–13·5 N sulphuric acid, using N-phenylanthranilic acid as indicator.[2] The reaction can also be traced potentiometrically; the results are still satisfactory even in titrations with 0·002 N vanadate (V) solution.[35]

The titration can also be carried out in at least 7 N hydrochloric acid solution, if iodine monochloride and chloroform are used as indicator.

Organic compounds, e.g. alcohols, phenols[36] or sugars,[5] do not interfere, whereas they do cause too high results in titrations with potassium dichromate or cerium (IV) sulphate.[36]

Iron (III) ions have to be reduced to iron (II) ions before the determination, preferably in a Jones reductor. One titrates with sodium or ammonium vanadate (V) in sulphuric acid solution, using as indicators diphenylamine, diphenylbenzidine, diphenylamine sulphonic acid[37] or compounds of phthalocyanine sulphonic acid with copper.[27]

Iron (III) ions can be reduced with hypophosphite[38] or hydroxylamine.[39] Syrokomskii and Zhukova[40] used compounds of divalent chromium for the reduction of trivalent iron. After oxidizing the excess of divalent chromium by atmospheric oxygen, the iron (II) ions produced are titrated with 0·002 N vanadate solution. This method is applied in the analysis of technical acids and bases.

Iron can be determined under suitable conditions in rocks,[41] minerals,[41a] and in various mixtures, e.g. of iron (II), iron (III), manganese (III) and manganese (IV) oxides.[41b]

Of only little practical significance are determinations of iron, which are based on the reduction of iron (III) with oxalic or lactic acid in ultraviolet light,[42–43a,37] or in which the iron is precipitated as dibromooxinate and, after dissolution, titrated with vanadate (V).[21]

Copper. Copper (II) ions can only be titrated indirectly with vanadate (V). The divalent copper has to be first reduced with bismuth amalgam under a carbon dioxide atmosphere in a medium containing 50 per cent of concentrated hydrochloric acid.[16] The copper (I) ions formed are oxidized with potassium iron (III) sulphate; the equivalent amount of iron (II) ions thus produced is titrated with ammonium vanadate solution, using N-phenylanthranilic acid as indicator.

According to another method, copper (II) ions are reduced by sulphur dioxide;[44] the monovalent copper thus formed is precipitated with an excess of potassium thiocyanate and the excess is titrated with sodium vanadate.

A further indirect copper determination is based on the precipitation of copper (II) ions with ammonium thiocyanate and mercury (II) chloride solution;[13] the precipitate is then dissolved in concentrated hydrochloric acid and the thiocyanate ions thus liberated are titrated with sodium vanadate (V).

Copper ions can similarly be precipitated as dibromo-oxinate which after dissolution is determined with vanadate (V).[21]

All methods mentioned here are, however, only of little practical interest.

Lead and manganese. Lead can only be determined indirectly with sodium vanadate (V).[13] Lead (II) ions are precipitated with an excess of potassium iodide; the precipitate is filtered off, washed and dissolved. The iodide ions are titrated with sodium vanadate (V) in hydrochloric acid solution following the method described for the determination of iodide. Lead (IV) and manganese (IV) oxide are both determined similarly.[44]

Zinc and cobalt. The determination of zinc and cobalt equals in principle the method discussed for the determination of mercury and copper.[13] Zinc (II) ions and cobalt (II) ions are precipitated in the presence of ammonium thiocyanate and mercury (II) chloride as $Zn[Hg(SCN)_4]$ and $Co[Hg(SCN)_4]$; the precipitates are redissolved in hydrochloric acid and the thiocyanate ions are titrated with sodium vanadate.

Another method is based on the amperometric titration of cobalt (II) ions with vanadate (V) in 0·05 N sulphuric acid solution containing 0·5 per cent EDTA.[44b]

Other metals. Antimony,[45,45a] nickel,[46,47] calcium,[23] titanium and molybdenum,[48] osmium,[49] palladium[50,51] and platinum[52] can be determined indirectly with sodium or ammonium vanadate (V).

Iodide. The quantitative oxidation of iodides with sodium vanadate (V) proceeds following the equation

$$I^- + 2\,VO_3^- + 8\,H^+ \rightleftharpoons I^+ + 2\,VO^{2+} + 4\,H_2O$$

in at least 7·5 N hydrochloric acid medium at room temperature. A 0·02 M iodine monochloride solution in the presence of chloroform serves as indicator, diphenylamine can also be used instead.[53] This determination can be carried out in the presence of chloride and bromide.

Thiocyanate, sulphide, thiosulphate and sulphite. These ions can be titrated directly with sodium vanadate (V) in at least 7·5 N hydrochloric acid solution at room temperature.[7] The reaction needs to be catalysed by iodine monochloride which simultaneously serves as indicator (in a chloroform layer).

In the literature, there are also described indirect determinations of thiocyanate,[54] sulphide,[44,55] thiosulphate,[56,57] hydrogen sulphite and sulphite.[56] In these determinations, an excess of vanadate (V) or vanadic acid (V) is added to the sample and the unconsumed amount is back-titrated with a standard solution of Mohr's salt; or, vice versa, the vanadyl ions produced can also be titrated with standard permanganate solution.

More details are to be found in the literature cited.

Dithionate, trithionate, tetrathionate and pentathionate. These substances are determined only indirectly.[4,56] Their oxidation with 0·2 N vanadic (V) acid solution proceeds at elevated or even boiling temperature. Thereupon, either the unconsumed vanadic (V) acid is back-titrated with standard iron (II) sulphate solution or the vanadyl ions produced are titrated with standard permanganate solution.

Hypophosphite and phosphite. The indirect determination of hypophosphite and phosphite is based on their oxidation by sodium vanadate (V) in the presence of silver (I) sulphate as catalyst.[58] The excess of vanadate (V) is back-titrated with Mohr's salt against *N*-phenylanthranilic acid as indicator.

According to another method,[59] phosphite is oxidized with iron (III) sulphate. The reaction is catalysed by palladium (II) chloride. The iron (II) ions produced are titrated with vanadate (V) against diphenylbenzidine.

Other substances. Of other substances can be determined indirectly: phosphate,[60] persulphate,[44] hexacyanoferrate (*II*),[44a] hydrogen peroxide,[44] selenium (*IV*) oxide,[44] compounds of chromium, calcium, silver and others.[2,16,60]

Hydrazine and its derivatives. The oxidimetric determination of hydrazine[7,16,61] proceeds at room temperature in strongly hydrochloric acid or sulphuric acid solution according to the equation

$$N_2H_4 + 4\,VO_3^- + 12\,H^+ \rightarrow N_2 + 4\,VO^{2+} + 8\,H_2O$$

Iodine monochloride in a chloroform layer serves as indicator; it also acts as catalyst. In sulphuric acid solution, *N*-phenylanthranilic acid is used as indicator.

Sodium vanadate (V) can also be used for the determination of:

Semicarbazide chloride, benzalazine, benzalsemicarbazone, amino guanidonium chloride, o-oxybenzalsemicarbazone, p-methoxybenzalsemicarbazone, chloralhydrazine, o-chlorobenzal semicarbazone and *methylethylketone semicarbazone.*[62] All the hydrazine groups of the compounds mentioned are quantitatively oxidized to nitrogen by vanadate involving the exchange of four electrons. The reactions proceed in hydrochloric acid solution under the catalytic action of iodine monochloride. During the oxidation of the above mentioned substances, ammonia and carbon dioxide can also be produced besides nitrogen. Iodine monochloride in a layer of chloroform is used to indicate the titration end-point. This method enables a fairly accurate determination (up to 20 mg) of the substances mentioned.

The *hydrazide of isonicotinic acid* is determined indirectly.[63] The excess of vanadate is back-titrated with Mohr's salt using N-phenylanthranilic acid as indicator. Lactose, glucose and starch do not interfere.

Hydroquinone and metol. Hydroquinone[64-66] and metol[65] are titrated with sodium vanadate (V) in sulphuric acid solution at room temperature under the catalytic action of oxalic acid; diphenylbenzidine or N-phenylanthranilic acid are used as indicator.

Other determinations. Most organic compounds react with vanadate (V) solutions very slowly. Therefore the determinations are carried out indirectly. The excess of vanadate (V) is back-titrated with Mohr's salt or with iron (II) sulphate solution. In this way, there have been determined, e.g.:

Amidopyrine,[67] *analgine,*[67a] *dibromohydroxyquinoline* and *dibromooxinates,*[21] then in hydrochloric or sulphuric acid solution *hydroxylamine,*[66,68] *tannin,*[66] *oxalic acid,*[6,6a] *tartaric acid,*[16,69,70] *citric acid, malic acid, lactic acid, malonic acid,*[16,69,71] various *α-amino acids,*[71a] *organic pharmaceutical preparates,*[71b] *formate,*[44] *glycerol,*[72.73] *indigo,*[74] *phthalocyanine,*[75] p-*phenylenediamine*[75a] and others.[76,76a]

Littler and Watters[77] as well as Jones and Watters[77a] studied the reaction of kinetics of the *oxidations of alcohols, ketones* and *cyclohexanols,* whereas West and Skoog[73] investigated the kinetics of the *oxidation of glycerol.*

A general survey of titrations with pentavalent vanadium is given in the references cited.[78-81]

REFERENCES

1. WILLARD, H. H. and MANALO, G. D., *Ind. Engng. Chem., Anal. Ed.,* **19**, 462 (1947).
2. SYROKOMSKII, V. S. and ANTROPOV, L. I., *Zavodskaya Lab.,* **9**, 818 (1940).
3. SMITH, G. F. and BANICK, JR., W. M., *Talanta,* **2**, 348 (1959).
4. LANG, R. and KURTENACKER, H., *Z. anal. Chem.,* **123**, 81 (1942).
5. BISHOP, E. and CRAWFORD, A. B., *Analyst,* **75**, 273 (1950).
6. WEST, D. M. and SKOOG, D. A., *Analyt. Chim. Acta,* **12**, 301 (1955).
6a. JONES, J. R. and WATTERS, W. A., *J. Chem. Soc.,* 4757 (1961).
7. SINGH, B. and SINGH, R., *Analyt. Chim. Acta,* **10**, 408 (1954).
8. GAUDEFROY, G., *Bull. Soc. Chim. France,* 1222 (1954).
9. RAO, B. K. and RAO, G. G., *Z. anal. Chem.,* **157**, 98 (1957).
10. TSUBAKI, I., *J. Chem. Soc. Japan,* **66**, 10 (1945).

COMPOUNDS OF PENTAVALENT VANADIUM

11. RAO, B. K., GOWDA, H. S. and RAO, G. G., *Z. anal. Chem.*, **156**, 180 (1957).
12. SURANOVA, Z. P. and OLENOVICH, N. L., *Zbornik Khim. Fakult. Odessk. Gosud. Univ. im. Mechnikova*, **4**, 73 (1954).
13. SINGH, B. and SINGH, S., *Analyt. Chim. Acta*, **13**, 405 (1955).
14. RAO, B. K., RAO, J. G. and RAO, G. G., *Z. anal. Chem.*, **157**, 181 (1957).
14a. MURTY, B. V. S. R. and RAO, G. G., *Talanta*, **8**, 426 (1961).
14b. MURTY, B. V. S. R. and RAO, G. G., *Talanta*, **8**, 438 (1961).
15. LANG, R. and GOTTLIEB, S., *Z. anal. Chem.*, **104**, 1 (1936).
16. SYROKOMSKII, V. S. and KLIMENKO, J. V., *Zavodskaya Lab.*, **7**, 1093 (1938).
17. ANKUDIMOVA, E. V., *Trudy Kommis. Anal. Khim.*, *Akad. Nauk SSSR Otdel. Khim. Nauk*, **5**, 8,197 (1954).
18. ANKUDIMOVA, E. V. and PETRASHEN, V. I., *Trudy Novocherkas. Politekh. Inst.*, **41**, 3 (1956).
19. SAITO, K., *J. Chem. Soc. Japan, Ind. Chem. Sect.*, **54**, 8 (1951).
20. RAO, G. G. and SURYANARAYANA, M., *Z. anal. Chem.*, **168**, 177 (1959).
21. NAZARENKO, V. A., VINKOVETSKAYA, S. J., *Zhur. Anal. Khim.*, **11**, 572 (1956).
22. STEPIN, V. V. and SILAEVA, E. V., *Zavodskaya Lab.*, **19**, 409 (1953).
23. SYROKOMSKII, V. S. and KLIMENKO, J. V., *Zavodskaya Lab.*, **9**, 1077 (1940).
24. SYROKOMSKII, V. S. and STEPIN, V. V., *J. Am. Chem. Soc.*, **58**, 928 (1936).
25. RAO, P. V., MURTY, B. V. S. R. and RAO, G. G., *Z. anal. Chem.*, **147**, 161 (1955).
26. SASTRI, M. N. and RAO, G. G., *Current Sci.*, **18**, 402 (1949).
26a. RAO, U. V., *Z. anal. Chem.*, **177**, 190 (1960).
26b. URBANSKI, S., *Chem. analit. (Warszawa)*, **5**, 687 (1960).
26c. UDALCOVA, N. I., *Zhur. Anal. Khim.*, **17**, 476 (1962).
27. RAO, G. G. and SASTRI, T. P., *Z. anal. Chem.*, **167**, 1 (1959).
28. ZVENIGORODSKAYA, V. M. and RIANICHEVA, M. I., *Zhur. Anal. Khim.*, **14**, 457 (1959).
29. UPOR, E., FEKETE, L. and NAGY, G., *Magyar Kémikusok Lapja*, **13**, 305 (1958).
30. MORACHEVSKII, J. V. and CHERKOVNITSKAYA, I. A., *Zhur. Anal. Khim.*, **13**, 337 (1958).
31. ESKEVICH, V. F. and KOMAROVA, L. A., *Zhur. Anal. Khim.*, **15**, 84 (1960).
32. RAO, G. G., RAO, P. V. and VENKATAMMA, N. C., *Z. anal. Chem.*, **150**, 178 (1956).
33. RAO, P. V. and RAO, G. G., *Z. anal. Chem.*, **160**, 190 (1958).
34. TSUBAKI, I., *Japan Analyst*, **4**, 77 (1955).
35. STEPIN, V. V., *Zavodskaya Lab.*, **8**, 262 (1939).
36. RAO, G. G., RAMANJANEYULU, J. V. S. and SASTRI, M. N., *Current Sci.*, **18**, 72, 169 (1949).
37. RAO, G. G., ARAVAMUDAN, G. and VENKATAMMA, N. C., *Z. anal. Chem.*, **146**, 161 (1955).
38. SASTRI, M. N. and RADHAKRISHNAMURTI, CH., *Z. anal. Chem.*, **147**, 16 (1955).
39. RAO, G. G. and SOMIDEVAMMA, G., *Z. anal. Chem.*, **165**, 432 (1959).
40. SYROKOMSKII, V. S. and ZHUKOVA, K. N., *Zavodskaya Lab.*, **11**, 373 (1945).
41. IKEGAMI, T., KAMMORI, O. and MORITA, S., *Tetsu-to-Hagane (J. Iron Steel Inst. Japan)*, **39**, 1350 (1953).
41a. WILSON, A. D., *Analyst*, **85**, 823 (1960).
41b. WATANABE, S., *Nippon Kinzoku Gakkai-Si (J. Japan Inst. Metals)*, **24**, 401 (1960).

42. SASTRI, M. N. and KALIDAS, CH., *Z. anal. Chem.*, **149**, 181 (1956).
43. SASTRI, M. N. and SARMA, L. S., *Z. anorg. Chem.*, **281**, 221 (1955).
43a. RAO, G. G. and RAO, V. M., *Current Sci.*, **13**, 317 (1944); *Proc. Nat. Inst. Sci. India*, **12**, 217 (1946).
44. SINGH, B. and SINGH, R., *Analyt. Chim. Acta*, **11**, 412 (1954).
44a. RAO, G. G. and RAO, V. B., *Current Sci.*, **18**, 124 (1949).
44b. VARMA, A., *J. Sci. Ind. Res. (India)*, **B 21**, 142 (1962).
45. RAO, K. B., *Z. anal. Chem.*, **165**, 195 (1959).
45a. NARUSHKEVICHIUS, L. and DAUKSHAS, K., *Nauch. Trudy Vys. Ucheb. Zaved. Lit. SSR., Khim. i Khim. Technol. N.* **1**, 11 (1961); *Ref. Zhur. Khim.*, 20 G 81 (1963).
46. SYROKOMSKII, V. S. and GUBELBANK, S. M., *Zhur. Anal. Khim.*, **6**, 207 (1951).
47. VERDI-ZADE, A. A., *Trudy Azerb. Zaoch. Pedagog. Inst.*, **4**, 77 (1957).
48. RÍUS, A. and DÍAZ-FLORES, C. A., *An. Real. Soc. Españ. Fisica Quim.*, **46B**, 289 (1950).
49. SYROKOMSKII, V. S., *Doklady Akad. Nauk. U.S.S.R.*, **46**, 307 (1945).
50. SYROKOMSKII, V. S. and GUBELBANK, S. M., *Zhur. Anal. Khim.*, **4**, 146 (1949).
51. SYROKOMSKII, V. S. and GUBELBANK, S. M., *Zhur. Anal. Khim.*, **4**, 203 (1949).
52. SYROKOMSKII, V. S. and PROSHENKOVA, N. N., *Zhur. Anal. Khim.*, **1**, 83 (1946).
53. SIERRA, F. and ASENSI MORA, G., *An. Real. Soc. Españ. Fisica Quim.*, **52B**, 13 (1958).
54. RAO, B. K., MURTHI, R. V. V. S. and RAO, G. G., *Z. anal. Chem.*, **157**, 178 (1957).
55. RAO, B. K. and RAO, G. G., *Analyt. Chim. Acta*, **13**, 313 (1955).
56. LANG, R. and KURTENACKER, H., *Z. anal. Chem.*, **123**, 169 (1942).
57. GOWDA, H. S., RAO, B. K. and RAO, G. G., *Analyt. Chim. Acta*, **12**, 504 (1955).
58. RAO, G. G. and GOWDA, H. S., *Z. anal. Chem.*, **146**, 167 (1955).
59. RAO, G. G. and SOMIDEVAMMA, G., *Z. anal. Chem.*, **164**, 391 (1958).
60. SYROKOMSKII, V. S. and STEPIN, V. V., *Zavodskaya Lab.*, **5**, 144 (1936).
61. SASTRI, M. N., RAO, V. B. and RAO, G. G., *Current Sci.*, **19**, 90 (1950).
62. SINGH, B. and SAHOTA, S. S., *Analyt. Chim. Acta*, **17**, 285 (1957).
63. GOWDA, H. S. and RAO, G. G., *Z. anal. Chem.* **165**, 36 (1959).
64. RAO, G. G., RAO, B. V. and SASTRI, M. N., *Current Sci.*, **18**, 381 (1949).
65. RAO, G. G. and SASTRI, T. P., *Z. anal. Chem.*, **151**, 415 (1956).
66. SURANOVA, Z. P., *Zbornik Khim. Fakult. Odessk. Univ.*, **3**, 73 (1953).
67. CHERKASOV, V. M. and PETROVA, V. A., *Zhur. Anal. Khim.*, **5**, 305 (1950).
67a. NOSENKOVA, N. G., *Trudy Sverdlovsk. Selskokhoz. Inst.*, **7**, 369 (1960).
68. GOWDA, H. S., RAO, B. K. and RAO, G. G., *Z. anal. Chem.*, **160**, 117 (1958).
69. SURANOVA, Z. P., *Zbornik Khim. Fakult. Odessk. Univ.*, **4**, 65 (1954).
70. RAO, G. G. and GOWDA, H. S., *Current Sci.*, **21**, 188 (1952).
71. WEST, D. M. and SKOOG, D. A., *Anal. Chemistry*, **31**, 583 (1959).
71a. RAO, S. P., RATHI, H. S. and GAUR, J. N., *Analyt. Chim. Acta*, **25**, 136 (1961).
71b. EROMINA, Z. I. and GUREVICH, V. G., *Farm. Zhur. (Kiev)*, **15**, 6 (1960).
72. WEST, D. M. and SKOOG, D. A., *Anal. Chemistry*, **31**, 586 (1959).
73. WEST, D. M. and SKOOG, D. A., *J. Amer. Chem. Soc.*, **82**, 280 (1960).
74. RAO, G. G. and SASTRI, M. N., *Current Sci.*, **21**, 189 (1952).

75. RAO, G. G. and SASTRI, T. P., *Z. anal. Chem.*, **169**, 11 (1959).
75a. SASTRI, T. P., *Z. anal. Chem.*, **196**, 349 (1963).
76. RAO, G. G. and MURTY, B. V. S. R., *Z. anal. Chem.*, **174**, 44 (1960).
76a. PANWAR, K. S. and GAUR, J. N., *J. electroanal. Chem.*, **3**, 348 (1962).
77. LITTLER, J. S. and WATTERS, W. A., *J. Chem. Soc.*, 1299, 3014, 4046 (1959).
77a. JONES, J. R. and WATTERS, W. A., *J. Chem. Soc.*, 1629, 2068 (1962).
78. SYROKOMSKII, V. S. and KLIMENKO, J. V., *Vanadometria*, Metalurgizdat, Sverdlovsk, 1950.
79. VULTERIN, J., *Chemie*, **10**, 869 (1958).
80. MACDONALD, A. M. G., *Ind. Chemist*, **32**, 280, 332 (1956).
81. BERKA, A., VULTERIN, J. and ZÝKA, J., *Chemist-Analyst*, **51**, 24 (1962).

CHAPTER 13

IRON (III) SALTS

IRON (III) salts are suitable for the oxidimetric determination of strong reducing agents such as the lower valency states of titanium, tungsten, molybdenum, tin and others. Iron (III) salts are, however, mostly used for volumetric determinations of titanium (III) salts, or, in indirect reductimetric titrations, for back-titrating the excess of the reducing agent. This can be done, e.g. using thiocyanate as indicator for the visual detection of the end-point, which is often impossible in other determinations.

STANDARD SOLUTION

Either iron (III) chloride, iron (III) sulphate or iron (III) ammonium sulphate are used for the titrations.

The standard solution can also be prepared by dissolving pure iron in a mixture of hydrochloric and nitric acid, and boiling until free from nitrogen oxide; the residue is then dissolved in sulphuric acid and water. The iron (III) solutions are stable.

Standardization

The standardization of the solution is carried out by one of the well-known methods for determining iron (III) ions, eventually by direct titration with thiosulphate.[1,19]

INDICATOR

Thiocyanate,[6,15,20] methylene blue[4,6] or indigo carmine[10] have been recommended as indicators for the visual titration, especially in the determination of trivalent titanium. In the titration of vanadium (II) salts, safranine or neutral red is used.[32]

Amongst electrometric methods, there is also recommended (apart from potentiometric titration) the amperometric titration, in which a dropping mercury electrode[2] or a rotating platinum electrode[3] is used.

REVIEW OF DETERMINATIONS

Titanium. After the sample has been dissolved and reduced to the trivalent state, using a metal reductor or amalgam, the titanium (III) ions are titrated visually or electrometrically[35] with standard iron (III) solution,[34,37] generally in an inert atmosphere. In this way, e.g. titanium in titan white,[36] in steels[3] and ores,[22] in the presence of uranium,[4] or titanium (III) in the presence of titanium (II) compounds[5] were determined.

Because titanium (IV) can be selectively reduced with zinc amalgam in tartaric acid solution, in the presence of the highest valency states of niobium and tantalum, it is possible to determine titanium in the presence of the two other substances.[6]

Recently, Konstantinov described a rapid method for the determination of titanium in ferrotitanium.[15]

Japanese authors reduced titanium (IV) salts at elevated temperatures (80°C) with aluminium, and titrated the titanium (III) ions thus produced in air with standard iron (III) solution at 50°C.[17] Small amounts of chromium, vanadium and tin do not interfere with the determination.

Molybdenum. The reddish-brown complex of trivalent molybdenum which is produced on reducing molybdate (VI) in an acid medium containing a sufficient amount of chloride, can be titrated potentiometrically[7] or amperometrically[8] with iron (III) solution ($Mo^{III} \rightarrow Mo^V$). The amperometric method is also used for the determination of phosphorus in steels after its transformation to phosphomolybdate.[8] The titration with standard iron (III) solution has also proved satisfactory for investigations of the composition of various compounds which contain molybdenum in lower valency states.[9]

In one of the most recent publications[18] it has been shown that pentavalent molybdenum compounds can also be titrated at elevated temperatures (98–100°C) either potentiometrically or visually, using rhodanine 6 G as indicator.[50]

The titration of molybdenum in citric or tartaric acid medium, containing 0·5–1 N HCl, using methylene blue or thionine as indicator, has also been reported.[18a]

Tin. In the titration of tin (II) using indigo carmine as indicator,[10] a clearly visible colour change from yellow to indigo appears. The decolorization of the iron (III) thiocyanate complex[20] or of methylene blue[21] can likewise be used for the indication of the end-point. The titration is carried out in an inert atmosphere.

Uranium. The reaction between uranium (IV) and iron (III) salts proceeds rapidly and quantitatively only at elevated temperatures.[11,38,39] The end-point is determined either with thiocyanate, photometrically,[11a] or, more precisely, potentiometrically;[12,25] the equivalence point can also be seen with the aid of a fluorescent indicator. If the titration is done at normal temperature, it is necessary to add an excess of reagent and to catalyse the reaction with osmium (VIII) oxide; the excess is then determined iodimetrically.[28]

Standard iron (III) solution can also be used for the titration of: *molybdenum (III)*,[23,24] *molybdenum (V)*[40] and *vanadium (II)*[32,44] after reduction in a cadmium reductor,[13] or *tungsten (III)*[41] after reduction with zinc amalgam.[14] The titration of *niobium (III) compounds* may be done also in the presence of an excess of tantalum.[42]

The potentiometric titration of *ascorbic acid* with iron (III) solution against thiocyanate, Variamine Blue or sulphosalicylic acid as indicator[26,46] is used for the indirect determination of *platinum (IV) salts* which have been reduced beforehand by an excess of ascorbic acid.[16]

Copper (I) salts. Copper (I) can be titrated in the presence of copper (II) salts, ammonium salts and various organic compounds.[27]

In recent publications, some titrations are termed "ferrimetric determination" which, in fact, are not titrations with iron (III) salts but which are based on the action of an excess of these, and titrating the iron (II) produced with standard vanadate (V) solution. *Arsenic (III)* and *antimony (III)*[29] (OsO_4 or ICl as catalyst) or *hydroxylamine*[30] ($CuSO_4$ as catalyst) are determined in this way.

The determination of *perrhenate* which is reduced with bismuth amalgam is based on a similar principle; after the oxidation with an excess of iron (III) solution, the iron (II) ions thus formed are titrated with cerium (IV) solution.

Finally, the possibility of titrating *indigosols* with iron (III) sulphate should be mentioned, which was studied by Matrka and co-workers,[31] and also the titration of nitroso *phenylhydroxylamine*[33] and *phytine*.[45]

The influence of complexing agents such as *o*-phenanthroline or 2,2-dipyridyl on the redox potential of the system Fe^{3+}/Fe^{2+} has been used in a very elegant way for a selective and sensitive determination of cobalt by titration with iron (III) solution; this method can be applied to the determination of very different materials.[47-49] 1×10^{-3} per cent cobalt in sodium can thus be determined with a 0·001 M $FeCl_3$ solution.[43]

A recent communication of this series deals with the titration of *uranium (IV)* in the presence of an excess of *o*-phenanthroline at pH 2.[49a]

REFERENCES

1. KOLTHOFF, I. M. and TOMÍČEK, O., *Pharm. Weekbl.*, **61**, 1205 (1924)
2. STRUBL, R., *Collection (Czech. Chem. Comm.)*, **10**, 475 (1938).
3. GALLAI, Z. A. and PESHKOVA, V. M., *Vestnik Moskov. Univ., Ser. Fiz.-Matem.*, **7**, 73 (1954); cit.: *Anal. Abstr.*, **6**, 2687 (1956).
4. YOSHIMURA, T., *J. Chem. Soc. Japan, Pure Chem. Sect.*, **72**, 1937 (1951).
5. MARTIN, H. and STEDEFEDER, J., *Liebigs Ann. Chem.*, **618**, 17 (1958).
6. TOMÍČEK, O., SPURNÝ, K., JERMAN, L. and HOLEČEK, V., *Collection (Czech. Chem. Comm.)*, **18**, 757 (1953).
7. DOLEŽAL, J., MOLDAN, B. and ZÝKA, J., *Collection (Czech. Chem. Comm.)*, **24**, 3769 (1959).
8. CHLEBOVSKÝ, T., *Hutnické Listy*, **16**, 252 (1958).
9. RIUS, A. and IRANZO, I. R., *An. Fisica Quim.*, **42**, 645, 761 (1926).
10. SCHLÜTTING, W., *Z. anal. Chem.*, **70**, 55 (1927).
11. *Analytical Chemistry of the Manhattan Project*, C. J. RODEN Ed., McGraw-Hill, New York 1950, p. 70.

11a. FLORENCE, T. M., *Analyt. Chim. Acta*, **23**, 282 (1960).
12. CELLINI, F. and LÓPEZ, A., *An. Real. Soc. Españ. Fisica Quim.*, **52B**, 163 (1956).
13. ERDEY, L. and MÁZOR, L., *Z. anal. Chem.*, **140**, 67 (1956); **143**, 228 (1954).
14. VLASÁK, F., *Collection (Czech. Chem. Comm.)*, **10**, 278 (1938).
15. KONSTANTINOV, E. F., *Zavodskaya Lab.*, **25**, 134 (1959).
16. MAKSIMYUK, E. A., *Izvest. Sekt. Plat. i Drug. Blagorod. Metallov*, **30**, 180 (1955).
17. ONO, K., HOKOYAMA, E. and IWAMATSU, H., *Japan Analyst*, **8**, 796 (1959).
18. RAO, G. G., SEETHARANARAJA, S. and SURYANARAYANE, M., *Z. anal. Chem.*, **169**, 248 (1959).
18a. RAJU SAGI, S. and RAO, G. G., *Acta Chim. Acad. Sci. Hung.*, **38**, 89 (1963).
19. JELLINEK, K. and WINOGRADOFF, L., *Z. anorg. Chem.*, **129**, 15 (1923).
20. JELLINEK, K. and KÜHN, W., *Z. anorg. Chem.*, **138**, 81 (1924).
21. RUSSO, C. and SENSI, G., *Gazz. Chim. Ital.*, **44**, 9 (1937).
22. KHU-CZI-FAN, *Khvasjue sicze* Nr. 12555 (1957).
23. RIUS, A. and MARTIN, H., *An. Fisica Quim.*, **43**, 897 (1947).
24. RIUS, A. and DIAZ-FLORES, C. A., *An. Real. Soc. Españ. Fisica Quim.*, **46B**, 289 (1950).
25. CELLINI, R. F. and LÓPEZ, J. A., *An. Real. Soc. Españ. Fisica Quim.*, **52B**, 163 (1956).
26. BARCZA, L., *Acta pharm. Hung.*, **26**, 152 (1956).
27. FREHDEN, O. and CONN, I., *Rev. chim.*, **8**, 600 (1957); *Ref. Zhur. Khim.*, **10**, No. 22110 (1958).
28. DESAI, M. W. and MURTHY, T. K. S., *Analyst*, **83**, 126 (1958).
29. RAO, B. K. and RAO, G. G., *Z. anal. Chem.*, **157**, 96 (1957).
30. RAO, B. K. and RAO, G. G., *Z. anal. Chem.*, **157**, 100 (1957).
31. MATRKA, M., NAVRÁTIL, F. and FILIPI, J., *Chem. Průmysl*, **7**, 343 (1957).
32. MATRKA, M. and SÁGNER, Z., *Chem. Průmysl*, **9**, 526 (1959).
33. SHEKA, J. A. and KARLYSHEVA, K. F., *Ukrain. Khim. Zhur.*, **25**, 656 (1959).
34. CALLAN, T. and HORROBIN, S., *J. Soc. Chem. Ind.*, **47**, 329 (1928).
35. GIUFFRÉ, L. and CAPIZZI, F. M., *Ann. Chim. (Roma)*, **50**, 1150 (1960).
36. YU-HENG-SUI, *Hua Hsüeh Tung Pao*, **3**, 164 (1960); *Chem. Abstr.*, **55**, 21968 (1961).
37. DESPHANDE, G. M. and NATARAJAN, R., *J. Amer. Pharm. Assoc.*, **47**, 633 (1958).
38. WEISS, G. and BLUM, P., *Bull. Soc. Chim. France*, **735** (1947).
39. NESSLE, G. J., U.S. Atomic Energy Commission, Document MDDC-1123, New York 1947.
40. SEETHARANARAJA, S., RAO, G. G. and SURYANARAYANE, M., *Z. anal. Chem.*, **169**, 248 (1959).
41. TREADWELL, W. D. and NIERIKER, N., *Helv. Chim. Acta*, **24**, 1067 (1949).
42. TREADWELL, W. D., *Helv. Chim. Acta*, **5**, 806 (1922).
43. VYDRA, F. and PŘIBIL, R., *Z. anal. Chem.*, **188**, 273 (1962).
44. NEMODRUK, A. A., ORESHKO, V. F., *Izvest. Vys. Ucheb. Zaved. Khim. i Khim. Technol.*, **3**, 316 (1960); *Ref. Zhur. Khim.*, 88371 (1960); *Anal. Abstr.* **8**, 3805 (1961).
45. SCHORMÜLLER, J., HÖHNE, R. and WÜRDIG, G., *Deutsche Lebensmittel-Rdsch.*, **52**, 213 (1956).
46. NASCIMENTO, R., *Rev. Soc. Brasil. Quim.*, **16**, 165 (1947).
47. VYDRA, F. and PŘIBIL, R., *Talanta*, **3**, 103 (1959).

48. VYDRA, F. and PŘIBIL, R., *Collection (Czech. Chem. Comm.)*, **26**, 2169 (1961).
49. VYDRA, F. and PŘIBIL, R., *Talanta*, **8**, 824 (1961).
49a. VYDRA, F. and PŘIBIL, R., *Talanta*, **9**, 1009 (1962).
50. SEETHARANARAJA, S. and RAO, G. G., *Z. anal. Chem.*, **188**, 164 (1962).

CHAPTER 14

COMPOUNDS OF TRIVALENT COBALT

COMPOUNDS of trivalent cobalt rank amongst the most powerful oxidants. This is shown by the high redox potential of the system

$$Co^{3+} + e^- \rightleftharpoons Co^{2+}$$

which is stated to be 1·82 V in 4 N HNO_3 solution.[1-3] Trivalent cobalt has nevertheless been used only in one single case[4] as a titrant; this is supposedly due to the difficulty of preparing the reagent and to the complicated storage of its solution.

STANDARD SOLUTION

Compounds of trivalent cobalt can be prepared in the form of complexes (commonly as carbonates) by oxidation of cobalt (II) salts[5-9] or in a non-complex form by oxidation with most strongly oxidizing agents,[10-12] or electrolytically.[13-15,18] For volumetric purposes, the preparation of cobalt (III) sulphate by electrolytic oxidation of a cobalt (II) salt[4,13] and dissolving the reaction product in 18 N H_2SO_4 is recommended. The usually 0·1 M solutions have to be stored at $-7°C$; they are then stable up to two months.[4]

Standardization

Cobalt (III) solutions are standardized with a standard solution of Mohr's salt using a potentiometric end-point.[4] The normality of the solution can also be determined indirectly by adding a known volume of standard iron (II) solution and back-titrating the excess with permanganate.[6] Finally, the normality can also be determined iodimetrically.[7-9]

INDICATOR

The equivalence point of the titration with cobalt (III) has been found photometrically.
The colour-intensity of the excess of reagent is followed up at 610 mμ. The change in colour of the compound to be titrated, e.g. in the cerium determination at 400 mμ, can likewise be followed up.

REVIEW OF DETERMINATIONS

The oxidation of some substances with compounds of trivalent cobalt, as e.g. iodide,[16] oxalate,[13] iron (II), chromium (III) and manganese (II),[2] alcohols, aldehydes and organic acids[13-17] has already been investigated. As an oxidimetric reagent, however, cobalt (III) solutions have only been used in the determination of *cerium (III) salts* (using the catalytic action of silver ions), *hexacyanoferrate (II)* and *iron (II) ions* (always in sulphuric acid solutions[4]). Suitable conditions for the quantitative oxidation of arsenic (III), chromium (III) and manganese (II), of hydrogen peroxide, bromide, chloride, iodide, of some organic acids and alcohols,[4] with cobalt (III) solution have so far not yet been established.

REFERENCES

1. JAHNS, S., *Z. anorg. Chem.*, **60**, 292 (1908).
2. NOYES, A. A. and DEAHL, T. J., *J. Amer. Chem. Soc.*, **59**, 1337 (1937).
3. SWETSER, P. B. and BRICKER, C. E., *Anal. Chemistry*, **25**, 253 (1952).
4. BRICKER, C. E. and LOEFFLER, L. J., *Anal. Chemistry*, **27**, 1419 (1955).
5. JOB, A., *Ann. Chim. phys.*, **20**, 214 (1909).
6. WILLARD, H. H. and HALL, D., *J. Amer. Chem. Soc.*, **44**, 2237 (1922).
7. METZI, A., *Z. anal. Chem.*, **53**, 537 (1914).
8. MORI, M. and SHIBATA, M., *J. Chem. Soc. Japan, Pure Chem. Sect.*, **75**, 1044 (1954); *Chem. Abstr.*, **49**, 6023 (1955).
9. HOFMANN-BANG, N. and WULFF, I., *Acta Chem. Scand.*, **9**, 1230 (1955).
10. FICHTER, H. and WOLFMANN, H., *Helv. Chim. Acta*, **9**, 1093 (1926).
11. BRUNNER, E., *Helv. Chim. Acta*, **12**, 208 (1929).
12. KITHASHIMA, S., *Bull. Inst. Phys. Chem. Res. (Tokyo)*, **70**, 1035 (1928).
13. SWANN, S. and XANTHAKOS, T. S., *J. Amer. Chem. Soc.*, **53**, 400 (1931).
14. BOMMER, V. H., *Z. anorg. Chem.*, **246**, 275 (1941).
15. KYUNO, M. E. and SHIBATA, M., *J. Chem. Soc. Japan, Pure Chem. Sect.*, **77**, 1434 (1956); *Chem, Abstr.*, **52**, 2628 (1958).
16. SCHELL, C. and MARKGRAF, H., *Trans. Amer. Electrochem. Soc.*, **45**, 161 (1924).
17. BAWN, C. E. G. and WRITE, A. G., *J. Chem. Soc.*, 331, 339, 343 (1951).
18. ROSSENINSKY, D. R. and HIGGINSON, W. C. E., *J. Chem. Soc.*, 31 (1960).

CHAPTER 15

HYDROGEN PEROXIDE

HYDROGEN PEROXIDE can react both as oxidant and as reductant, depending on the substance with which it reacts and on the medium in which the reaction takes place. For example, hydrogen peroxide oxidizes arsenite, but reduces hypohalites and hexacyanoferrate (III).

Erdey and Buzás[1] presumed that the following equilibrium is established between the hydrogen peroxide molecule, acting as donor, and the hydrogen peroxide anion in alkaline solutions:

$$H_2O_2 \rightleftharpoons H^+ + OOH^-$$

In the presence of a suitable hydrogen acceptor (i.e. a substance which can be reduced) hydrogen peroxide is decomposed with the evolution of oxygen; the equilibrium is shifted in favour of the formation of molecular hydrogen until all the hydrogen peroxide is consumed (hydrogen peroxide acts as reductant). In the presence of an oxygen acceptor (i.e. a substance which can be oxidized) the oxygen atom of the hydrogen peroxide anion reacts; here, the hydrogen peroxide molecule is completely dissociated and the total active oxygen of the peroxide is used for the oxidation (hydrogen peroxide acts as an oxidant).

Hydrogen peroxide has so far been described as a titrant in very few publications.

STANDARD SOLUTION

A 0·1 N hydrogen peroxide solution is prepared by diluting the 30 per cent commercially available solution.[1,2] In some publications, a reagent is recommended, made of hydrogen peroxide and titanium (IV) salts.[3,11]

Standardization

The normality is determined by titrating with permanganate or iodimetrically,[3] in some cases with standard arsenite solution, using lucigenin as indicator.[1]

Hydrogen peroxide solutions, stored in dark bottles, change their titre after 8 hr.[2] To increase their stability, it is recommended that the solutions be acidified with sulphuric acid. Other stabilizing reagents usually interfere with most of the determinations.[1,5] The reagent in the form of a complex compound with titanium (IV) is stable for about 14 days.[3]

INDICATOR

The well-known reaction of titanium (IV) ions with hydrogen peroxide[8] (the slightest excess of the reagent forms a yellow colour) has been used in earlier work for the determination of the equivalence point in the titration of tin (II) ions. The equivalence point is also indicated by iodide and starch solution or by iron (II) and thiocyanate solution (a slight excess of reagent produces through oxidation of the iron (II) ions the well-known red colour).[9]

Some titrations can be carried out against lucigenin as indicator.[1] Lucigenin is a chemifluorescent indicator (dimethyldiacridilium nitrate) which fluoresces itself in strong alkaline solutions (pH 13); in less alkaline solutions the substance only fluoresces in the presence of hydrogen peroxide.[4]

Nickel (II) salt can also be used as indicator in titrations with hydrogen peroxide;[6] here, a pale green precipitate of nickel (II) hydroxide forms in alkaline solution which is coloured black by an excess of hydrogen peroxide, owing to the formation of nickel (III) hydroxide. It can also be titrated potentiometrically[2,3] or photometrically in certain cases.

REVIEW OF DETERMINATIONS

Newer determination methods, including the less accurate visual titration of cerium (IV), are summarized in Table 3.

TABLE 3. SURVEY OF TITRATIONS WITH H_2O_2

Substance to be determined		Optimal conditions	Indicator
$Fe(CN)_6^{3-}$	$\to Fe(CN)_6^{4-}$	0·1 N NaOH, 80°C	Lucigenin[1]
AsO_3^{3-}	$\to AsO_4^{3-}$	0·1 N NaOH, 80°C	Lucigenin[1]
ClO^-	$\to Cl^-$	0·1 N NaOH, 80°C	Lucigenin[1]
BrO^-	$\to Br^-$	0·1 N NaOH	Lucigenin[1]
IO^-	$\to I^-$	0·1 N NaOH	Lucigenin; $Ni(OH)_2$[6]
Ce^{4+}	$\to Ce^{3+}$	10% H_2SO_4, 50°C	Photometry, potentiometry,[3] visual indication (decolorization of the solution)[7] amperometry[12]
Sn^{2+}	$\to Sn^{4+}$	Conc. HCl	Potentiometry;[2] KI + starch; thiocyanate and iron (II) salt;[9] complex formation with titanium (IV) salt[8]

The titration of *hypobromites* proceeds best, this can be applied to the indirect determination of *chromium (III)*,[1] *arsenic (III)*, *ammonium salts*, *sulphides* (sulphur in steel), *sulphites* and *thiosulphites*. The substance to be determined is oxidized with an excess of hypobromite, and the excess is back-titrated with hydrogen peroxide.[6]

A method has recently been described for the determination of very dilute peroxide solutions (10^{-5} M) in which the sample is reduced in alkaline solution with an excess of arsenic (III) solution, and the unconsumed amount is titrated amperometrically with hydrogen peroxide.[10] In one of the latest publications, the indirect determination of uranium with a peroxide solution, stabilized with Ti^{4+}, has been described.[11]

REFERENCES

1. ERDEY, L. and BUZÁS, I., *Acta Chim. Acad. Sci. Hung.*, **6**, 77 (1955).
2. VULTERIN, J. and ZÝKA, J., *Chem. Listy*, **50**, 311 (1956).
3. BECK, M. T., *Acta Chim. Acad. Sci. Hung.*, **5**, 209 (1955).
4. ERDEY, L., *Acta Chim. Acad. Sci. Hung.*, **3**, 95 (1953).
5. ERDEY, L., *Acta Chim. Acad. Sci. Hung.*, **3**, 81 (1953).
6. ERDEY, L. and INCZÉDY, I., *Z. anal. Chem.*, **166**, 410 (1959).
7. JOB, A., *Compt. Rendu*, **128**, 101 (1899).
8. JELLINEK, K. and KREBS, P., *Z. anorg. Chem.*, **130**, 276 (1923).
9. JELLINEK, K. and KÜHN, W., *Z. anorg. Chem.*, **138**, 84 (1924).
10. KOLTHOFF, J. M., MEEHAN, E. J., BRUCKENSTEIN, S. and MINATO, H., *Microchem. J.*, **4**, 33 (1960).
11. FACSKO, GH., *Rev. Chim. (Bucuresti)*, **11**, 236 (1960); *Chem. Abstr.*, **57**, 7903 (1962).
12. ZHDANOV, A. K. and RYAZANOVA, T. M., *Uzbekh. Khim. Zhur.*, 18 (1962); *Anal. Abstr.*, **11**, 917 (1964).

CHAPTER 16

CHROMIUM (II) SALTS

DIVALENT chromium compounds exhibit a strong reducing action. Their solutions are therefore used as active reductants in the determination of many inorganic substances. Dimroth and Frister introduced chromium (II) compounds for the first time in analytical chemistry in 1922 (determination of α, α'-dipyridyl).[1a]

The normal potential of the redox system

$$Cr^{2+} \rightleftharpoons Cr^{3+} + e^-$$

is stated to be $- 0\cdot412$ V[1]. Grube and Schlecht[1] investigated the redox potential in various media and found that it changes only very little. Thus they found the redox potential to be in a neutral solution containing acetate $- 0\cdot403$ V, in $0\cdot003$ N sulphuric acid medium $- 0\cdot412$ and in a weakly acid chloride containing solution $- 0\cdot398$ V. Other authors arrived at similar results.[2-3a]

Chromium (II) chloride, chromium (II) sulphate or chromium (II) acetate are used as standard solutions for the titrations.

STANDARD SOLUTION

Standard chromium (II) solutions are usually prepared by reducing potassium dichromate (VI) or chromium (III) alum either electrolytically[1,2,4-7] or with zinc,[8-11] or in some cases with reductors.[12-17]

For the preparation of chromium (II) sulphate by electrolytical reduction,[4] dissolve 40 g potassium dichromate in 200 ml water, add 15 ml of concentrated H_2SO_4, reduce by passing sulphur dioxide; expel the excess of sulphur dioxide by passing air through the solution; dilute then to 250 ml, electrolyse whilst stirring constantly at a potential of 16 V and at a current strength of $0\cdot02$ A/cm².

The two lead electrodes are usually so positioned that the cathode forms a lead cylinder and the anode a lead " jacket ". The reduction is completed within about 7–12 hr.

The electrolytic preparation of a chromium (II) chloride solution is carried out similarly:

Dissolve 40 g of potassium dichromate (VI) in 200 ml of concentrated HCl, dilute to 250 ml and reduce electrolytically for 3 hr at 8 V and 0·02 A/cm^2. A carbon electrode serves as anode in this case.

The reduced solution is forced into the storage container of the burette by means of carbon dioxide and diluted to 250 ml.

In the reduction with zinc[8] one also uses potassium dichromate (VI) to which pure concentrated hydrochloric acid has been added and boiled until the evolution of chlorine ceases. The solution is cooled down; then the reduction is carried out with zinc in a glass flask fitted with a Bunsen valve. The reduction takes several hours and is complete when the solution shows a clear blue colour. The reduced solution is forced through a glass filter into another flask by means of an oxygen-free hydrogen stream which passes through a wash bottle (bubbler) containing chromium (II) chloride solution; this flask is filled with an excess of sodium acetate solution which has been boiled and through which hydrogen gas has been passed beforehand. Red chromium acetate precipitates instantaneously from the solution; it is thoroughly washed with water. The water used for this washing has to be boiled out and then cooled in a carbon dioxide atmosphere. It is recommended to heat the last portion of the wash-water with dilute nitric acid and to test with silver nitrate for the absence of chloride. This is especially necessary if a chromium (II) sulphate solution is to be prepared.

To the washed chromium (II) acetate add sufficient 2 per cent hydrochloric or sulphuric acid so that only a small part of the substance remains undissolved. The clear blue solution is forced with a stream of hydrogen into a storage container of the burette, previously filled with hydrogen. The undissolved part of chromium (II) acetate is dissolved by adding some millilitres of acid. The total solution is brought to the required volume by boiled water.

Chromium (II) sulphate solutions can also be prepared from chromium (III) alum, $KCr(SO_4)_2 \cdot 12\,H_2O$.[18] Hatfield[19] and Postis[20] describe the preparation of chromium (II) acetate. In all cases mentioned above, 0·1 N solutions are used. Very recently, Lux and Illman[20a] found that solutions of divalent chromium can easily be prepared by simply dissolving metallic chromium in hydrochloric, sulphuric or perchloric acid. The chromium used has to be, however, very pure. If less pure chromium is used for preparing the solutions (the authors mention chromium supplied by MERCK which contains 99 per cent chromium and 0·34 per cent iron), when it is dissolved, mostly trivalent chromium is produced. Accordingly the failures formerly obtained with this method can be explained.

Chromium (II) solutions are easily oxidized by atmospheric oxygen and have therefore to be stored in an inert atmosphere. If these conditions are strictly followed, the titre of the solution hardly alters;[4,5,8] it is sufficient to check the normality once or twice a month. It was found[4,8] that less stable solutions result from reduction with zinc and which have not been purified via the chromium (II) acetate.

The solutions to be titrated have likewise to be freed of oxygen by boiling them in a stream of carbon dioxide or by prior reduction. Apart from oxidation by atmospheric oxygen, chromium (II) solutions can also be ruined spontaneously by hydrogen ions to form hydrogen molecules:

$$2\,Cr^{2+} + 2\,H^+ \rightarrow 2\,Cr^{3+} + H_2$$

This reaction proceeds however very slowly, especially if no catalytically active substances are present, such as platinum, gold or some forms of silicic acid.[23] Other authors[7] found that arsenic sulphides, antimony sulphides as well as copper and palladium act also catalytically. Asmanov[2] observed that chromium (II) compounds in the absence of platinum are less stable in hydrochloric acid than in sulphuric acid solution. The spontaneous decomposition of the solution increases with increasing hydrochloric acid concentration and also with an increasing chloride ion concentration. In 4 N sulphuric acid, no formation of hydrogen was noticed.

For titrations with chromium (II) solutions all kinds of apparatus can be used which completely exclude atmospheric oxygen.[5,8] An automatic storage burette is the most suitable;[8] contact of the solution with greased stopcocks before it enters the graduated tube thus can be avoided; errors are then reduced significantly.

Standardization

Copper (II) sulphate and potassium dichromate (VI) are mostly used as standard substances for the standardization of the solution. Copper (II) sulphate has to be purified very carefully and, particularly to be freed of iron.[8,21]

It is more advantageous to standardize the chromium (II) sulphate solution against potassium dichromate (VI) solution:[5]

Dissolve the weighed amount of the standard substance in 10–20 ml of sulphuric acid, dilute to 100 ml with distilled water, transfer to the titration apparatus and pass CO_2 through the solution for at least 15 min; then titrate potentiometrically with the chromium (II) sulphate solution.

If a chromium (II) chloride solution has to be standardized, iron (II) ammonium sulphate or iron (II) sulphate is oxidized by potassium dichromate and the iron (III) sulphate thus formed is titrated with the chromium (II) chloride solution to be standardized.[15] This method has been examined by the authors of this book and has been found to be both simple and accurate.

Chromium (II) solutions can also be standardized with bromate, potassium permanganate, iodate, potassium periodate and cerium (IV) sulphate.[4,22]

INDICATOR

For chromometric determinations, the potentiometric end-point indication is the most suitable; a platinum indicator and a calomel or mercury (I) sulphate reference electrode are used. The titration can also be carried out visually,[23a] i.e. using indicators such as N-phenylanthranilic acid, ferroin, diphenylbenzidine, p-ethoxychrisoidine, neutral red, methyl red, etc.

REVIEW OF DETERMINATIONS

Tin. Tin (IV) chloride[9,24] is titrated in the presence of catalysts— antimony (V) chloride or better bismuth (III) chloride (in about the same concentration as the tin (IV) chloride to be titrated)—after adding 20 ml of concentrated hydrochloric acid and 30 g of crystallized calcium chloride.

Dilute the solution to about 100 ml, boil for 5 min and titrate with chromium (II) chloride solution whilst passing CO_2.

The temperature of 90–100°C and the high chloride concentration bring about a quantitative titration. Bismuth (III) and antimony (V) ions act as catalysts. If antimony (V) chloride is used, this is first reduced to Sb^{3+} and then Sn^{4+} is reduced to Sn^{2+}. Using bismuth (III) chloride as catalyst, the tetravalent tin is reduced first and then the trivalent bismuth, both to the metal.

The determination of tin in the presence of other cations is described below.

Antimony. The determination of antimony proceeds easily at elevated temperatures in 20 per cent hydrochloric acid solution in the presence of calcium chloride.[9] This determination has also been used for the microtitration.[10] Good results are achieved in solutions containing about 2×10^{-4} gm-atoms of pentavalent antimony per litre. The titration is carried out at 80–85°C with 0·4 N chromium (II) chloride solution.

Shatko[10a] described the indirect determination of tri- and pentavalent antimony.

Copper. The reaction between copper (II) and chromium (II) ions proceeds quantitatively in 2–20 per cent hydrochloric acid or in 2–15 per cent sulphuric acid, or also in acetic solution in the presence of sodium or ammonium chloride.[8,21,26–27a]

The divalent copper is first reduced to the monovalent state and then to the metal. In hydrochloric or in acetic acid solution containing chloride ions, the divalent copper is only reduced to the monovalent state. In this case, the chloride complex of monovalent copper is formed and so its valency stabilized.

In the potentiometric titration in hydrochloric acid medium[8,21] atmospheric oxygen is removed either by boiling the solution for 5 min in a carbon dioxide atmosphere, or by prior reduction with chromium (II) chloride solution, added to the hot solution. The copper (I) ions thus formed consume the oxygen in the solution very rapidly, and all the copper is oxidized potentiometrically either with potassium bromate or with potassium permanganate. Then it is titrated with 0·1 N chromium (II) chloride solution at 80°C. The results obtained are very precise.

This determination is not affected by lead, cadmium, bismuth, silver and tungsten (VI) salts. In the presence of mercury (II) and arsenic (III) salts, sodium chloride has to be added. Hexavalent molybdenum is reduced to pentavalent molybdenum[28] with the simultaneous reduction of copper.

Copper can also be determined in the presence of iron,[29] tin (IV) and gold.[8,21] Antimony (III), antimony (V) and arsenic (V) ions do interfere with the determination.[8,21]

In sulphuric acid solution, the determination is not affected by lead, silver, aluminium, cobalt, nickel salts and small amounts of antimony[26] as well as arsenic (V), mercury (II) and iron. Arsenic (III) has to be oxidized with potassium dichromate.

The determination of copper in the presence of iron gives satisfactory results if the excess of iron is not more than the 25-fold.[30] Molybdenum interferes.[28]

Domange[31] as well as Bard and Petropoulos[31a] studied the determination of copper.

Silver and gold. The potentiometric determination of silver in 2–20 per cent sulphuric acid at elevated temperatures is described in the literature;[32] the silver ions are reduced to the metal. Brennecke found that the results in 1·6–5·6 per cent sulphuric acid are always too high by 1 per cent.[33]

Gold (III) ions are reduced to the metal in hot hydrochloric acid solution by chromium (II) chloride. The precipitated gold catalyses the reaction between chromium (II) and hydrogen ions; this brings about a noticeable consumption of excess of the titrant. If copper (II) salt is added to the sample solution, the reaction

$$Cr^{2+} + Cu^{2+} \rightarrow Cr^{3+} + Cu^{+}$$

$$3\,Cu^{+} + Au^{3+} \rightarrow 3\,Cu^{2+} + Au$$

proceeds so rapidly, that the undesired catalytic effect does not come into action. The determinations of gold alone, or of gold in the presence of copper are based on this principle.[8,21,32]

If gold has to be determined alone, a precisely measured amount of copper (II) sulphate is added and then it is titrated in the same way. Lead, cadmium, bismuth and iron ions do not interfere with this determination.

Mercury. The determination of mercury[21,34] is carried out in 3–5 per cent hydrochloric acid or in 5 per cent hydrochloric acid in the presence of bismuth (III), iron (III) or copper (II) chloride. If it is titrated in acetic acid solution ammonium and bismuth (III) chloride has to be added.

The titration error[33] can amount to + 3 per cent and in the presence of bismuth (III) chloride to about + 1·5 per cent. The reason is to be found probably in the reaction between chromium (II) and hydrogen ions. The mechanism of the catalysis has not, however, so far been successfully clarified.

The titration can also be carried out in 3 per cent sulphuric acid solution,[33] but then the titrant has to be added only slowly. The titration error averages only 0·5 per cent.

Bismuth. The titration with chromium (II) sulphate or chloride can be done in hydrochloric acid solution at room temperature[35,36] as well as at elevated temperature.[37] The acid concentration should not exceed 6 N, otherwise the potential jump is indistinct. The oxygen is removed by passing carbon dioxide through the solution for 15–20 min. This time can be shortened if the solution is reduced beforehand (see copper) and then oxidized with potassium bromate.

Cadmium (II), lead (II), sulphate, chloride and small amounts of acetate ions do not interfere with the determination.

The titration can also be carried out in an acetic acid solution containing chloride and tartrate.[37] Other authors[24] worked out a method for determining bismuth, which is carried out under the same conditions as described for the tin.

Iron. The determination of iron is carried out potentiometrically at room temperature in sulphuric acid solution.[38,39]

The presence of potassium thiocyanate acts favourably upon the reaction procedure.[12] Zintl and Schloffer[26] recommend for the titration of iron in sulphuric acid solution the same procedure as for the determination of copper. The oxygen is also removed from the solution by prior reduction. The reaction between the iron (II) and oxygen proceeds, however, more slowly, so that the following oxidation with permanganate or dichromate can only take place after some minutes.

Iron can thus be determined in the presence of hexavalent molybdenum, pentavalent arsenic, titanium and copper. Lead, aluminium, cobalt and manganese do not interfere with this determination. The determination can be carried out in 10 per cent hydrochloric acid solution in the presence of a catalyst (bismuth (III) chloride).

Cooke and co-workers[40] as well as Syrokomskii and Zhukova[41] studied the titration of iron.

Cobalt. Přibil and Švestka[42] developed a method based on complex formation of trivalent cobalt with the disodium salt of ethylenediamine tetraacetic acid (EDTA). Cobalt (III) EDTA is produced by oxidation of cobalt (II) salts with cerium (IV) sulphate solution, in the presence of an excess of EDTA. After the cerium (IV) ions have been removed, the trivalent cobalt is determined potentiometrically with chromium (II) chloride.

This method has been used for the analysis of special steels.

Molybdenum. The reaction between chromium (II) and molybdate (VI) ions proceeds quantitatively in hydrochloric or sulphuric acid solution at 80–100°C. The hexavalent molybdenum is first reduced to the pentavalent and then to the trivalent state; this manifests itself by two potential jumps in the potentiometric titration.[43]

The titration can also be done in the presence of oxalic acid, but the solution must then contain phosphoric or silicic acid.[11]

Molybdenum can also be determined in a mixture with iron;[44,45] copper, titanium, vanadium, tungsten and iron salts do not interfere.[28]

Molybdenum can furthermore be determined amperometrically[46] using a rotating platinum micro-anode in hydrochloric, sulphuric or phosphoric acid solution; manganese, zinc, aluminium, chromium, cobalt, titanium, tungsten and iron do not interfere with this determination. This method has been applied to the analysis of ferromolybdenum and other steels.

Molybdenum in molybdenite can be determined potentiometrically in 6 N hydrochloric acid solution with chromium (II) chloride in the presence of potassium thiocyanate; the latter causes a sharp potential change and, at the same time, a colour change at the equivalence point.

Höltje and Geyer,[48] Goryushina and Cherkashina[49] and also other authors[49a,b,c] deal with the reduction of hexavalent molybdenum by chromium (II) sulphite.

Tungsten. Flatt and Sommer[28,50] worked out a potentiometric method for determining hexavalent tungsten in hydrochloric acid solution at 70–90°C. The rather small but distinct potential jump corresponds with the reduction to pentavalent tungsten. The determination is not affected by iron, copper, molybdenum (VI) and dichromate.

Determinations of tungsten in tungstenite and Scheelite[47] with chromium (II) compounds are carried out potentiometrically in 10 N hydrochloric acid solution. Molybdenum (up to 0·1 per cent) does not interfere. Other publications deal also with the determination of tungsten.[49b,51-53]

Uranium. Flatt and Sommer[50] studied the conditions for a quantitative reduction of uranium compounds with chromium (II) salts. Hexavalent uranium is reduced to the tetravalent state in 2–6 N hydrochloric or 1–6 N sulphuric acid solution. The titration proceeds best in the heat and in the presence of iron (II) ions which accelerate the equilibrium establishment. Shamy and co-workers[49b,54] as well as other authors[54a,b] deal with a similar topic.

Chromium (II) acetate has likewise been used for the reduction of hexavalent uranium.[55]

Acidify the solution of uranyl nitrate in 30 ml ethanol with 0·5 ml of concentrated sulphuric acid and titrate with 0·02 N chromium (II) acetate in dioxan. In this way, copper and vanadium can also be determined side by side.

Uranyl acetate can be determined visually with chromium (II) sulphate[25] against neutral red, methyl red, phenosafranine or *p*-ethoxychrysoidine as indicators. The titration can only be carried out in the presence of an excess of sulphuric acid, when methyl red or *p*-ethoxychrysoidine is used. Neutral red and phenosafranine are not reliable.

CHROMIUM (II) SALTS

Dichromate (VI) and chromate (VI). Dichromate is reduced to chromium (III) ions by chromium (II) salts in sulphuric acid solution.[56] This reduction proceeds quantitatively at room temperature in 3–5 per cent sulphuric acid. At lower acid concentrations the potential adjusts itself badly, whereas at higher acid concentrations the dichromate (especially at boiling temperature when atmospheric oxygen is removed) may be decomposed. In the latter case the results are always unprecise.[27] The standard solution has to be added slowly. If chromium (II) chloride is used for the titration, iron (II) ammonium sulphate has to be added to the dichromate solution; the trivalent iron produced by oxidation is then titrated with chromium (II) chloride.[13,35]

The polarimetric determination of chromate (VI)[57] and dichromate (VI) with standard chromium (II) sulphate solution is carried out in 4 N sulphuric acid medium. Chromate (VI) and molybdate (VI) can be determined together in this way.

Chromate (VI)[58] can be determined indirectly when methylene blue is used:

Add the chromate to a 0·7 per cent solution of methylene blue, previously reduced by chromium (II) sulphate in 10 N sulphuric acid solution; wait for 2 min and then titrate with chromium (II) sulphate.

Vanadate (V). The reaction between vanadate (V) and chromium (II) salts proceeds under the same conditions as described for the determination of dichromate.[56] Here, the vanadate (V) is first reduced to the vanadyl (IV) ion and then to the vanadous ion, using a potentiometric end-point indication.

Muraki[59] studied the potentiometric titration of vanadate (V) with chromium (II) sulphate under the catalytic action of iodine monochloride.

Vanadate (V) can be determined by visual titration with chromium (II) sulphate[25] in 10 N sulphuric acid using N-phenylanthranilic acid, diphenylamine sulphonic acid, diphenylamine or diphenylbenzidine as indicators; the pentavalent vanadium is reduced to the tetravalent state. This determination can be carried out in the presence of copper (II), iron (III), cerium (IV) and chromium (VI) salts.

Titanium. Titanium (IV) salts are reduced quantitatively to the trivalent state in hydrochloric acid solution containing sufficient calcium or sodium chloride.[60] The titration is carried out potentiometrically at 90°C. Here it is important to add the standard chromium (II) solution only slowly in order to be able to see the relatively little potential jump. For that reason, Lingane[38] criticizes this method as being unsuitable and recommends the titration be done in 4 N sulphuric acid solution using mercury indicator electrodes; this reaction proceeds at room temperature and the end-point is represented by a distinct potential change.

Thallium. Thallium (III) ions are reduced quantitatively to the monovalent state by chromium (II) salts. This has been applied for the potentiometric determination of trivalent thallium.[60a] The determination is carried out in 2–3 N hydrochloric acid at 30–35°C in a carbon dioxide atmosphere. One can also titrate visually using methyl orange as an irreversible indicator. The potentiometric determination, however, gives more accurate results. The determination of trivalent thallium in the presence of other ions has not been investigated so far.

Titration of mixtures. Of all the determinations with standard chromium (II) solution in analytical practice, those titrations have proved advantageous which make it possible to determine several cations in the presence of each other. A number of such titrations can be carried out with standard chromium (II) chloride, sulphate or acetate solutions using a potentiometric end-point indication. Most of these determinations are applied to the analysis of steels, alloys, minerals, ores and various raw materials. In the further review, the most important determinations of this kind are mentioned. Detailed descriptions are to be found in the original literature cited.

Copper and silver. The determination is carried out in ammonium chloride, sodium acetate or acetic acid solution. First, the copper (II) ions are reduced to copper (I) and then the silver (I) ions to the metal. This method is suited to the rapid determination of copper and silver in alloys.[21] The titration can also be done in sulphuric acid solution[32,37] but the determination of silver then shows a positive error.

Copper–iron, copper–iron–dichromate. The determination is best carried out in sulphuric acid solution under the same conditions as described for the iron.[26,29,30] This method has been applied to the determination of copper and iron in sulphides and pyrites.[26] Other mixtures that can be titrated are: *copper and gold,*[34] *copper and silver,*[32] *copper, gold and mercury,*[32] *copper and mercury,*[34] *copper and antimony,*[24,61] *copper and selenium,*[27] *copper and molybdenum,*[28] *copper and titanium,*[60] *copper and vanadium,*[55] *copper and tungsten.*[28]

Tin in mixture with other substances. Tin and *copper,* tin and *antimony,* tin and *bismuth,* tin and *iron,* tin, *copper* and *antimony,* tin, *copper* and *bismuth,* tin, *iron* and *bismuth*[9,24,61] can be determined potentiometrically. The methods have been applied[24] for rapid determination of various alloys, e.g. bronzes, bearing metals and others.

Iron–titanium. The mixture of these ions is titrated potentiometrically in 4 N sulphuric acid at room temperature.[38] First, the trivalent iron is reduced when a platinum indicator electrode is used; then, after having changed the platinum for a mercury electrode, the tetravalent titanium is reduced.

Another determination is based on the titration of iron and titanium[60,62] in hydrochloric acid solution containing calcium chloride at 90°C. In the analysis of ferrotitanium, the titration is carried out in 2 N hydrochloric or 2 N sulphuric acid in the presence of ammonium chloride at a constant temperature of 85°C. Flatt and Sommer[28] recommend the use of 8 per cent hydrochloric acid or 10 per cent sulphuric acid at 90°C. The successive determination of *iron, copper* and *titanium* in ferrotitanium can be carried out using various electrodes.[63] The tungsten electrode is the most suitable. When titrating in sulphuric or hydrochloric acid solution, 3 potential jumps can be noticed corresponding to the reductions Fe^{3+} to Fe^{2+}, Cu^{2+} to Cu^+ and Ti^{4+} to Ti^{3+}.

Iron and titanium can be determined with chromium (II) sulphate solution also in titanium-containing clay.[64]

Slavík,[5] Martinchenko and Shimko[65] dealt with the determination of iron and titanium.

Iron–vanadium. The process of the potentiometric determination of iron (III) and vanadium (V) depends on the titration conditions.[28,56,66] If the iron concentration is low, the titration is done in 10 per cent sulphuric acid; if it is higher, in 20 per cent sulphuric acid whilst cooling. The method is used in the analysis of ferrovanadium.

Iron–molybdenum. The reduction of iron (III) and molybdenum (VI) proceeds in 5–30 per cent sulphuric acid at 90°C,[28] or in concentrated hydrochloric acid in the presence of calcium chloride, likewise at 90°C.[44,45] Slavík[5] also studied this problem.

Iron–molybdenum–chromium. The titration with chromium (II) sulphate is carried out in 5–8 per cent sulphuric acid at 85°C using a potentiometric end-point.[45] Successively are reduced: chromium (VI) to chromium (III), iron (III) to iron (II) and molybdenum (VI) to molybdenum (V). This method is recommended for the analysis of steel.

Iron–tungsten. For the determination of a mixture of iron (III) and tungsten (VI) ions, the same conditions are suitable as described for the determination of tungsten alone.[28,50] Muraki[67] mentioned that the titration can be done in 10 N hydrochloric acid at room temperature. The determination is affected by phosphoric acid and fluoride.

Other mixtures which can be determined potentiometrically are: *iron, tungsten and molybdenum;*[28] *iron and chromium;*[26,45,56] *iron, vanadium and chromium.*[56]

Molybdenum–iron–copper. Busev and Gyn[68] studied the potentiometric titration of this mixture with chromium (II) chloride using a tungsten indicator electrode. The reduction proceeds best in hydrochloric acid medium at 70°C. This method has been applied to the analysis of ferromolybdenum.

Furthermore can be determined simultaneously:

Molybdenum and tungsten;[28,69,70] *molybdenum and vanadium;*[28,71] *molybdenum and chromium;*[45,57,71] *molybdenum, chromium and vanadium;*[71] *molybdenum and titanium.*[28]

Uranium–vanadium–iron–chromium. Minczewski and co-worker[72,73,55] studied the reduction of these substances in various mixtures with chromium (II) sulphate and chromium (II) acetate. A mixture of chromium

(VI), vanadium (V), iron (III) and uranium (VI) can be determined by titration with standard chromium (II) sulphate solution in 5 per cent sulphuric acid at room temperature.[72] In the same way there have been determined potentiometrically in 5 per cent sulphuric acid solution the following mixtures:

Uranium and vanadium; iron, chromium and uranium; vanadium and iron. These ions can also be titrated individually. This method is used for the determination of ores which have been first dissolved in sulphuric and hydrochloric acid.

Nikolayeva[74] also worked about the titration of uranium.

Other mixtures. The following mixtures can likewise be determined with standard chromium (II) solution using a potentiometric end-point indication:

Mercury and bismuth; mercury and iron;[34] *selenium and tellurium;*[75] *vanadium and titanium;*[5] *tungsten and chromium.*[28] Tungsten in tungstenite and molybdenum in molybdenite can also be determined with chromium (II) chloride.[76]

Other determinations. Of the various other inorganic compounds the following can be titrated with standard chromium (II) solution either directly or indirectly:

Hydrogen peroxide,[58,77] *sodium peroxide,*[58] *potassium disulphate,*[58,77,78] *oxygen* in water and gases,[77,79,79a] *potassium periodate, iron (III) salts,*[58] *permanganate,*[57] *nitrate,*[80,81] *selenite (IV),*[82] *hexacyanoferrate (III),*[14,59] *chlorate,*[83] *arsenic,*[84] *ruthenium,*[86] *rhenium,*[85,a,b,c] *osmium,*[14] *plutonium.*[86b,c]

Azo, nitro, nitroso compounds and quinones. Jucker[87] and other authors[58] worked out an electrometric method for the determination of azo, nitro, nitroso compounds and quinones with chromium (II) sulphate.

The reaction can be illustrated schematically by the following equations:

$$\text{>}-N=N-R + 4\,Cr^{2+} + 4\,H^+ \rightarrow \text{>}-NH_2 + R-NH_2 + 4\,Cr^{3+}$$

$$R-NO_2 + 6\,Cr^{2+} + 6\,H^+ \rightarrow R-NH_2 + 6\,Cr^{3+} + 2\,H_2O$$

$$R-NO + 4\,Cr^{2+} + 4\,H^+ \rightarrow R-NH_2 + 4\,Cr^{3+} + H_2O$$

$$\underset{O}{\overset{O}{\|}}\bigcirc\underset{\|}{} + 2\,Cr^{2+} + 2\,H^+ \rightarrow \underset{OH}{\overset{OH}{|}}\bigcirc\underset{|}{} + 2\,Cr^{3+}$$

Dissolve the sample (50–500 mg) in water, dilute sulphuric acid, dimethylformamide or pyridine, dilute to 50 ml with 1 N sulphuric acid and titrate potentiometrically with 0·1 N chromium (II) sulphate solution using a platinum indicator and a calomel reference electrode. In some cases it is necessary to work in buffered solution of sodium acetate and acetic acid (pH 2), or of pyridine and sulphuric acid (pH 2, pH 7).

With this method there have been determined after dissolving in 1 N sulphuric acid:

p-*Nitrosophenol*, o-*nitrosodimethylaniline*, p-*nitrosodimethylaniline*, a mixture (1:1) of o- *and* p-*nitrosodimethylaniline*, 1-*nitroso-2-naphthol*, 3-*nitroso-2-naphthol* and a mixture of the two (2:1), *anthraquinone-2-sulphonic acid, anthraquinone-2,7-disulphonic acid;* after dissolving in water: p-*quinone*.

In a solution buffered with sodium acetate acetic acid (pH 2): o- *and* p-*nitrophenol* and their mixture (2:1), and in a pyridine and sulphuric acid solution (pH 2 and pH 7) m-*nitrophenol* may be determined.

Picric acid, trinitrotoluene, tetracene and aminoazobenzene can also be determined similarly.

Nitro-, nitroso-, and *acetylene compounds* can be determined indirectly, after adding an excess of chromium (II) chloride by potentiometric back-titration with 0·1 N iron (III) ammonium sulphate solution.[88–89a] Anthraquinone is titrated directly with chromium (II) chloride. *Azoxy-compounds, azoxybenzene, hydrazobenzene* and *dicarboxylic acid* were also titrated with chromium (II) chloride solution.[89a]

Diazosalts. They are reduced by chromium (II) to hydrazine compounds.[90] The reaction proceeds in 0·05 N hydrochloric acid or in neutral solution. The excess of reducing agent is back-titrated potentiometrically with standard iron (III) ammonium sulphate solution.

The determination of azo, nitro and nitroso compounds as well as of quinone and sugars can also be carried out by visual titration.[91] One titrates them either directly with chromium (II) sulphate using neutral red, phenosafranine and *p*-ethoxychrysoidine as indicators, or one back-titrates the added excess of chromium (II) sulphate with iron (III) ammonium sulphate using potassium thiocyanate as indicator.

The determination of *sugar* is based on the well-known reaction with Fehling's solution, where the unconsumed copper (II) sulphate is titrated with standard chromium (II) sulphate solution.

Terent'ev and Goryacheva[92] as well as Belcher and Bhatty[93] studied analogous determinations of nitro compounds and other substances.

Organic dyes. Various dyes (especially azo dyes) can also be titrated with standard chromium (II) sulphate solution after dissolution in water and ethanol or in hydrochloric acid; atmospheric oxygen has to be removed beforehand by passing carbon dioxide into the solution.[91,94] In this way, methyl red, methyl orange, Bismarck brown, Congo red, methylene blue, crystal violet, eosin, malachite green, indigo (after its sulphonation) and others can be determined.

A comparative study has recently been made of the reduction of *triphenyl methane, indigoide* and *quinone imine dyes* with chromium (II) sulphate or titanium (III) chloride.[94a] For the volumetric determination, titration with chromium (II) sulphate solution is recommended. The following dyes of the single groups have been investigated:

Rhodamine, rosaniline, methyl violet, aniline blue, isatine and safranine.

Acetoxime and diacetyldioxime. Acetoxime and diacetyldioxime are quantitatively reduced by a solution of divalent chromium at pH 0·8–11·7 and 7–12·4. After the reaction is completed, the solution is acidified with hydrochloric acid and titrated with 0·1 N iron (III) ammonium sulphate solution against potassium thiocyanate as indicator.[95]

Chloroform, carbon tetrachloride,[96] *picrolonic acid* and *picrolonates*[96a] have been similarly determined by indirect titration.

The determinations of copper (II), iron (III), antimony (V), titanium (IV), iodine monochloride, bromine and iodine with chromium (II) chloride solution in N,N-dimethylformamide have been recently described,[97] but have only little practical significance.

A general review of titrations with divalent chromium is also to be found in the literature.[98–100]

REFERENCES

1. GRUBE, G. and SCHLECHT, L., *Z. Elektrochem.*, **32**, 178 (1926).
1a. DIMROTH, O. and FRISTER, F., *Ber. dt. chem. Ges.*, **55**, 3693 (1922).
2. ASMANOV, A., *Z. anorg. Chem.*, **160**, 209 (1927).
3. JABLCZYNSKI, K., *Z. phys. Chem.*, **64**, 748 (1908).
3a. FORBES, G. S. and RICHTER, H. W., *J. Amer. Chem. Soc.*, **34**, 1140 (1917).
4. FLATT, R. and SOMMER, F., *Helv. Chim. Acta*, **25**, 684 (1942).
5. SLAVÍK, J., *Chem. Technik*, **6**, 528 (1954); *Chem. Průmysl*, **4**, 412 (1954).
6. TRAUBE, W. and GOODSON, A., *Ber. dt. chem. Ges.*, **49**, 1679 (1916).
7. TRAUBE, W., BURMEISTER, E. and STAHN, R., *Z. anorg. Chem.*, **147**, 50 (1925).
8. ZINTL, E. and RIENÄCKER, G., *Z. anorg. Chem.*, **161**, 374 (1927).
9. BRINTZINGER, H. and RODIS, F., *Z. anorg. Chem.*, **166**, 53 (1927).
10. TOURKY, A. R. and MOUSA, A. A., *J. Chem. Soc.*, 759 (1948).
10a. SHATKO, P. P., *Zhur. Anal. Khim.*, **12**, 201 (1957).
11. TOURKY, A. R. and EL-SHAMY, H. K., *Analyst*, **68**, 40 (1943).
12. THORNTON, W. M. and SADUSK, JR., J. F., *Ind. Engng. Chem. Anal. Ed.*, **4**, 240 (1932).
13. STONE, H. W. and BEESON, C., *Ind. Engng. Chem. Anal. Ed.*, **8**, 188 (1936).
14. CROWELL, W. R. and BAUMBACH, H. L., *J. Amer. Chem. Soc.*, **57**, 2607 (1935).
15. LINGANE, J. J. and PECSOK, R. L., *Anal. Chemistry*, **20**, 425 (1948).
16. MURAKI, I., *J. Chem. Soc. Japan, Pure Chem. Sect.* (*Nippon Kagaku Zassi*), **71**, 407 (1950).

17. TREADWELL, W. D. and NIERIKER, R., *Helv. Chim. Acta*, **24**, 1067 (1941).
18. TANDON, J. P. and MEHROTRA, R. C., *Z. anal. Chem.*, **158**, 20 (1957).
19. HATFIELD, M. R., *Inorg. Syntheses*, **3**, 148 (1950).
20. POSTIS, J. DE, *Bull. Soc. Chim. France*, 283 (1952).
20a. LUX, H. and ILLMANN, H., *Chem. Ber.*, **91**, 2143 (1958).
21. RIENÄCKER, G., Doctoral Thesis, Munich, 1926.
22. TANDON, J. P. and MEHROTRA, R. C., *Z. anal. Chem.*, **159**, 353 (1958).
23. DÖRING, TH., *J. prakt. Chem.*, **66**, 65 (1902).
23a. TANDON, J. P. and MEHROTRA, R. C., *Z. anal. Chem.*, **187**, 410 (1962).
24. BRINTZINGER, H. and RODIS, F., *Z. Elektrochem. angew. phys. Chem.*, **54**, 246 (1928).
25. TANDON, J. P. and MEHROTRA, R. C., *Z. anal. Chem.*, **164**, 314 (1958).
26. ZINTL, E. and SCHLOFFER, F., *Z. angew. Chem.*, **41**, 956 (1928).
27. HÖLEMAN, H., *Z. anorg. Chem.*, **220**, 33 (1934).
27a. MALIK, W. U. and ABUBAKER, K. M., *Analyt. Chim. Acta*, **23**, 518 (1960).
28. FLATT, R. and SOMMER, F., *Helv. Chim. Acta*, **27**, 1522 (1944).
29. BUEHRER, TH. and SCHUPP, O., *Ind. Engng. Chem.*, **18**, 121 (1926).
30. RIENÄCKER, G. and JERSCHKEWICZ, H. G., *Z. anal. Chem.*, **133**, 47 (1951).
31. DOMANGE, L., *Ann. Chim. Anal. Appl.* (4), **25**, 5 (1943).
31a. BARD, A. J. and PETROPOULOS, A. G., *Analyt. Chim. Acta*, **27**, 44 (1962).
32. ZINTL, E., RIENÄCKER, G. and SCHLOFFER, F., *Z. anorg. Chem.*, **168**, 97 (1928).
33. BRENNECKE, E., *Neuere Massanalytische Methoden*, 3rd Ed., p. 197, Ferdinand Enke, Stuttgart 1951.
34. ZINTL, R. and RIENÄCKER, G., *Z. anorg. Chem.*, **161**, 385 (1927).
35. BUSEV, A. I., *Doklady Akad. Nauk SSSR*, **74**, 55 (1950).
36. BUSEV, A. I., *Zhur. Anal. Khim.*, **6**, 178 (1951).
37. SCHLOFFER, F., Doctoral Thesis, Munich, 1928.
38. LINGANE, J. J., *Anal. Chemistry*, **20**, 797 (1948).
39. LINGANE, J. J., *Anal. Chemistry*, **20**, 285 (1948).
40. COOKE, W. D., HAZEL, F. and McNABB, W. M., *Anal. Chemistry*, **21**, 643, 1011 (1949).
41. SYROKOMSKII, V. S. and ZHUKOVA, K. N., *Zavodskaya Lab.*, **11**, 373 (1945).
42. PŘIBIL, R. and ŠVESTKA, L., *Chem. Listy*, **44**, 30 (1950).
43. BRINTZINGER, H. and OSCHATZ, F., *Z. anorg. Chem.*, **165**, 221 (1927).
44. BRINTZINGER, H. and SCHIEFERDECKER, W., *Z. anal. Chem.*, **78**, 110 (1929).
45. BRINTZINGER, H. and ROST, B., *Z. anal. Chem.*, **115**, 241, 250 (1939).
46. PESHKOVA, V. M., GALLAI, Z. A. and ALEXEYEVA, N. N., *Khim. Redkikh Elementov*, **3**, 119 (1957).
47. KAO, S. S., TAI, S. K. and CHENG, S. H., *Hua Hsüeh Hsüeh Pao*, **22**, 328 (1956).
48. HÖLTJE, R. and GEYER, R., *Z. anorg. Chem.*, **246**, 243 (1941).
49. GORYUSHINA, V. G., CHERKASHINA, T. V., *Zavodskaya Lab.*, **14**, 255 (1948).
49a. BUSEV, A. I. and LI GYN, *Vestnik Moskov. Univ.; Khim.*, **2**, 73 (1960).
49b. EL-SHAMY, H. K. and BARAKAT, M. F., *Egypt. j. Chem.*, **2**, 191 (1959).
49c. BUSEV, A. I. and LI GYN, *Vestnik Moskov. Univ., Ser., Mat., Mekhan, Astron., Fiz., Khim.*, **14**, 187 (1959).
50. FLATT, R. and SOMMER, F., *Helv. Chim. Acta*, **27**, 1518 (1944)
51. EL WAKKAD, S. E. S. and RIZK, H. A. M., *Analyst*, **77**, 161 (1952).
52. CHERNIKHOV, J. A., GORYUSHINA, V. G., *Zavodskaya Lab.*, **11**, 137 (1945).

53. CHERNIKHOV, J. A., GORYUSHINA, V. G., *Zavodskaya Lab.*, **12**, 397 (1946).
54. EL-SHAMY, H. K. and EL-DIN ZAYAN, S., *Analyst*, **80**, 65 (1955).
54a. GALLAI, Z. A., KALENCHUK, G. E., *Zhur. Anal. Khim.*, **16**, 63 (1961).
54b. SINGER, E., *Chem. Průmysl*, **12**, 307 (1962).
55. MINCZEWSKI, J., KOLYGA, S. and WODKIEWICZ, I., *Nukleonika Spec.*, No. 3, 62 (1958).
56. ZINTL, E. and ZAIMIS, P., *Z. angew. Chem.*, **40**, 1286 (1927).
57. GALLAI, Z. A., *Nauch. Doklady Vys. Shkoly (Moskva), Khim. i. Khim. Technol.*, **3**, 498 (1958).
58. TANDON, J. P. and MEHROTRA, R. C., *Z. anal. Chem.*, **162**, 31 (1958).
59. MURAKI, I., *J. Chem. Soc. Japan, Pure Chem. Sect. (Nippon Kagaku Zassi)*, **76**, 201 (1955).
60. BRINTZINGER, H. and SCHIEFERDECKER, W., *Z. anal. Chem.*, **76**, 277 (1929).
60a. MAJUMDAR, R. and BHATNAGAR, M. L., *Analyt. Chim. Acta*, **25**, 203 (1961).
61. LINGANE, J. J. and AUERBACH, C., *Anal. Chem.*, **23**, 986 (1951).
62. BRINTZINGER, H. and ROST, B., *Z. anal. Chem.*, **117**, 1 (1939).
63. BUSEV, A. I. and LI GYN, *Zavodskaya Lab.*, **25**, 30 (1959).
64. GOTTFRIED, J., *Chem. Průmysl*, **8**, 176 (1958).
65. MARTYNCHENKO, I. and SHIMKO, A., *Zavodskaya Lab.*, **5**, 1297 (1936).
66. BRINTZINGER, H. and ROST, B., *Z. anal. Chem.*, **117**, 4 (1939).
67. MURAKI, I., *J. Chem. Soc. Japan, Pure Chem. Sect. (Nippon Kagaku Zassi)*, **76**, 193 (1955).
68. BUSEV, A. I. and LI GYN, *Zhur. Anal. Khim.*, **13**, 519 (1958).
69. GORYUSHINA, V. G. and CHERKASHINA, T. V., *Zavodskaya Lab.*, **14**, 873 (1948).
70. SUÁREZ ACOSTA, R., *An. Real. Soc. Españ. Fisica Quim.*, **54B**, 285 (1958).
71. ZANKO, A. M. and SHLYAKMAN, M. J., *Zavodskaya Lab.*, **3**, 777 (1934).
72. MINCZEWSKI, J. and KOLYGA, S., *Chem. analit. (Warszawa)*, **3**, 467 (1958).
73. MINCZEWSKI, J. and KOLYGA, S., *Chem. analit. (Warszawa)*, **3**, 463 (1958).
74. NIKOLAYEVA, E. R., *Vestnik Moskov. Univ., Ser., Mat., Mekhan., Astron., Fiz., Khim.*. **13**, 105 (1958).
75. LINGANE, J. J. and NIEDRACH, L., *J. Amer. Chem. Soc.*, **70**, 1997 (1948).
76. KAO, S. S., TAI, S. K. and CHENG, S. H., *Acta Chim. Sinica*, **22**, 327 (1956).
77. TANDON, J. P. and MEHROTRA, R. C., *Z. anal. Chem.*, **159**, 422 (1958).
78. MURAKI, I., *J. Chem. Soc. Japan, Pure Chem. Sect. (Nippon Kagaku Zassi)*, **76**, 196 (1955).
79. STONE, H. W. and EICHELBERGER, R. L., *Anal. Chemistry*, **23**, 868 (1951).
79a. BÜCHNER, K., *Glas-Instrumenten-Technik*, **3**, 191 (1959).
80. LINGANE, J. J. and PECSOK, R. L., *Anal. Chemistry*, **21**, 622 (1949).
81. WIERCINSKI, J., *Przemysl chem.*, **17**, 57 (1933).
82. MURAKI, I., *J. Chem. Soc. Japan, Pure Chem. Sect. (Nippon Kagaku Zassi)*, **76**, 198 (1955).
83. SOMEYA, K., *Z. anorg. Chem.*, **160**, 355 (1927).
84. SHATKO, P. P., VASINA, N. T., PODOLSKAYA, V. I., MALKINA, L. A. and PONOMAREVA, T. F., *Zhur. Anal. Khim.*, **14**, 358 (1959).
85. TRIBALAT, S., *Ann. Chimie*, **4**, 289 (1949).
85a. RYABCHIKOV, D. I., ZARINSKII, V. A. and NAZARENKO, I. I., *Zhur. Anal. Khim.*, **14**, 737 (1959).
85b. RYABCHIKOV, D. I., ZARINSKII, V. A. and NAZARENKO, I. I., *Zhur. Anal. Khim.*, **15**, 752 (1960).
85c. ZARINSKII, V. A. and FROLKINA, V. A., *Zhur. Anal. Khim.*, **17**, 75 (1962).

86. ZINTL, E. and ZAIMIS, P., *Ber. dt. chem. Ges.*, **60**, 842 (1927).
86a. FUDGE, A. J., WOOD, A. J. and BANHAM, M. F., At. Energy Research Estab. (Gt. Brit.) R 3264, 13 (1960); *Chem. Abstr.*, **54**, 19301 (1960).
86b. HELBIG, W., *Z. anal. Chem.*, **182**, 85 (1961).
87. JUCKER, H., *Analyt. Chim. Acta*, **16**, 210 (1957).
88. BOTTEI, R. S. and FURMAN, N. H., *Anal. Chemistry*, **27**, 1182 (1955).
89. FURMAN, N. H. and BOTTEI, R. S., *Anal. Chemistry*, **29**, 121 (1957).
89a. BOTTEI, R. S., *Analyt. Chim. Acta*, **30**, 6 (1964).
90. BOTTEI, R. S. and FURMAN, N. H., *Anal. Chemistry*, **29**, 119 (1957).
91. TANDON, J. P., *Z. anal. Chem.*, **167**, 184 (1959).
92. TERENT'EV, A. P. and GORYACHEVA, G. S., *Uchennye Zapisky Moskov. Gosudarst. Univ.*, **3**, 227 (1934).
93. BELCHER, R. and BHATTY, M. K., *Analyst*, **81**, 124 (1956).
94. TANDON, J. P. and MEHROTRA, R. C., *Z. anal. Chem.*, **158**, 189 (1957).
94a. TANDON, J. P., *Z. anal. Chem.*, **188**, 161 (1962).
95. KIBA, T. and YAMAZAKI, Y., *J. Chem. Soc. Japan, Pure Chem. Sect.*, **74**, 808 (1953).
96. KIBA, T. and TERADA, K., *J. Chem. Soc. Japan, Pure Chem. Sect.*, **75**, 196 (1954).
96a. DWORZAK, R., KRAUSE, H. and FRIEDRICH, P., *Liebigs Ann. Chem.*, **653**, 12 (1962).
97. HINTON, J. F. and TOMLINSON, H. M., *Anal. Chemistry*, **33**, 1502 (1961).
98. PALLAUD, R., *Chim. analytique*, **33**, 181 (1951).
99. ZANKO, A. M. and STEFANOVSKI, W. F., *Zavodskaya Lab.*, **7**, 17 (1933).
100. BUSEV, A. I., *Primenenie Soyed. Dvukhvalent. Khroma v Anal. Khim.*, Akad. Nauk. SSSR, Moskva 1960.

CHAPTER 17

TIN (II) CHLORIDE

TIN (II) chloride is a powerful reducing agent; it was suggested over one hundred years ago as a titrant for the direct titration of trivalent iron.[1] Most of the " stannometric methods " (titrations with tin (II) chloride solution) were developed, however, more recently, after suitable indicators had been found.

The normal potential of the redox system

$$Sn^{2+} \rightleftharpoons Sn^{4+} + 2\,e^-$$

is + 0·15 V.[1a] It is dependent on the hydrogen ion concentration.

The reduction of the various substances proceeds usually in hydrochloric acid solution, at room or at elevated temperature in an inert atmosphere.

STANDARD SOLUTION

A tin (II) chloride solution has to be prepared with exclusion of atmospheric oxygen.[2]

Procedure. Transfer 80 ml of concentrated hydrochloric acid for each litre of final solution to a bottle, add 4–5 g calcium carbonate to achieve a brisk formation of carbon dioxide. This expels completely the air from the solution, as well as from the bottle. Rapidly weigh crystalline tin (II) chloride (12 g/l.), dissolve in the acid and dilute to 1 l. with boiled-out distilled water.

The whole procedure must be carried out whilst carbon dioxide is evolved continuously.

Besides tin (II) chloride, tin (II) sulphate is also used as a reagent for some determinations. The preparation of the standard solution is the same as for the tin (II) chloride solution. Tin (II) chloride solutions are easily oxidized by atmospheric oxygen and therefore need to be stored in an inert atmosphere. For this purpose it is convenient to use an apparatus similar to that used in titrations with strong reducing agents (titanium (III) chloride, chromium (II) chloride or chromium (II) sulphate). Such an apparatus has been suggested by Szabó and Sugár.[2] But even when atmospheric oxygen is excluded, the normality of the standard solution changes. There is, however, no necessity for a daily check on the normality.

Standardization

Tin (II) chloride is standardized by titration with potassium bromate in strong hydrochloric acid solution using rubrophen as indicator.

Potassium iodate can also be used as standard substance; after acidifying it with concentrated hydrochloric acid, it is titrated with the tin (II) chloride solution,[3] using starch solution as indicator. If the titration is carried out quickly, an atmosphere of carbon dioxide is not absolutely necessary.

INDICATOR

In titrations with tin (II) chloride, visual indication of the end-point is often used. Cacotheline[4, 4a] has been suggested as indicator for most of the determinations. Potassium thiocyanate, ammonium molybdate (VI), diphenylamine and starch also serve as indicators.

All determinations can also be done potentiometrically.

REVIEW OF DETERMINATIONS

Copper. Copper (II) ions are determined with tin (II) chloride[5] in hydrochloric acid medium and in 10 per cent potassium thiocyanate solution. The optimal acidity in the titration is between 0·17 and 0·68 N; a 2 per cent ammonium molybdate (VI) solution is used as indicator. Copper (I) thiocyanate is formed during the titration. The pink solution changes to red–brown at the equivalence point. The titration is done in an atmosphere of carbon dioxide, because otherwise the results are too high.

Copper (II) ions can also be titrated coulometrically.[5a]

Iron. The reaction between iron (III) and tin (II) ions has been investigated extensively.[2, 4, 6–11] The direct titration of iron (II) ions[2, 9] proceeds very rapidly at 60–75°C. The best concentration of hydrochloric acid at the beginning of the titration is 0·5–1·6 N. In the presence of ammonium chloride, the trivalent iron is reduced to the divalent state at the temperature mentioned above. Two indicators are used for the detection of the equivalence point.

Add first 2–3 drops of 0·1 N potassium thiocyanate solution; a yellow colour forms after the greater part of the iron (III) ions has been reduced. Then add 1–2 drops of 0·1 M Na_2HPO_4 solution and 5–6 drops of saturated ammonium molybdate (VI) solution, and titrate with 0·1 N tin (II) chloride solution to the equivalence point; this is marked by a characteristic change of colour from green to light blue. The titration is carried out in a carbon dioxide atmosphere.

The determination is not affected by lead, silver, aluminium and manganese compounds, nor by trivalent or pentavalent arsenic, pentavalent antimony and tungsten. Larger amounts of copper and chromium interfere owing to their colour. If iron is present as sulphate, a sufficient amount of chloride has to be added. This method is suited for the determination of iron in various minerals and ores.

A saturated cacotheline solution can also be used as indicator in the determination of iron.[12]

Titrate in hydrochloric acid solution at elevated temperature to a violet colour. If cacotheline is combined with other indicators, e.g. diphenylamine, even a mixture of iron (III) and dichromate or of iron (III) and vanadium (V) can easily be determined.[4] Another indicator used in the determination of iron (III) is 2-hydroxyphenoxazone-(3).[13]

The potentiometric determination of iron (III) ions with tin (II) chloride proceeds under the same conditions as given for the visual titration. Feil[13a] recommended for the determination of iron in ores the use of a AgI reference electrode.

The reducing action of tin (II) chloride has also been applied to the determination of a mixture of iron (II) and iron (III) salts.[14] In a solution buffered with ammonium tartrate, the iron (III) ions are first titrated with tin (II) chloride and then the iron (II) ions are titrated with permanganate.

Iron and chromium compounds can likewise be titrated potentiometrically side by side.[15]

Chromium. Hexavalent chromium is reduced to the trivalent state by tin (II) chloride. The reaction proceeds in hydrochloric acid solution sufficiently rapidly and quantitatively at room temperature.

In the visual determination,[2] the potassium dichromate (VI) is determined in 3–3·5 N hydrochloric acid solution with tin (II) chloride against diphenylamine as indicator. The solution assumes a green colour at the equivalence point. To obtain the necessary carbon dioxide atmosphere, add very small pieces of marble to the solution. The method is suitable for the determination of chromium in ferrochrome.[2a]

The indirect determination of dichromate is less suitable in which the excess of tin (II) chloride is back-titrated with iodine solution.[16]

The potentiometric determination of hexavalent chromium with tin (II) chloride solution[10,11,17] is carried out in the same medium as has been already described for the visual titration.

Vanadium. Vanadate (V) is reduced in acid solution to vanadyl ions by tin (II) chloride. The visual titration of vanadate (V)[2] in 3–3·5 N hydrochloric acid solution using diphenylamine as indicator is based on this principle. An adequate amount of calcium carbonate (about 2 g) has to be added to the solution in order to produce sufficient carbon dioxide throughout the whole titration. It is titrated at room temperature with 0·1 N tin (II) chloride solution to the change of colour from violet to light green. Diphenylamine may be replaced by ammonium molybdate (VI) used as an external indicator.[18]

Vanadate (V) can likewise be titrated coulometrically with tin (II) chloride in an acidified bromide solution.[19]

The indirect determination of vanadate (V) in which the excess of tin (II) chloride is back-titrated with iodine is only of little significance.[20]

Mercury. Compounds of divalent mercury are reduced to metallic mercury by tin (II) ions. Tin (II) sulphate is used as titrant.[10] The titration is carried out potentiometrically in 1 M sulphuric acid solution at 75°C in an inert atmosphere.

Gold and platinum. The potentiometric determination of trivalent gold compounds with tin (II) chloride proceeds in 2 N hydrochloric acid solution in an inert atmosphere.[10a,b] The trivalent gold is reduced to the metal at room temperature. The titration can also be carried out coulometrically.[19]

Platinum (IV) compounds are likewise reduced to the metal. The reduction with tin (II) chloride solution proceeds quantitatively in 4 per cent hydrochloric acid solution at 75°C in an inert atmosphere.[10,10a] The determination is affected by nitric acid.

Müller and Stein[10c] described the reduction of platinum (IV) to platinum (II); if an excess of tin (II) chloride is used, metallic platinum is formed, which turns to a black precipitate on boiling. The error of this determination is 1–2 per cent.

Rhenium and palladium. Turkiewicz[28] and later Hölemann[29] described the potentiometric determination of rhenium (VII). The titration was carried out at 80°C in 15 per cent HCl. The reaction proceeds according to the equation

$$ReO_4^- + Sn^{2+} + 8 H^+ \rightleftharpoons Re^{5+} + Sn^{4+} + 4 H_2O$$

The determination is not affected by the presence of rhenium (IV). If more than 100 mg of rhenium are applied to the analysis, the error does not exceed ± 0·5 per cent.

The potentiometric titration of *palladium (II)* is characterized by a high potential change near the equivalence point; the reaction, however, is not quantitative and therefore not suitable for analytical purposes.

Molybdenum. Compounds of hexavalent molybdenum are reduced to the pentavalent state by tin (II) chloride. The reaction proceeds quantitatively only in concentrated hydrochloric acid.[21] If the titration is done in 5 N hydrochloric acid solution, the temperature has to be raised to 75°C; in 8 N hydrochloric acid solution, one titrates at room temperature.

The titration of molybdenum compounds with tin (II) chloride solution was used for the determination of this element in steels, ores and minerals;[31–33] it has also been modified for the indirect determination of phosphorus.[34]

Antimony. Antimony (V) compounds can be titrated with tin (II) chloride in concentrated hydrochloric acid solution, using a potentiometric end-point.[22]

Iodine and bromine. The reduction of iodine to iodide with tin (II) chloride solution is applied to the volumetric determination of iodine.[2,10]

Acidify the sample to be analysed with 2 N hydrochloric acid and titrate with 0·1 N tin (II) chloride solution. The equivalence point is detected either with starch[2] or potentiometrically.[10]

It is not necessary to titrate in an inert atmosphere.

Iodine and bromine are determined by coulometric titration in 3–4 N sodium bromide and 0·2 N hydrochloric acid solution.[23]

Iodate. In a solution acidified with hydrochloric acid, potassium iodate is reduced by tin (II) chloride to iodide. The determination is carried out with 0·1 N tin (II) chloride solution at room temperature using starch solution as indicator.[2]

Bromate. Potassium bromate can be determined with standard tin (II) chloride solution in three ways:[2]

1. Add an excess of potassium bromide to the acidified bromate solution and titrate the liberated bromine immediately with 0·1 N tin (II) chloride solution. The very high initial concentration of bromine is a disadvantage of this determination and often causes negative errors.

2. Titrate the bromate with tin (II) chloride solution under the same conditions as given above for the determination of iodate. Bromine is liberated during the reaction so that the end-point of the titration is detected by discoloration of the solution.

3. Add a small amount of potassium iodide to the solution; then it is possible to detect the end-point of the titration using starch solution. The reduction proceeds otherwise as in the two cases mentioned before. The titration is carried out in hydrochloric acid solution with 0·1 N tin (II) chloride solution.

Hexacyanoferrate (III). The visual determination of potassium hexacyanoferrate (III) proceeds in 7 per cent hydrochloric acid solution in an atmosphere of carbon dioxide.[2] Titrate at room temperature with 0·1 N tin (II) chloride solution until the intermediately formed green colour of the solution changes to clear blue; thus, no indicator is necessary.

The determination is not affected by the presence of: mercury (II), bismuth (III), cadmium (II), arsenic (III), arsenic (V), aluminium, zinc, calcium, and magnesium ions, nor by phosphate, acetate, fluoride, thiocyanate, chlorate (V), perchlorate, tetraborate and oxalate. In the presence of iodide, starch solution has to be used as indicator.

Hexacyanoferrate (III) can be titrated potentiometrically in 10 per cent hydrochloric acid solution at room temperature.[10]

Nitrate. The indirect determination of nitrate is based on its reduction with 20 per cent iron (II) ammonium sulphate solution.[24] The iron (III) ions thus produced are then titrated potentiometrically in hydrochloric acid solution with tin (II) chloride in carbon dioxide atmosphere.

Murakami[25] recommends the reduction of nitrate or nitric acid with an excess of 0·2 N tin (II) chloride solution. The reaction proceeds in concentrated hydrochloric acid in an inert atmosphere at boiling temperature; the nitrate is reduced to hydroxylamine. The excess of tin (II) chloride is back-titrated with 0·2 N iron (III) chloride solution.

Permanganate. In 1 M sulphuric acid solution, permanganate is reduced by tin (II) sulphate to manganese (II) salt.[10] The determination is carried out potentiometrically at 18°C.

Silicic acid. The direct titration of silicomolybdic acid with tin (II) solution in sulphuric acid solution is applied to the determination of silicic acid (dissolved) in water.[26]

Organic compounds. Only a few determinations of organic compounds can be found in the literature; Belen'kii and Sokolov[35] described the determination of α-*nitroso*-β-*naphthol*, according to the reaction

$$RNO + 2 Sn^{2+} + 4 H^+ \rightarrow RNH_2 + 2 Sn^{4+} + H_2O$$

which is carried out in hydrochloric acid solution at room temperature and in an inert atmosphere; the authors report on the similar determination of *nitrosodiphenylamine, dinitroresorcinol*, etc., and on the possibility of the general determination of nitro-groups in aromatic compounds.[27] The determination of p-*nitraniline* and α-*naphthylamine* was described later.[36]

Erdey and co-workers[37] used tin (II) chloride solution to study the reactions of *Variamine Blue* as redox indicator, and Okáč and Šimek[38] for the reduction of *nickel (IV) dimethylglyoximate.*

REFERENCES

1. FRESENIUS, R., *Z. anal. Chem.*, **1**, 26, 32 (1862).
1a. HUEY, C. S. and TARTAR, H. V., *J. Am. Chem. Soc.*, **56**, 2585 (1934).
2. SZABÓ, Z. G. and SUGÁR, E., *Analyt. Chim. Acta*, **6**, 293 (1952).
2a. SUGÁR, E. and KOVÁCS, K. M., *Magyar Kém. Lapja*, **17**, 428 (1962).
3. KOLTHOFF, I. M., BELCHER, R., STENGER, V. A. and MATSUYAMA, G., *Volumetric Analysis,* III, Titration Methods: Oxidation–Reduction Reactions, p. 622, Interscience, New York 1957.
4. SZARVAS, P., LANTOS, J., *Magyar Kém. Folyóirat*, **65**, 145 (1959).
4a. SZARVAS, P. and LANTOS, J., *Talanta*, **10**, 477 (1963).
5. RAO, B. K. S. and LADDHA, G. S., *Analyt. Chim. Acta*, **20**, 528 (1959).
5a. LINGANE, J. J., *Analyt. Chim. Acta*, **21**, 227 (1959).
6. NOYES, A. A., *Z. phys. Chem.*, **16**, 546 (1895).
7. WEISS, J., *J. Chem. Soc.*, 309 (1944).
8. CORIN, M. N., *J. Am. Chem. Soc.*, 1787 (1936).
9. SZABÓ, Z. and SUGÁR, E., *Anal. Chemistry*, **22**, 361 (1950); *Magyar Kém. Folyóirat*, **58**, 1 (1952).
10. MÜLLER, E. and GÖRNE, J., *Z. anal. Chem.*, **73**, 385 (1928).
10a. MÜLLER, E. and BENNEWITZ, R., *Z. anorg. Chem.*, **179**, 113 (1929).
10b. MÜLLER, E. and STEIN, W., *Z. Elektrochem.*, **36**, 376 (1930).
10c. MÜLLER, E. and STEIN, W., *Z. Elektrochem.*, **36**, 220 (1930).
11. HOSTETTER, J. C. and ROBERTS, H. S., *J. Amer. Chem. Soc.*, **41**, 1337 (1919).
12. KUKHMENT, M. L. and GENGRINOVICH, A. I., *Zavodskaya Lab.*, **11**, 267 (1945).
13. MUSHA, S. and KITAGAWA, T., *J. Chem. Soc. Japan, Pure Chem. Sect. (Nippon Kagaku Zassi)*, **76**, 1289 (1955).
13a. FEIL, E., *Angew. Chem.*, **49**, 606 (1936).
14. NEUMANN, B. and MEYER, G., *Z. anal. Chem.*, **129**, 229 (1949).

15. MÜLLER, E. and HAASE, G., *Z. anal. Chem.*, **91**, 241 (1933).
16. YOUNG, S. W., *J. Am. Chem. Soc.*, 809 (1897).
17. TRZEBIATOWSKI, W., *Z. anal. Chem.*, **82**, 45 (1930); *Rochniki Chem.*, **10**, 411 (1930).
18. WARYNSKI, M. and MDIVANI, *Bull. Soc. Chim. France* (4), **3**, 626 (1908).
19. BARD, J. A. and LINGANE, J. J., *Analyt. Chim. Acta*, **20**, 581 (1959).
20. KARANTASSIS, T., *Compt. Rendu.*, **224**, 1564 (1947).
21. HÖLTJE, R. and GEYER, R., *Z. anorg. Chem.*, **246**, 243 (1941).
22. DAUKSHAS, K., NARUSHKEVICHIUS, L., *Uchennye Zapiski Vilnus. Univ., Ser. Mat., Fiz., Khim.*, **7**, 161 (1957).
23. BARD, J. A. and LINGANE, J. J., *Analyt. Chim. Acta*, **20**, 463 (1959).
24. SZABÓ, Z. G., BARTHA, L. G. and SIMON-FIALA, J., *Acta chim. Acad. Sci. Hung.*, **3**, 231 (1953).
25. MURAKAMI, T., *Japan Analyst*, **7**, 766 (1958).
26. TAKAHASHI, T. and MIYAKE, S., *Talanta*, **4**, 1 (1960).
27. REICHEL, J. and DOBRESCU, F., *Bull. stiint. si tekn. Inst. Politekn. Timisoara*, **4**, 247 (1959).
28. TURKIEWICZ, E., *Rochniki Chem.*, **12**, 589 (1932).
29. HÖLEMANN, H., *Z. anorg. Chem.*, **217**, 105 (1934); **220**, 33 (1934).
30. MÜLLER, E. and STEIN, W., *Z. Elektrochem.*, **40**, 133 (1934).
31. SOSNOVSKII, B. A., *Zavodskaya Lab.*, **3**, 696 (1934); *Z. anal. Chem.*, **103**, 44 (1935).
32. FOGEL'SON, JE. I. and KALMYKOVA, N. V., *Zavodskaya Lab.*, **5**, 148 (1936); cit.: *Chem. Abstr.*, **30**, 4782 (1936).
33. KRÜLL, F., *Centr. Mineral. Geol.*, A 331 (1934); *Chem. Abstr.*, **23**, 1739 (1935).
34. VASIL'EV, D. V., *Zhur. Priklad. Khim.*, **14**, 689 (1941).
35. BELEN'KII, L. I. and SOKOLOV, I. I., *Za Rekonstr. Textil. Prom.*, **14**, 34 (1935); *Khimie a Industrie*, **35**, 917 (1935); *Chem. Abstr.*, **30**, 4789 (1936).
36. BELEN'KII, L. I. and SOKOLOV, I. I., *Prom. Org. Khim.*, **1**, 618 (1936).
37. ERDEY, L., ZÁLAY, E. and BODOR, E., *Acta Chim. Hung.*, **3**, 231 (1953).
38. OKÁČ, A. and ŠIMEK, M., *Chem. Listy*, **52**, 2285 (1958).

CHAPTER 18

SODIUM ARSENITE

STANDARD arsenite solutions are prepared by dissolution of arsenic (III) oxide (which is easily obtainable in a pure state) in alkali hydroxide, followed by neutralization of the solution. These solutions are used in the standardization of some standard solutions, e.g. iodine solution.

A known amount of arsenite solution is titrated with the unknown solution which has to be standardized.

Standard arsenite solution is also used as a weak reducing agent, especially for the titration of free halogen or of hypohalites or for the determination of manganese after its transformation to permanganate. The formal redox potential of the system As^V/As^{III} depends on the pH of the solution;[30,31] at pH 7 it is stated to be 0·316 V and increases to about 0·520 V at pH 3.[30]

Very recently, Szekeres and co-workers tried to replace the iodimetric determination of the excess of halogen, hypohalite or bromate (V) in various indirect titrations by the "arsenometric titration". By this means, the costly consumption of potassium iodide, especially in large series of analysis, can be avoided.

STANDARD SOLUTION

For the preparation of 0·1 N sodium arsenite solution, dissolve 4·945 g of pure dried arsenic (III) oxide in 60 ml of 1 N sodium hydroxide. Neutralize the solution with 50–60 ml of 1 N hydrochloric or sulphuric acid and dilute to 1 l. The solution should be neutral to lithmus paper.

The solution thus prepared serves as standard and is stable for several months. Strongly acid or alkaline solutions change their normality after some weeks.[1]

INDICATOR

In the reaction of arsenite solution with halogen or hypohalite, easily destructible dyes, e.g. methyl red, methyl orange, indigo, quinoline yellow,[2] brasilin,[3] or luminescent indicators such as luminol[4] serve as indicators; some other examples are given on page 128.

A more precise detection of the end-point can be achieved by one of the well-known electrometric methods, e.g. by potentiometric titration.

REVIEW OF DETERMINATIONS

Amongst the determinations of hypohalites with arsenite, the method of Penot ranks as one of the oldest methods of titration. It has been applied to the determination of hypohalites, the equivalence point being detected with iodide–starch paper. Recently, many papers have been published about direct and indirect determinations of *hypohalites* or of *chlorine* and *iodine* with arsenite solution.[3,5-7,18-23a,25,26,29,32-38] The method is recommended for the determination of iodine on the ultra-micro scale.[39] *Chlorites* can be titrated in the presence of osmium (VIII) oxide as catalyst.[27]

Szekeres and co-workers[8-10] replaced the iodimetric determination by arsenometric titration in the determination of the excess of *bromine*, liberated from a mixture of bromate–bromide in acid medium in the oxidation of oxalate, nitrite, hexacyanoferrate (II) and ammonium salts.

Gleu and Katthän[11] determined *periodate* by reduction with arsenite in dilute sulphuric acid solution. The reaction was catalysed by ruthenium (VIII) oxide and ferroin was used as indicator; periodate is reduced to iodate (V).

The reduction of SO_5^{2-} -*ions* (Caro acid),[40] *cerium (IV)* (ICl as catalyst)[41] and of *hexacyanoferrate (III)* (OsO_4 as catalyst)[42] were also recommended for analytical purposes.

Gold (III) compounds can be reduced at 90°C to the monovalent state in a solution containing bromide; potentiometric titration is used.[28]

In steel analysis, arsenite is useful as a reductant,[12] e.g. for the selective reduction of *dichromate* in the presence of a large amount of vanadate (V); an excess of reagent is added and the unconsumed part back-titrated with permanganate.[13,43] This method is suited for the rapid analysis of steels. The iron is masked with fluoride or phosphate to obtain a better end-point in the visual titration.[14,24]

The titration of manganese after its transformation to *permanganate* is suitable for determinations of manganese in ores, steel, ferromanganese and in titanium, etc. The reaction does not proceed exactly according to the reaction scheme to be expected, but is empirical; it yields, however, well reproducible results and is successfully applied in steel laboratories; a visual[15,17,24] or an electrometric end-point is used.[44-46]

In one of the suggested methods, arsenite is used as titrant together with nitrite; in acid solution, the equivalence point is indicated by the change of colour from violet to yellow–brown.[16] As in all titrations of permanganate, it is necessary to standardize the titrant against a steel sample with a known concentration of manganese; in one of the most recent publications, a mixture of 0·1 N arsenite and 0·1 N sodium azide has been recommended as standard solution for this titration.[28]

The titration of *manganese (III)* compounds was used for the indirect determination of PbO_2.[47]

REFERENCES

1. KOLTHOFF, I. M., *Z. anal. Chem.*, **60**, 393 (1921).
2. KOLTHOFF, I. M., STENGER, V. A., BELCHER, R. and MATSUYAMA, G., *Volumetric Analysis*, III, Titration Methods: Oxidation–Reduction Reactions, Interscience, New York 1957.
3. BITSKEI, J., *Acta Chim. Acad. Sci. Hung.*, **8**, 203 (1956).
4. ERDEY, L. and BUZÁS, L., *Acta Chim. Acad. Sci. Hung.*, **6**, 123 (1955).
5. SINN, V., *Chim. analytique*, **29**, 58 (1947).
6. YOUNG, J. H. and DAS GUPTA, R. N., *Analyst*, **74**, 367 (1949).
7. BELCHER, R., *Analyt. Chim. Acta*, **5**, 27 (1951).
8. SZEKERES, L. and MOLNÁR, L. G., *Magyar Kém. Folyóirat*, **64**, 96 (1958).
9. SZEKERES, L. and ZERGÉNYI-BALÁS, G., *Z. anal. Chem.*, **163**, 359 (1958).
10. KELLNER, A., SZABÓ, CH. and SZEKERES, L., *Z. anal. Chem.*, **157**, 13 (1957).
11. GLEU, K. and KATTHÄN, W., *Chem. Ber.*, **86**, 1077 (1953).
12. KOLTHOFF, I. M. and SANDELL, E. B., *Ind. Engng. Chem., Anal. Ed.*, **2**, 140 (1930).
13. SZABÓ, Z. G. and CSÁNYI, L., *Anal. Chemistry*, **21**, 1144 (1949).
14. Standard Methods of Analysis of Iron, Steel and Ferro-Alloys, United Steel Co. Ltd., Sheffield 1951.
15. SANDELL, E. B., KOLTHOFF, I. M. and LINGANE, J. J., *Ind. Engng. Chem., Anal. Ed.*, **7**, 256 (1935).
16. HILLSON, H. D., *Ind. Engng. Chem., Anal. Ed.*, **16**, 560 (1944).
17. HALLER, J. and LISTEK, S. S., *Anal. Chemistry*, **20**, 637 (1948).
18. MAHAN, W. A., *Water Sewage*, **96**, 171 (1949).
19. MARKS, H. C. and BANNISTER, G. L., GLASE, J. R. and HERIGEL, E., *Anal. Chemistry*, **19**, 200 (1947).
20. MARKS, H. C. and CLASS, J. R., *J. Amer. Water Works Assoc.*, **64**, 1227 (1941).
21. CSÁNYI, L. and SOLYMOSI, F., *Acta Chim. Acad. Sci. Hung.*, **17**, 69 (1958).
22. ZHDANOV, A. K., GHADEYEV, V. A. and JAKOVENKO, G. D., *Zhur. Anal. Khim.*, **14**, 367 (1957).
23. CORBETT, J. A., *Analyst*, **83**, 53 (1958).
23a. ZAVAROV, G. V., *Zavodskaya Lab.*, **24**, 681 (1958).
24. BAPAT, M. G. and SHARMA, B., *Z. anal. Chem.*, **157**, 258 (1957).
25. KISS, S. A., *Magyar Kém. Lapja*, **14**, 497 (1959).
26. NORKUS, P. K. and PROKOPCHIK, J., *Zhur. Anal. Khim.*, **26**, 323 (1961).
27. SZEBELLEDY, L. and VICZIAN, B., *Oesterr. Chemiker-Ztg.*, **41**, 431 (1938).
28. JIMENO, S. A. and IGLESIES CASTANO, J. M., *An. Real. Soc. Españ. Fisica Quim.*, **57B**, 691 (1961); *Chem. Abstr.*, **57**, 4024 (1962).
29. NORKUS, P. K., *Zhur. Anal. Khim.*, **18**, 884 (1963).
30. FURMAN, N. H. and MILLER, C. O., *J. Amer. Chem. Soc.*, **59**, 152 (1937).
31. BOCK, R. and GREINER, G., *Z. anorg. Chem.*, **295**, 61 (1958).
32. TREADWELL, W. D., *Helv. Chim. Acta*, **4**, 396 (1921).
32a. MÜLLER, E. and DIETMANN, H., *Z. anal. Chem.*, **73**, 138 (1928).
33. SCHLEICHER, A. and TOUSSAINT, L., *Z. anal. Chem.*, **65**, 399 (1925).
34. BLAKELY, J. D., PRESTON, J. M. and SCHOLEFIELD, F., *J. Soc. Dyers Colourists*, **46**, 230 (1930); *Chem. Abstr.*, **24**, 5253 (1930).
35. GAUCHMANN, M. S. and STEFANOVSKII, V. F., *Zavodskaya Lab.*, **9**, 493 (1940).
36. GALLUS-OLENDER, J., *Chem. analit. Warszawa*, **3**, 859 (1958); *Z. anal. Chem.*, **171**, 458 (1960).
37. SCHLEICHER, A. and WESLY, W., *Z. anal. Chem.*, **65**, 406 (1925).

38. KOLTHOFF, I. M. and LAUR, A., *Z. anal. Chem.*, **73**, 177 (1928).
39. BISHOP, E., *Mikrochim. Acta*, 619 (1956).
40. MÜLLER, E. and HOLDER, G., *Z. anal. Chem.*, **84**, 410 (1931).
41. LANG, R. and ZWEŘINA, J., *Z. anal. Chem.*, **91**, 5 (1933).
42. SOLYMOSI, F., *Acta Chim. Acad. Sci. Hung.*, **16**, 267 (1958).
43. ZINTH, E. and ZAIMIS, P., *Angew. Chem.*, **40**, 1286 (1927).
44. HALL, W. T. and CARLSON, C. E., *J. Amer. Chem. Soc.*, **45**, 1615 (1933).
45. CHLOPIN, N. J., *Z. anal. Chem.*, **102**, 263 (1935).
46. AVRUNINA, A. M. and ZAN'KO, A.M., *Zavodskaya Lab.*, **7**, 1238 (1938).
47. LANG, R. and ZWEŘINA, J., *Z. anal. Chem.*, **93**, 248 (1933).

CHAPTER 19

MERCURY (I) NITRATE AND MERCURY (I) PERCHLORATE

FROM the normal potential of the redox system $Hg_2^{2+}/2\ Hg^{2+}$ which is 0·906 V, it follows[1] that solutions of mercury (I) salts cannot be regarded as strong reducing agents; (they are even applied to oxidimetric determinations in some cases). Only strong oxidants such as permanganate, cerium (IV) compounds (under the catalytic action of $AuCl_3$), bromine, hypobromite or gold (III) compounds can be titrated directly with mercury (I) salts.

If sufficient thiocyanate is added to mercury (I) solution, the reaction proceeds quantitatively according to the following equation:

$$Hg_2^{2+} + 4\ SCN^- \rightarrow [Hg(SCN)_4]^{2-} + Hg$$

In the presence of a reducible substance, metallic mercury is transformed into the complex form:

$$Hg + 4\ SCN^- - 2\ e^- \rightarrow [Hg(SCN)_4]^{2-}$$

The total reaction of the titration with mercury (I) salt in the presence of thiocyanate is represented by the following equation:

$$Hg_2^{2+} + 8\ SCN^- - 2\ e^- \rightarrow 2[Hg(SCN)_4]^{2-}$$

Because the complex $[Hg(SCN)_4]^{2-}$ is only very little dissociated, the potential of the system Hg^{2+}/Hg is considerably more negative;[2] in 1 M thiocyanate solution it is 0·12 V. This makes it possible to carry out reductimetric titrations with mercury (I) salts in the presence of thiocyanate, not only for systems with high redox potentials but also for many other substances, provided, however, that they do not react with thiocyanate. Substances which do react with thiocyanate are usually reduced with iron (II) salts and the iron (III) ions thus formed are then titrated with mercury (I) salt in the presence of thiocyanate. Some substances can be determined by oxidation with an excess of hexacyanoferrate (III) in alkaline solution; the excess is back-titrated with mercury (I) salt in the presence of iodide:

$$Hg_2^{2+} + 4\,I^- \to [HgI_4]^{2-} + Hg$$

$$Hg + 2[Fe(CN)_6]^{3-} + 4\,I^- \to 2[Fe(CN)_6]^{4-} + [HgI_4]^{2-}$$

or $\quad 2[Fe(CN)_6]^{3-} + Hg_2^{2+} + 8\,I^- \to 2[Fe(CN)_6]^{4-} + 2[HgI_4]^{2-}$

STANDARD SOLUTION

For the preparation of about 0·1 N mercury (I) nitrate solution dissolve 30 g of $Hg_2(NO_3)_2 \cdot 2\,H_2O$ in 500 ml of hot water, containing such an amount of nitric acid that the final solution will be between 0·01–0·05 N with regard to nitric acid.

Because the $Hg_2(NO_3)_2 \cdot 2\,H_2O$ may contain nitrogen oxides, boil the solution for 3 hr, then filter off and dilute with water.

For the preparation of about 0·1 N mercury (I) perchlorate solution, dissolve 10·89 g of HgO (red modification) in 10 g of $HClO_4$, warm the mercury (II) perchlorate thus formed with metallic mercury in a porcelain dish on a water bath for 3 hr. Add some more $HClO_4$ to avoid a possible hydrolysis and dilute to 1 l. with distilled water.

In the literature it is recommended that metallic mercury be added to the standard solutions in order to stabilize them,[3] but even without this addition they are stable for an unlimited time.

Standardization

Mercury (I) salt solutions are standardized in the presence of thiocyanate by visual titration of iron (III) solution,[4] or of iodine, produced by reaction of dichromate or potassium iodate with iodide; potentiometric titration of potassium hexacyanoferrate (III) can also be applied for this purpose.[5,6,25]

In the literature, there has also been recommended a gravimetric method in which the Hg_2^{2+} is precipitated as Hg_2Cl_2, but this is not suitable for standardization because Hg_2Cl_2 begins to sublime at 100°C.

INDICATOR

In the titration of cerium (IV) salts, N-phenylanthranilic acid is used as redox indicator, in the determination of copper, benzidine or o-dianisidine, in the determination of hexacyanoferrate (III) in alkaline solution, barium diphenylamine sulphonate.

Iron (III) salt is titrated to the disappearance of the red colour formed with thiocyanate (in the same way as in the determination of copper (II) ions in the presence of SCN^- after the addition of Fe^{2+}).

In the titration of permanganate, the disappearance of its own colour serves as end-point.

In the determination of iodine, the end-point of the titration is detected by the decolorization of the blue iodine starch solution or by the turbidity of the solution caused by mercury (I) iodide, produced through the reaction of iodide by the first excess drop of the titrant. In the presence of bromide, mercury (I) bromide forms with the first excess drop of reagent; this has been used for the visual indication of the end-point in the titration of bromine, hypobromite, bromate and hypochlorite.

All systems so far mentioned and certain others can also be determined potentiometrically.

REVIEW OF DETERMINATIONS

Gold. Gold (III) salts are reduced to the metal in the potentiometric titration with mercury (I) compounds. The determinations are carried out in 1 N sulphuric acid; lead (II), copper (II), iron (III) ions, 0·5 N nitric acid and 0·1 N hydrochloric acid do not interfere with the titration.[2]

Cerium.[9,10] Cerium (IV) salts are reduced to the trivalent state by mercury (I) perchlorate. In the direct titration in 0·5–6 N sulphuric acid solution, gold (III) chloride is added as catalyst in order to ensure a sufficiently rapid and quantitative reaction. The equivalence point is detected potentiometrically or visually using N-phenylanthranilic acid as indicator. The titration with mercury (I) nitrate yields results too low by 1–2 per cent.

Bromine, hypobromite, bromate,[11] *hypochlorite.*[12] Mercury (I) salts reduce bromine and hypobromite rapidly and quantitatively. In the determination of bromate and hypochlorite, an equivalent amount of bromine has to be liberated first by adding an excess of bromide. The equivalence point is detected either potentiometrically or by the appearance of the turbidity in the solution to be titrated due to mercury (I) bromide (see above). Chlorides do not interfere with this determination.

Manganese. Permanganate can be determined potentiometrically with mercury (I) salts in 1 N sulphuric acid solution (reduction to manganese (II)). Chromate and vanadate (V) do not interfere with the determination. The titration can also be carried out visually in 3 N sulphuric acid solution using gold (III) chloride as catalyst.[13] In the presence of diphosphate or fluoride as complex forming reagents, permanganate is transformed to the trivalent state. This can be applied to the direct potentiometric titration. Because the complex formed with ammonium fluoride is colourless, the titration can also be done visually (in 0·5 N sulphuric acid); in this case, the end-point is detected by the decolorization of the permanganate;[14] a catalyst is not necessary.

Iron. Bradbury and Edwards[7] found that the reducing power of mercury (I) salt is increased in the presence of thiocyanate so that iron (III) ions can be titrated directly to the disappearance of the red coloured iron (III) thiocyanate complex. The conditions for this reaction have been thoroughly studied in further publications.[8,15–20] It has been observed that the reaction proceeds quantitatively in sulphuric or nitric acid (up to 2·5 N) solution. Hydrochloric acid affects the reaction; its inhibiting action can be partially avoided by applying a higher concentration of thiocyanate; using a great excess of thiocyanate, the titration can even still be carried out in 3 N hydrochloric acid solution.[21]

Ce^{4+}, Sn^{2+}, Ti^{3+}, Tl^{3+}, Cu^{2+}, Co^{2+}, Bi^{3+} and most of the anions affect the titration. Iron (III) can be determined in the presence of Ce^{4+}, $Cr_2O_7^{2-}$ and VO_3^- after reduction of these ions with an excess of thiocyanate.[22]

The interference of ions which produce a colour with thiocyanate, can be avoided by potentiometric titration in about 1 N sulphuric acid.[18] Under these conditions, the determination is not affected by Co^{2+}, Cr^{3+} and Bi^{3+}.

The mercurimetric titration has proved suitable for the determination of iron in ores,[3,33] silicates,[21] ashes, slags,[4,22a] in pharmaceutical products[23] and mixtures of forage.[35]

Iodine. The mercurimetric determination of iodine is carried out in acid medium in the presence of an excess of thiocyanate. The equivalence point can be detected by the disappearance of the blue colour (after addition of starch solution) or by the turbidity of mercury (I) iodide, produced by the first excess drop of the titrant. The titration can also be carried out potentiometrically.[2] In *iodimetric determinations*, monovalent mercury can be used instead of thiosulphate.

Molybdenum. The titration of molybdate (VI) with mercury (I) perchlorate is carried out potentiometrically in 1·0 or 1·5 N sulphuric acid in the presence of thiocyanate and iodide. Tungsten does not interfere. Mercury (I) nitrate cannot be used because of the great potential fluctuations during the titration.[24]

Copper. Copper (II) salts can be determined in the presence of thiocyanate by mercurimetric titration potentiometrically as well as visually.

For the visual indication, add some drops of a solution of Mohr's salt to the sulphuric acid solution and titrate to the disappearance of the coloured iron (III) thiocyanate (the trivalent iron is produced by oxidation of the divalent iron with the divalent copper).[26]

The redox indicators benzidine and *o*-dianisidine are also used for visual indication of the end-point.[19]

In the potentiometric titration of copper (II) ions in the presence of iron (III) ions, both ions are determined simultaneously. If the determination is carried out in a solution containing sodium acetate and ammonium fluoride (in the presence of thiocyanate) only one potential jump appears in the curve, corresponding to the quantitative reduction of divalent copper. The iron present does not influence the accuracy of the determination. This method is used for the determination of copper in ores.

Permanganate persulphate, vanadate (V), cerium (IV) salts, dichromate,[27,28] *chlorate,*[27] *hypochlorite,*[12] *hydrogen peroxide*[27] *and copper (II) salts.*[28-30] These substances are determined by reduction with an excess of iron (II) salt and by titration of the iron (III) ions formed with mercury (I) nitrate. The determination of nitrates after their reduction with Mohr's salt at boiling temperature is based on the same principle.[31]

Mercury (I) salts[32] *and hydroxylamine.*[31] Both compounds are oxidized by an excess of iron (III) salt. The unconsumed amount of iron (III) is then back-titrated with mercury (I) nitrate solution.

Hexacyanoferrate (III). Hexacyanoferrate (III) is titrated potentiometrically with mercury (I) salt solution in acid medium in the presence of thiocyanate.[25,34] In 1–5 N NaOH solution, the end-point of the titration with mercury (I) perchlorate in the presence of iodide (see p. 132) can be detected potentiometrically as well as visually against diphenylamine sulphonate (disappearance of the red colour).[5] The titration is not affected by 5 per cent nitrate, chloride and sulphate ions.

Hydrazine, hydrogen peroxide, arsenic (III) and chromium (III) salts. These compounds are determined indirectly by oxidation with an excess of hexacyanoferrate (III) in alkaline solution.[6] The unconsumed excess is back-titrated with mercury (I) perchlorate solution.

REFERENCES

1. JANDER, G. and co-worker, *Neuere Massanalytische Methoden*, p. 120, Ferdinand Enke, Stuttgart 1956.
2. TARAYAN, V. M., *Merkuroreduktometria*, Izd. Erevansk. Univ. Erevan, 1958.
3. FINKELSHTEIN, D. N. and KRYUCHKOVA, G. N., *Zavodskaya Lab.*, **21**, 403 (1955).
4. BABUCHKIN, S. A. and POGREBINSKAYA, M. L., *Zavodskaya Lab.*, **14**, 1182 (1948).
5. BURRIEL, F. M., LUCENA CONDE, F. and ARRIBAS, S. J., *Analyt. Chim. Acta*, **10**, 301 (1954).
6. BURRIEL, F. M., LUCENA CONDE, F. and ARRIBAS, S. J., *Analyt. Chim. Acta*, **11**, 214 (1954).
7. BRADBURY, F. R. and EDWARDS, E. G., *J. Soc. Chem. Ind. (London)*, **59**, 96 (1940).
8. BELCHER, R. and WEST, T. S., *Analyt. Chim. Acta*, **5**, 260 (1951).
9. TARAYAN, V. M. and OVSEPYAN, E. N., *Zavodskaya Lab.*, **18**, 1066 (1952).
10. TARAYAN, V. M. and EKIMYAN, M. G., *Nauch. Trudy Erevansk. Univ.*, **44**, 87 (1954); *Ref. Zhur. Khim.*, **8**, 247 (1955).
11. TARAYAN, V. M. and OVSEPYAN, E. N., *Nauch. Trudy Erevansk. Univ.*, **44**, 77 (1954); *Ref. Zhur. Khim.*, **8**, 258 (1955).
12. TARAYAN, V. M. and MELIKSETYAN, A. P., *Nauch. Trudy Erevansk. Univ.*, **60**, 73 (1957); *Chem. Abstr.*, **53**, 11109 (1959).
13. TARAYAN, V. M., *Nauch. Trudy Erevansk. Univ.*, **44**, 65 (1954); *Ref. Zhur. Khim.*, **8**, 250 (1955).

14. TARAYAN, V. M. and EKIMYAN, M. G., *Izvest. Akad. Nauk Armensk. SSR, Khim. Nauk* No. 2, 105 (1957).
15. PUGH, W., *J. Chem. Soc.*, 588 (1945).
16. BELCHER, R. and WEST, T. S., *Analyt. Chim. Acta*, **5**, 268 (1951).
17. BELCHER, R. and WEST, T. S., *Analyt. Chim. Acta*, **5**, 472 (1951).
18. BELCHER, R. and WEST, T. S., *Analyt. Chim. Acta*, **7**, 470 (1952).
19. MATSUO, E., *J. Chem. Soc. Japan, Ind. Chem. Sect.*, **58**, 962 (1955); *Anal. Abstr.* **4N 1**, Abstr. N. 113 (1957).
20. FLASCHKA, H., *Mikrochemie*, **35**, 473 (1950).
21. TARAYAN, V. M. and EKIMYAN, M. G., *Zavodskaya Lab.*, **21**, 304 (1955).
22. TARAYAN, V. M. and ARUTYUNYAN, A. A., *Nauch. Trudy Erevansk. Univ.*, **36**, 53 (1952).
22a. TROFIMOVA, S. G., *Izvest. Akad. Nauk Kazakh. SSR, Ser. Met. Obogachkchen. i Ogneuporov*, 55 (1961); *Chem. Abstr.*, **56**, 8001 (1962).
23. BRADBURY, F. R., CHATTERGEE, K. C. and EDWARDS, E. G., *Quart. J. Pharmacol.*, **13**, 297 (1940).
24. TARAYAN, V. M. and OVSEPYAN, E. N., *Zavodskaya Lab.*, **17**, 526 (1951).
25. TARAYAN, V. M. and ARUTYUNYAN, A. A., *Izvest. Akad. Nauk Armensk. SSR*, **3**, 651 (1950).
26. TARAYAN, V. M., ARUTYUNYAN, A. A., *Zavodskaya Lab.*, **19**, 900 (1953).
27. BELCHER, R. and WEST, T. S., *Analyt. Chim. Acta*, **5**, 360 (1951).
28. TARAYAN, V. M., *Izvest. Akad. Nauk Armensk. SSR*, **3**, 677 (1950).
29. BELCHER, R. and WEST, T. S., *Analyt. Chim. Acta*, **5**, 364 (1951).
30. BURRIEL, F. and LUCENA CONDE, F., *An. Real. Soc. Españ. Fisica Quim.*, **47B**, 257 (1951).
31. BELCHER, R. and WEST, T. S., *Analyt. Chim. Acta*, **5**, 546 (1951).
32. BELCHER, R. and WEST, T. S., *Analyt. Chim. Acta*, **5**, 474 (1951).
33. TRUSOV, J. P., *Zhur. Anal. Khim.*, **14**, 139 (1959).
34. LUCENA CONDE, F. and BELLIDO, I. S., *Talanta*, **1**, 305 (1958).
35. HIRSJÄRVI, V. P., SALOVIUS, B. and UOSUKAINEN, M., *Mikrochim. Acta*, 534 (1960).

CHAPTER 20

COMPOUNDS OF MONOVALENT COPPER

COPPER (I) compounds are effective reducing agents, owing to the normal potential of the redox system

$$Cu^+ \rightleftharpoons Cu^{2+} + e^-$$

which has the value + 0·15N.[1] The disadvantage of these compounds is their lack of stability against atmospheric oxygen. Standard copper (I) solutions therefore need to be stored, and the titrations must be carried out, in an inert atmosphere.

STANDARD SOLUTION

Solutions prepared by dissolving copper (I) chloride in hydrochloric acid are only very unstable,[2] even if stored in containers specially constructed for this purpose, and which at the same time make titrations in an inert atmosphere possible.[3] It is better to prepare the standard copper (I) solution from a 0·1 N copper (II) chloride or copper (II) sulphate storage solution which is 2 M with respect to HCl:

The copper (II) storage solution is allowed to pass from the container through a sliver reductor (7 ml/min) whereby the divalent copper is transformed quantitatively to the monovalent state.[4,10] By this means it is possible to prepare the standard copper (I) solution immediately it is required, the titration must then proceed immediately.

Standardization

Copper (I) solutions can be standardized, e.g. with dichromate or cerium (IV) sulphate in an inert atmosphere, preferably by potentiometric titration. The copper (II) storage solution from which the copper (I) solution is prepared by means of a silver reductor, can be standardized compleximetrically using pyrocatechol violet as indicator.[5]

Belcher and co-workers[11] attempted to increase the reducing action of the system Cu^{II}/Cu^{I} by the presence of EDTA, but did not get positive results.

INDICATOR

In all determinations described here, the equivalence point can be detected potentiometrically. The visual indication has proved satisfactory in the titration of cerium (IV) salts using ferroin,[4] and of iron (III) salts using thymolindophenol.[10]

REVIEW OF DETERMINATIONS

Müller and Tänzler[2] thoroughly investigated the use of copper (I) salts in acid medium as standard solution. In their publications they describe the titration of *platinum (IV)* and *gold (III) salts* ($PtCl_6^{2-} \rightarrow PtCl_4^{2-}$; $Au^{3+} \rightarrow Au$).

Determinations of platinum have been carried out by other authors in a similar way[3,6,12] and revised recently;[15] they used this method for the analysis of alloys containing platinum, rhodium[7] or palladium.[13,14]

Iridium (IV) compounds can also be determined with standard copper (I) solutions.[8,9]

Flaschka[10] used copper (I) solutions, freshly prepared when needed from copper (II) salt, by means of a silver reductor, for the determination of trivalent iron using thymolindophenol as indicator. In the most recent publications, this method of preparing the standard solution has been extensively studied.[4] Conditions have been established for the potentiometric determination of *cerium(IV) salts*, *dichromate (VI)*, *vanadate(V)*, *hexacyanoferrate (III)*, *iodate (V)* and *permanganate* in acid solution. A mixture consisting of some of the above compounds can indeed be determined potentiometrically by successive reductions, but the results show noticeable errors.[4]

REFERENCES

1. LATIMER, W. M., *The Oxidation States of the Elements and Their Potentials in Aqueous Solutions*, Prentice Hall, New York 1952.
2. MÜLLER, E. and TÄNZLER, K., *Z. anal. Chem.*, **89**, 339 (1932).
3. GRINBERG, A. A. and GOLBRAYKH, Z. E., *Zhur. Obsch. Khim.*, **14**, 808 (1944).
4. SUCHOMELOVÁ, L., DOLEŽAL, J. and ZÝKA, J., *J. Electroanal. Chem.*, **1**, 403 (1960).
5. DOLEŽAL, J., DRAHOŇOVSKÝ, J. and ZÝKA, J., *Collection (Czech. Chem. Comm.)*, **24**, 3649 (1959).
6. GRINBERG, A. A., MAKSIMYUK, E. A. and PTITSIN, B. V., *Doklady Akad. Nauk SSSR*, **51**, 687 (1946).
7. RYABCHIKOV, D. I., *Izvest. Sektora Platiny i Drugikh Blagorodnykh Metallov*, **22**, 20 (1948).
8. RYABCHIKOV, D. I. and NERSESOVA, S. V., *Izvest. Sektora Platiny i Drugikh Blagorodnykh Metallov*, **18**, 100 (1945).
9. GRINBERG, A. A. and MAKSIMYUK, E. A., *Izvest. Sektora Platiny i Drugikh Blagorodnykh Metallov*, **20**, 149 (1947).

10. FLASCHKA, H., *Mikrochim. Acta*, 15 (1951).
11. BELCHER, R., GIBBONS, D. and WEST, T. S., *Analyt. Chim. Acta*, **12**, 107 (1955).
12. RYABCHIKOV, D. I., *Zhur. Anal. Khim.*, **1**, 47 (1946).
13. PCHENITSYN, N. K., PROKOF'EVA, I. V. and BUKANOVA, A. E., *Zhur. Anal. Khim.*, **16**, 611 (1961).
14. PCHENITSYN, N. K., GINZBURG, S. J. and PROKOF'EVA, I. V., *Zhur. Anal. Khim.*, **17**, 343 (1962).
15. PCHENITSYN, N. K., PROKOF'EVA, I. V. and BUKANOVA, A. E., *Zhur. Anal. Khim.*, **16**, 605 (1961).

CHAPTER 21

COMPOUNDS OF PENTAVALENT AND TRIVALENT MOLYBDENUM

IN VOLUMETRIC determinations of molybdenum compounds, molybdate (VI) is usually transformed by reduction into compounds of penta- or trivalent molybdenum, which then can be titrated with various oxidizing titrants.

Attempts have been made to use molybdenum compounds of the lower valency states as reductimetric reagents. These are more stable than the respective tungsten compounds and can be more easily prepared; their reducing action is, however, weaker.

The redox potentials of the systems Mo^{VI}/Mo^V and Mo^{VI}/Mo^{III} depend on the hydrogen ion concentration.

In 2 N hydrochloric acid Mo^{VI}/Mo^V is + 0·53 V, and Mo^{VI}/Mo^{III} + 0·1 V (red form of the trivalent molybdenum complex) or even −0·25 V (unstable green form of the trivalent molybdenum complex).[1]

STANDARD SOLUTION

Solutions of *pentavalent molybdenum* can be prepared by electrolysis of molybdates (VI) in strongly acid solution (8 N HCl). The emerald green solution thus prepared has, however, too high a redox potential. Therefore it is usually diluted with water so that a solution results which is about 2 N with regard to hydrochloric acid; the colour of the solution then changes to a red–brown tint. The solution has to be stored in an inert atmosphere,[2,3] but the titrations can be carried out in air.[2]

Standardization

The standardization is best done by potentiometric titration with cerium (IV) sulphate at elevated temperature.[2]

Solutions of *trivalent molybdenum* can be prepared by energetic reduction of strongly acid molybdate (VI) solutions, e.g. by electrolysis on a platinum cathode, or by zinc amalgam.[4-8] In solutions containing sufficient chloride ions, the more stable red form of the chloro complex of trivalent molybdenum is produced.[4,6,7] Only this form of the complex is used in volumetric analysis, because the green form of the complex with trivalent molybdenum is only little stable in solution.[1,4,6]

Standardization

The red solutions of the chloro complex of trivalent molybdenum can be standardized by titration with a solution of cerium (IV) salt or of dichromate (VI).[7] Not long ago Busev[6] investigated extensively the stability of the solutions. He found that the normality of the red complex compound in 2 N HCl solution, containing about 5 per cent potassium chloride, does not change for 14 days. Fidler[7] recommends a special container for storage and titration in an inert atmosphere.

INDICATOR

All further determinations are carried out potentiometrically. When the conditions of acidity are strictly followed, the end-point of the titration of iron (III) salt or of dichromate with solutions of pentavalent molybdenum compounds, can also be detected visually. The well-known molybdenum blue forms at the equivalence point.[2]

REVIEW OF DETERMINATIONS

Iron (III), cerium (IV) salts, iodate bromate, vanadate (V), dichromate as well as mixtures of iron (III) salt with dichromate or iron (III) salt with vanadate or cerium (IV) salt and vanadate, can be reliably titrated with solutions of *pentavalent molybdenum compounds*. The single ions in the mixture of cerium (IV) salt, vanadate (V) and iron (III) salt can likewise be determined quantitatively. For the details of the necessary conditions for these determinations the reader should consult the original publications.[2,12]

Solutions of *trivalent molybdenum compounds* are suitable for the same determinations[5-7a] and moreover, methods have been described using this standard solution also for the titration of *gold (III)* and *iridium (IV) salts*.[7] The determinations of *iodate* and *bromate* yield results with noticeable errors.[5]

Apart from these possibilities there have also been mentioned in the literature:

The reaction between molybdenum (III) and *copper (II)* salt which is suitable for the standardization of the standard molybdenum solution,[6] and the interesting titration of molybdate (VI) (Mo^{III} + 2 Mo^{VI} → 3 Mo)^V; Fidler[7] (later Busev[10]) discusses in his paper very extensively this reaction and the reducing action of trivalent molybdenum in volumetric analysis. He applied this titration with very good results to some practical analysis, e.g. for the determination of iron and chromium in chromium ores and in ferrochromium, and for the determination of vanadium and iron in ferrovanadium.

Only one publication exists in the literature dealing with the reduction of organic compounds using trivalent molybdenum, namely of *nitro* and *nitroso compounds* to amino compounds; it is said that *picric acid* or the well-known analytical reagent *Cupferron* can be determined in this way.[9]

Very recently, the preparation of $K_3Mo(CN)_8$ and its application to the volumetric determination of *cobalt* and *manganese*[11-13] have been described.

REFERENCES

1. FOERSTER, F., *Z. phys. Chem.*, **146**, 177 (1930).
2. TOURKY, A. R., FARAH, M. Y. and EL-SHAMY, H. K., *Analyst*, **73**, 258, 262, 266 (1948).
3. FURMAN, N. H. and MURRAY, W. N., *J. Amer. Chem. Soc.*, **58**, 1689 (1936).
4. DOLEŽAL, J., MOLDAN, B. and ZÝKA, J., *Collection (Czech. Chem. Comm.)*, **24**, 3769 (1959).
5. FARAH, M. Y. and MIKHAIL, S. Z., *Z. anal. Chem.*, **166**, 24 (1959).
6. BUSEV, A. I. and LI GIN, *Zhur. Anal. Khim.*, **14**, 668 (1959).
7. FIDLER, J., Diploma Thesis, Charles University, Prague 1959.
7a. MIKHAIL, S. Z. and STEWART, J. J., *Z. anal. Chem.*, **178**, 335 (1961).
8. WARDLAW, W. and WORMELL, R. L., *J. Chem. Soc.*, **130**, 1087 (1927).
9. GAPCHENKO, M. V., *Zavodskaya Lab.*, **10**, 245 (1941).
10. BUSEV, A. I. and LI GIN, *Zhur. Anal. Khim.*, **15**, 191 (1960).
11. KRATOCHVIL, B. and DIEHL, H., *Talanta*, **3**, 346 (1960).
12. MIKHAIL, S. Z. and STEWART, J. J., *Z. anal. Chem.*, **178**, 335 (1961).
13. MALIK, W. U. and IFTIKHAR ALI, S., *Talanta*, **8**, 737 (1961).

CHAPTER 22

COMPOUNDS OF PENTAVALENT AND TRIVALENT TUNGSTEN

It is evident from the redox potential of the system W^{VI}/W^V, which is + 0·25 V in 10·5 N hydrochloric acid solution,[8] that various strong oxidizing agents can be titrated potentiometrically with compounds of pentavalent tungsten.[1−3]   Geyer and Henze[9] have studied recently the redox potentials of tungsten compounds in hydrochloric acid medium.

Compounds of trivalent tungsten are, however, the more active reductants; the use of potassium chlorotungstate (III), $K_3W_2Cl_9$ solutions, have so far been described as titrants only in one publication.[4] This compound is similar in its reactions to titanium (III) compounds.

STANDARD SOLUTION

Solutions of pentavalent tungsten can be prepared by electrolytic reduction of tungstate (VI) in 10 N hydrochloric acid solution;[1,4] when stored in an inert atmosphere they are stable for 14 days.[1]

For the preparation of tungsten (III) solutions, the compound $K_3W_2Cl_9$ is used which itself is prepared from tungstate (VI) by electrolytic reduction,[6,7] or by reduction with tin[4,5]   This standard solution has also to be stored in an inert atmosphere.

Standardization

The standardization of solutions of pentavalent and trivalent tungsten compounds can be carried out potentiometrically against dichromate[4] in acid medium and in an inert atmosphere.

INDICATOR

All determinations mentioned here are best carried out potentiometrically. The end-point of the titration can also be detected by the appearance of the blue colour produced by the first excess drop of tungsten (V) solution. Also in titrations with standard tungsten (III) solutions, the end-point of the titration can likewise be detected by this blue colour, because here the first excess drop of the titrant reacts with the tungstate (VI) produced during the titration to form tungsten (V).

In the titration of cerium (IV) salts with chlorotungstate (III), ferroin can also be used as indicator; thiocyanate is used as indicator in the titration of iron (III) salts.

REVIEW OF DETERMINATIONS

Tungsten (V) solutions have been applied to the titration[1,2,3,8] of *copper (II)* ($Cu^{2+} \rightarrow Cu^+$), *iron (III)*, *hexacyanoferrate (III)*, *dichromate*, *vanadate* and also of mixtures such as iron (III) and vanadate (V),[3] always in strongly acid solution at elevated temperatures.

The reactions with cerium (IV) salt and bromate (V) only proceed quantitatively, if the standard solution is titrated with the sample solution.[3]

The titration of vanadate enables as little as 30 μg of vanadium to be determined accurately and is specially recommended.[2]

Titrations with *chlorotungstate (III)* can be carried out in sulphuric acid solution.[4] Apart from the determinations already described above, *permanganate* (in the presence of manganese (II) salts), *bromate* and *cerium (IV) salts* can likewise be determined. *Copper (II) salts* are titrated after addition of a sufficient amount of thiocyanate. Mixtures, such as copper (II) and iron (III) salts can also be easily titrated.

Very recently, the titration of *tungstate (VI)* with tungsten (III) solutions has been studied.[10]

REFERENCES

1. TOURKY, A. R., ISSA, I. M. and ARMIN, A. M., *Analyt. Chim. Acta*, **10**, 168 (1954).
2. TOURKY, A. R., ISSA, I. M. and DAESS, A. M., *Rec. Trav. Chim. Pays Bas*, **75**, 22 (1956).
3. TOURKY, A. M., ISSA, I. M. and DAESS, A. M., *Analyt. Chim. Acta*, **16**, 81 (1957).
4. UZEL, R. and PŘIBIL, R., *Chem. Listy*, **33**, 102 (1939).
5. OLSSON, O., *Ber.*, **46**, 566 (1913); *Ber.*, **47**, 917 (1914); *Z. anorg. Chem.*, **88**, 49 (1914).
6. COLLENBERG, O. and SANDVED, K., *Z. anorg. Chem.*, **130**, 1 (1923).
7. COLLENBERG, O. and GUTHE, A., *Z. anorg. Chem.*, **134**, 317 (1924).
8. COLLENBERG, O. and GUTHE, A., *Z. anorg. Chem.*, **136**, 252 (1924).
9. GEYER, R. and HENZE, G., *Z. anal. Chem.*, **177**, 185 (1960).
10. GEYER, R. and HENZE, G., *Wiss. techn. Hochsch. Chem. Leuna-Merseburg*, **3**, 261 (1960/61); *Chem. Abstr.*, **57**, 1542 (1961).

CHAPTER 23

URANIUM (IV) SULPHATE

URANIUM (IV) solutions (uranium (IV) sulphate $U(SO_4)_2$ is used) are medium strong reducing agents. They can be used for titrating strong oxidants.[1-4] The redox potential of the system

$$U^{4+} + 2H_2O \rightleftharpoons UO_2^{2+} + 4H^+ + 2e^-$$

is -0.33 V in 0·1 N acid medium[1] and changes with the acidity, [2,7,9] in 1–9 N sulphuric acid, the redox potential is about 0·6 V.[2]

According to Belcher and co-workers,[1] uranium (IV) solutions are stable in air; but they consider that there is no advantage over other titrants. Other authors found,[2,3,7,8] on the contrary, that the standard solutions are oxidized by atmospheric oxygen and have therefore to be stored in an inert atmosphere and in the dark. The stability of the solutions increases with increasing hydrogen ion concentration.

STANDARD SOLUTION

A 0·1 N standard solution is prepared by reduction of uranium (VI) salts, best in a silver reductor[1,6] or with zinc[2,4] or electrolytically;[5] here the solution to be reduced should be about 1 N hydrochloric acid or 0·2 N sulphuric acid.

Standardization

Add about 10 ml of 2 N hydrochloric acid to 20 ml of 0·1 N dichromate (VI) solution. Titrate this solution at 60°C potentiometrically with the uranium (IV) solution to be standardized.[1]

It is also possible to standardize the solution with permanganate.[1,10]

INDICATOR

Only iron (III) salts can be titrated directly using a visual end-point (thiocyanate as indicator).[4] Otherwise only indirect determinations are possible with this reagent. To the substance to be determined there is added either an excess of iron (II) salt, and the iron (III) thus produced is then titrated with standard uranium (IV) solution, or a known amount of uranium (IV) solution is added and the unconsumed titrant is back-titrated with iron (III) solution.

Potentiometric titration, however, is more suitable. This has also extended the use of uranium (IV) solutions to systems which otherwise could not be titrated.

REVIEW OF DETERMINATIONS

Vortmann and Binder[4] first used uranium (IV) salt for the titration of *iron (III)* and carried out also some indirect determinations. Later on, Belcher and co-workers[1] described the determination of *vanadate (V), dichromate, permanganate* and *cerium (IV) salts*. Recently, Issa and El-Sherif[2,3] used uranium (IV) solutions for potentiometric titration. Apart from the substances already mentioned, they also determined *bromate, hexacyanoferrate (III)* and *tellurate* and studied the conditions necessary for these determinations. All these titrations, which are carried out in acid solution at elevated temperature, proceed very slowly. It is therefore necessary to wait several minutes near the equivalence point for the potential to reach equilibrium.

According to other authors[11] *mixtures* can likewise be determined potentiometrically, e.g.: cerium (IV) and iron (III); vanadate (V) and iron (III); vanadate (V), cerium (IV) and iron (III); permanganate and iron (III); dichromate and iron (III) as well as permanganate, vanadate (V) and iron (III).

REFERENCES

1. BELCHER, R., GIBBONS, D. and WEST, T. S., *Anal. Chemistry*, 26, 1025 (1954).
2. ISSA, I. M. and EL-SHERIF, I. M., *Analyt. Chim. Acta*, 14, 466 (1956).
3. ISSA, I. M. and EL-SHERIF, I. M., *Analyt. Chim. Acta*, 14, 474 (1956).
4. VORTMANN, G. and BINDER, F., *Z. anal. Chem.*, 67, 269 (1925).
5. EL-SHAMY, H. K. and EL-DIN-ZAYAN, S., *J. Chem. Soc.*, 384 (1953).
6. WILLARD, H. H. and DIEHL, K., *Advanced Quantitative Analysis*, p. 68, van Nostrand. New York, 1944.
7. McCOY, H. N. and BUNZEL, H. H., *J. Amer. Chem. Soc.*, 31, 367 (1909).
8. GUSTAVSON, R. and KNUDSON, C. M., *J. Amer. Chem. Soc.*, 44, 2756 (1922).
9. LUTHER, R. and MICHIE, A. E., *Z. Elektrochem.*, 14, 826 (1903).
10. EWING, D. P. and ELDRIDGE, E. F., *J. Amer. Chem. Soc.*, 44, 1484 (1922).
11. TOURKY, A. R., ISSA, I. M. and DAESS, A. M., *Rec. Trav. Chim. Pays Bas*, 75, 22 (1956).

CHAPTER 24

VANADIUM (II) SULPHATE

THE properties of divalent vanadium have been investigated from many standpoints.[1-5] It has been found that divalent vanadium is distinguished by its strong reducing action and can therefore be used mainly in quantitative analysis. The oxidation of vanadium (II) ions does not, however, yield a uniform product of oxidation. According to the nature of the reducing substances and the reaction medium, various valency states of vanadium are formed;[6] most frequently they are compounds of trivalent vanadium but the oxidation may also proceed to the tetravalent and pentavalent state.

The normal potential of the redox system

$$V^{2+} \rightleftharpoons V^{3+} + e^-$$

is stated to be -0.255 V at 25°C.[7] This makes it possible to determine reductimetrically many inorganic, and especially organic substances.

Titrations with vanadium (II) sulphate are mostly carried out in sulphuric, hydrochloric, or acetic acid, or buffered solution at normal or elevated temperature in an inert atmosphere.

STANDARD SOLUTION

The standard solution is prepared by reducing acid solutions of ammonium meta-vanadate (V), vanadyl sulphate or vanadyl chloride either electrolytically[6,8] or with zinc amalgam[9-11] or zinc.[12] The latter two methods are the more suitable ones.

Dissolve an adequate amount of ammonium meta-vanadate (V) in 1 N sulphuric acid; transfer the solution thus prepared to a bottle which contains zinc amalgam[9,10] and from which the air has been removed with carbon dioxide. Shake the mixture until the violet colour formed does not intensify further. Filter the solution through a glass filter using vacuum and transfer to an automatic burette.

The standard solution has to be stored always in an atmosphere of carbon dioxide, otherwise the vanadium (II) is oxidized rapidly to higher valency states.

This lengthy reduction with zinc amalgam can be replaced by a simple reduction using zinc.[12]

Dissolve 12 g of ammonium meta-vanadate (V) in 1000 ml of 1 N sulphuric acid and add about 8 g of powdered zinc in small portions to the solution. The originally orange-coloured solution changes intermediately to blue, to green and assumes finally the violet colour of the vanadium (II) salt solution. The reduction is finished when there is no further change in the violet colour. Then filter the solution through a sintered glass as described above, by applying nitrogen pressure, and transfer to the burette.

In some determinations, also vanadium (II) chloride is used as titrant. The preparation and properties of this solution are very similar to the above-mentioned case.

Vanadium (II) solutions are very rapidly oxidized by atmospheric oxygen and need therefore to be stored in an atmosphere of hydrogen, nitrogen or carbon dioxide. For this purpose, an apparatus is proposed[12] as commonly used in titrations with titanium (III) solutions. In other cases, apparatus specially designed for this particular purpose are recommended.[11,13-15]

The reaction between vanadium (II) and hydrogen ions, which yields molecular hydrogen is also known:

$$2\,V^{2+} + 2\,H^+ \rightarrow 2\,V^{3+} + H_2$$

This reaction, which renders vanadium (II) solutions unstable, proceeds, however, very slowly and can be neglected if there are no catalysts present.

Ellis and Vogel[15] observed no change in 3 weeks with 1 N sulphuric acid solution and in 2 weeks with 1 N hydrochloric acid solution.

Standardization

The normality of the standard solution is determined by titration with iron (III) ammonium sulphate solution in 0·1 N sulphuric acid or with iron (III) chloride solution in 0·2 N hydrochloric acid, either potentiometrically or visually, using a 0·2 per cent solution of phenosafranine as indicator.[10,12,15]

The iodimetric standardization[15] is based on the oxidation of vanadium (II) ions with an excess of 0·1 N potassium iodate solution. After adding potassium iodide, the liberated iodine is titrated with 0·1 N thiosulphate using starch as indicator.

The standardization can also be carried out by titration with permanganate,[1,7,16] cerium (IV) or potassium dichromate (VI).[15]

Recently, neutral red, phenosafranine, methylene blue or gallocyanine have been used as visual indicators for the standardization of vanadium (II) solutions with iron (III) or copper (II) salts.[16a]

INDICATOR

In titrations with vanadium (II) sulphate or chloride, the equivalence point is detected either potentiometrically[6,12] or amperometrically,[11] or visually using phenosafranine, N-phenylanthranilic acid,[15,17] diphenylamine, thiocyanate[18] or neutral red[17] as indicators.

REVIEW OF DETERMINATIONS

Silver and copper. Silver ions are quantitatively reduced to metallic silver with vanadium (II) sulphate solution at 90–95°C in acetic acid solution.[6] The titration is carried out potentiometrically. Nitrate and acetate do not affect the determination.

Copper is determined in different ways in hydrochloric, acetic or sulphuric acid solution.[6,16a] When titrating in hydrochloric acid solution, the following reactions take place:

$$2\,Cu^{2+} + V^{2+} + H_2O \rightleftharpoons 2\,Cu^+ + VO^{2+} + 2\,H^+$$
$$VO^{2+} + V^{2+} + 2\,H^+ \rightleftharpoons 2\,V^{3+} + H_2O$$

In acetic acid–sodium acetate buffered solution, divalent copper is reduced to the metal. The reaction is similar to the above formulated one. The vanadyl (IV) ions produced are reduced to vanadium (III) ions.

In sulphuric acid solution, copper (II) ions are likewise reduced to the metal but no vanadyl (IV) ions are formed, so that the reaction simply proceeds as in the following equation:

$$Cu^{2+} + 2\,V^{2+} \rightarrow Cu + 2\,V^{3+}$$

All these titrations are carried out potentiometrically at temperatures near the boiling point.

Iron and vanadium. Iron (III) salts are titrated in hydrochloric, sulphuric or acetic acid solution.[6] In the potentiometric titration, two potential jumps appear which correspond to the reduction of iron (III) to iron (II) ions and to the reduction of vanadyl (IV)—produced during the reaction—to vanadium (III) ions. The determination is not affected by alkali metals, alkaline earth metals and cadmium. In sulphuric acid solution iron can be determined also in the presence of copper.

Vanadium (II) chloride can likewise be used for the titration in hydrochloric and acetic acid solution. In all cases, it is necessary to work at boiling temperature.

Iron (III) ions can be determined visually using phenosafranine[15] or thiocyanate[8,11] as indicators; this method is, however, less suitable.

The potentiometric titration of vanadyl (IV) compounds with vanadium (II) sulphate or vanadium (II) chloride, proceeds quantitatively at boiling temperature in hydrochloric or acetic acid solution to form vanadium

(III) ions. The potential jump becomes more distinct on the addition of chloride, sulphate or alkali metals.

Chromium. In the reduction of dichromate with vanadium (II)[6] in acid medium the reaction proceeds according to the following equations:

$$Cr_2O_7^{2-} + 2V^{2+} + 2H^+ \rightleftharpoons 2Cr^{3+} + 2VO_3^- + H_2O$$

$$2VO_3^- + V^{2+} + 6H^+ \rightleftharpoons 3VO^{2+} + 3H_2O$$

$$VO^{2+} + V^{2+} + 2H^+ \rightleftharpoons 2V^{3+} + H_2O$$

The titration with vanadium (II) sulphate is carried out either directly[11] against thiocyanate and iron (III) salt as indicator or indirectly[8] by back-titration of the added excess of vanadium (II) salt with iron (III) ammonium sulphate solution in the presence of thiocyanate.

Potentiometric titration is, however, more suitable. For the calculation of the result it is better to use the amount of titrant corresponding to the second potential jump ($VO_3^- \rightarrow VO^{2+}$).

Cerium and titanium. Cerium (IV) salts, or (after preliminary oxidation with peroxodisulphate) also cerium (III) salts, can be determined with vanadium (II) sulphate in sulphuric acid solution using diphenylamine as indicator.[18]

Titanium can be titrated potentiometrically[19,20] in sulphuric acid medium at 80°C, and also in the presence of iron.

Thallium and bismuth. Thallium and bismuth as well as mixtures of the two, can be titrated potentiometrically with vanadium (II) chloride or vanadium (II) sulphate solution in 6–10 N hydrochloric or sulphuric acid medium.[20a,b] Zinc (II), cadmium (II) and lead (II) ions do not interfere with the titration. The determination is, however, affected by iron (III) and copper (II) ions as well as by the presence of oxidizing agents, such as permanganate, dichromate, peroxosulphate, etc. The method has been successfully used for the analysis of technical materials.

Thallium and antimony. In a similar way, thallium (III) and antimony (V) salts as well as their mixtures can be titrated. At low concentrations of hydrochloric acid, thallium (III) ions are first reduced to the monovalent state; thereupon, the concentration of hydrochloric acid is increased to 10–12 N, when the antimony (V) ions are reduced to the trivalent state.[20c]

Other authors describe the determination of thallium (III).[20d]

Palladium. Gusev and Ketova[20e] worked out the conditions necessary for the quantitative reduction of palladium compounds with vanadium (II) chloride. Palladium (II) salts are reduced with 0·1 or 0·05 N vanadium (II) chloride in 0·2–2 N hydrochloric or sulphuric acid at room temperature to the metal. The potentiometric determination is not affected by the presence of nickel, cobalt, iron (III), gold (III) and silver salts. Copper and platinum interfere.

Uranium. Uranium (VI) can be determined visually with vanadium (II) sulphate solution.[21a] The reduction proceeds in sulphuric acid solution according to the equation:

$$UO_2^{2+} + 2 V^{2+} + 4 H^+ \rightarrow U^{4+} + 2 V^{3+} + 2 H_2O$$

p-ethoxy chrysoidine is used as indicator.

Other substances. Chlorate,[21,21a] bromate,[21a] nitrate,[15,21] peroxodisulphate,[21,21a] peroxide, perborate, tin (II),[21a] hydroxylamine[15] and *diazo phenols*[21b] are determined indirectly with vanadium (II) sulphate in acid solution. The excess of vanadium (II) ions is back-titrated with permanganate or with iron (III) salt solution.

Azobenzene and azo dyes. The volumetric determination of *azobenzene*[15] is based on its reaction with vanadium (II) ions which proceeds according to the equation:

$$C_6H_5N{=}NC_6H_5 + 2 V^{2+} + 2 H^+ \rightleftharpoons C_6H_5NH{-}NHC_6H_5 + 2 V^{3+}$$

The titration is carried out in 0·5–2 N hydrochloric acid in a nitrogen atmosphere with vanadium (II) chloride solution, using a 0·2 per cent solution of phenosafranine as indicator. Vanadium (II) sulphate can likewise be used for the determination of *azo dyes*.[10]

The titration proceeds quantitatively even at room temperature in sulphuric acid solution and with exclusion of atmospheric oxygen. This method is as suitable as the titration with titanium (III) solution; the " titanium number " (giving the ml of 0·1 N titanium (III) chloride solution consumed by 1 g of the dye) can be replaced here by the " vanadium number". Matrka and Ságner[10a] observed that determinations in which a visual end-point is used, yield erroneous results; they ascribe this to the fact that vanadium (III) ions produced during the titration reduce the azo dye to be determined and so diminish considerably the consumption of titrant. In such cases it is preferable to carry out the titration potentiometrically or with titanium (III) chloride.

Triphenylmethane dyes. The reduction of triphenylmethane dyes[22] can be illustrated by the example of crystal violet:

$$[(CH_3)_2N{\cdot}C_6H_4]_3 C^+ + 2 e^- + H^+ \rightarrow [(CH_3)_2N{\cdot}C_6H_4]_3 CH$$

Because the reduction products formed are usually insoluble in buffered solutions, it is recommended that sodium *m*-xylene sulphonate[22] be added instead of alcohol to the solution to be titrated. This hydrotropic solvent keeps all the dyestuff compounds in solution. The direct visual or potentiometric titration is carried out with 0·1 N vanadium (II) sulphate solution at 50°C in a nitrogen atmosphere. This determination yields more reliable results than other reductimetric methods.

Hydroxy-triphenylmethane dyes. The reductimetric determination of these dyes is carried out indirectly:[22a]

Dissolve the sample in a mixture of ethanol and water, acidify with concentrated sulphuric acid and add a known excess of 0·1 N vanadium (II) sulphate solution. Wait for 1 min, then cool and back-titrate with 0·1 N iron (III) ammonium sulphate solution. The end-point is detected visually, using a 0·2 per cent solution of safranine-T as indicator.

The mechanism of the reaction is demonstrated by the reduction of aurine:

$$(HO \cdot C_6H_4)_2 \cdot C = C_6H_4 = O + 2\ e^- + 2\ H^+ \rightarrow (HO \cdot C_6H_4)_3 \cdot CH$$

Eriochromazurol, chromium blue R, chromoxan brown 5R and naphtho-chromazurine can be determined similarly. The relative error is about ± 0·50 per cent.

Anthraquinone. The reduction of anthraquinone to anthrahydroquinone with standard vanadium (II) sulphate solution[12] proceeds at 50°C in acetic solution according to the scheme:

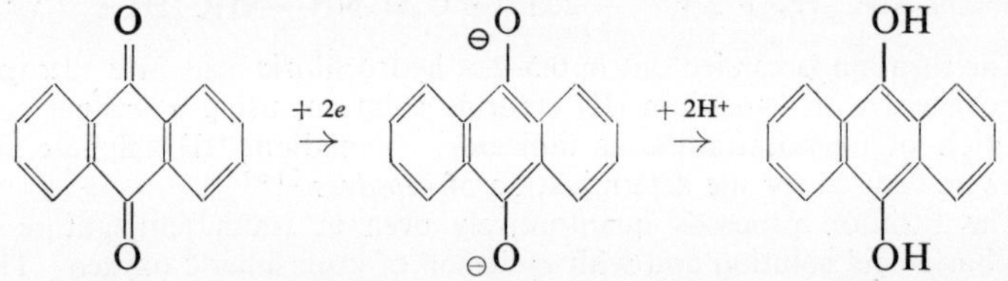

In sulphuric acid solution, the reduction proceeds to anthranol involving the exchange of four electrons. It is therefore necessary to choose conditions for the titration, so that the consumption of standard solution, which is considerably acid, is as small as possible lest the acidity of the solution to be titrated increases. If the concentration of sulphuric acid does not exceed 0·5–2 per cent until the end of titration, the desired course of reduction is not unfavourably influenced by the reaction.

Pyrazol anthrone. The following reaction takes place in the reduction of pyrazol anthrone with vanadium (II) sulphate solution:[23]

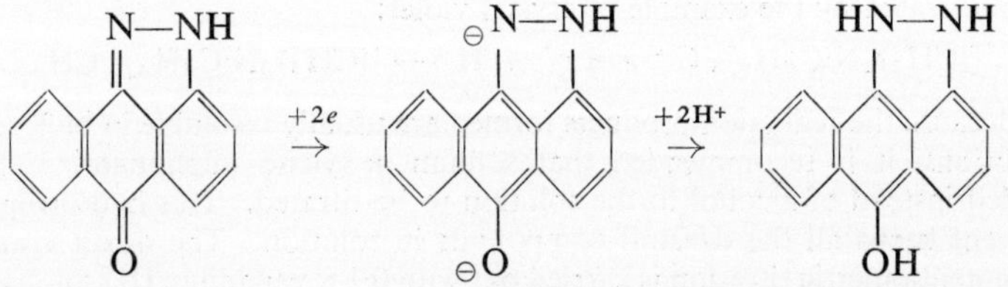

The determination is carried out in saturated sodium citrate solution after the sample has been dissolved in ethanol. This method is of great significance for the industrial analysis of this important dyestuff intermediate product.

Nitro compounds. Organic substances which contain one or several NO_2 groups, or also substances which can be transformed to nitro-compounds, can be determined with standard vanadium (II) sulphate solution.[9,15,24,25] Here, the reaction proceeds according to the equation:

$$R\text{—}NO_2 + 6H^+ + 6V^{2+} \rightarrow R\text{—}NH_2 + 6V^{3+} + 2H_2O$$

The method is based on the reaction between the nitro-compound and the vanadium (II) sulphate, the excess of which is back-titrated either with sodium chromate (VI) in concentrated sulphuric acid, using phenosafranine as indicator,[25] or with iron (III) ammonium sulphate[15] using phenosafranine or thiocyanate as indicator.[24] In all cases, the titrations are carried out in an inert atmosphere.

Sometimes reduction in acetate buffered solution (pH 4·5) is recommended whereby more reliable results can be achieved.[15] *2-Nitro*-m-*xylene* and *nitroguanidine* are determined by this method.[15] Other nitrocompounds such as *nitrobenzene, dinitrobenzene,* o-*nitro benzoic acid,* p-*nitraniline,* p-*nitrophenyl hydrazine, 2,4-dinitrophenylhydrazine* and *picric acid* are determined indirectly in sulphuric acid solution. The excess of vanadium (II) ions is titrated with sodium chromate (VI) against phenosafranine,[25] or with iron (III) ammonium sulphate solution in the presence of thiocyanate as indicator.[24]

Other substances[25] such as *formaldehyde, acetaldehyde, acetone, benzaldehyde, nitrobenzaldehyde, benzophenone, glucose, camphor* and *vanillin* are transformed into the respective nitro-compounds with the aid of 2,4-dinitrophenylhydrazine. The dinitrophenylhydrazones thus formed are reduced with an excess of vanadium (II) sulphate; the unconsumed amount is back-titrated with sodium chromate solution in the usual way. *Alcohols* and *phenols* are determined as 3,5-dinitro benzoates, *amines* and *aromatic hydrocarbons* as picrates. Other substances are transformed to the respective nitro-compounds in a similar way.

The literature quoted here gives a survey of the possible titrations with vanadium (II) sulphate.[26,27]

REFERENCES

1. Piccini, A. and Marino, L., *Z. anorg. Chem.*, **32**, 55 (1902).
2. Rutter, T. F., *Z. anorg. Chem.*, **52**, 368 (1907).
3. Connant, J. B. and Cutter, H. B., *J. Amer. Chem. Soc.*, **48**, 1016 (1926).
4. Meyer, J. and Aulich, M., *Z. anorg. Chem.*, **194**, 278 (1930).
5. Foerster, F. and Böttcher, F., *Z. phys. Chem.*, **A151**, 321 (1930).
6. Maass, K., *Z. anal. Chem.*, **97**, 241 (1934).
7. Jones, G. and Colvin, J. H., *J. Amer. Chem. Soc.*, **66**, 1573 (1944).
8. Banerjee, P. Ch., *J. Indian Chem. Soc.*, **12**, 198 (1935).
9. Gapchenko, M. V., Sheintsis, O. G., *Zavodskaya Lab.*, **9**, 562 (1940).
10. Gapchenko, M. V., *Zavodskaya Lab.*, **16**, 1126 (1950).
10a. Matrka, M. and Ságner, Z., *Chem. Průmysl*, **10**, 474 (1960).
11. Meites, L., *J. Chem. Educat.*, **27**, 458 (1950).
12. Matrka, M. and Ságner, Z., *Chem. Listy*, **51**, 68 (1957).
13. Matrka, M., Smetana, B. and Ságner, Z., *Chem. Průmysl*, **8**, 367 (1958).

14. KARSTEN, P., KIES, H. I. and BERGSHOEFF, G., *Chem. Wbl.*, **48**, 734 (1952).
15. ELLIS, C. M. and VOGEL, A. I., *Analyst*, **81**, 693 (1956).
16. ROSCOE, H. E., *Philos. Trans.*, **158**, 1 (1868).
16a. MITTAL, R. K., TANDON, J. P. and MEHROTRA, R. C., *Z. anal. Chem.*, **189**, 330 (1962).
17. MATRKA, M. and SÁGNER, Z., *Chem. Průmysl*, **9**, 526 (1959).
18. BANERJEE, P. CH., *J. Indian Chem. Soc.*, **15**, 475 (1938).
19. RUSSELL, A. S., *J. Chem. Soc.*, **128**, 497 (1926).
20. BANERJEE, P. CH., *J. Indian Chem. Soc.*, **19**, 30 (1952).
20a. KETOVA, L. A. and GUSEV, S. I., *Izvest. Vys. Ucheb. Zavedenii, Khim. Khim. Technol.*, **3**, 59 (1960).
20b. GUSEV, S. I. and KETOVA, L. A., *Zhur. Anal. Khim.*, **16**, 552 (1961).
20c. GUSEV, S. I. and KETOVA, L. A., *Zhur. Anal. Khim.*, **17**, 137 (1962).
20d. BHATNAGAR, R., BHATNAGAR, M. L. and MATHUR, N. K., *J. Electroanal. Chem.*, **4**, 182 (1962).
20e. GUSEV, S. I. and KETOVA, L. A., *Zhur. Anal. Khim.*, **17**, 1018 (1962).
21. BANERJEE, P. CH., *J. Indian Chem. Soc.*, **13**, 301 (1936).
21a. MITTAL, R. K., TANDON, J. P. and MEHROTRA, R. C., *Z. anal. Chem.*, **189**, 406 (1962).
21b. NEMODRUK, A. A. and ORESHKO, V. F., *Izvest. Vys. Ucheb. Zavedenii, Khim. i Khim. Technol.*, **3**, 316 (1960).
22. MATRKA, M. and SÁGNER, Z., *Chem. Průmysl*, **8**, 22 (1958).
22a. MATRKA, M. and SÁGNER, Z., *Chem. Průmysl*, **11**, 135 (1961).
23. MATRKA, M. and SÁGNER, Z., *Chem. Průmysl*, **7**, 484 (1957).
24. BANERJEE, P. CH., *J. Indian Chem. Soc.*, **19**, 35 (1942).
25. WITRY-SCHWACHTGEN, G., *Instr. Gr.-Ducal Luxembourg. Sect. Sci. Natur., Phys. et Math., Arch.*, **22**, 87 (1955).
26. VULTERIN, J., *Chem. Listy*, **53**, 384 (1959).
27. MACDONALD, A. M. G., *Ind. Chemist*, 332 (1956).

CHAPTER 25

VANADIUM (IV) SULPHATE AND VANADIUM (IV) ACETATE

TETRAVALENT vanadium in the form of vanadyl (IV) ions is distinguished by its reducing power in acid as well as in alkaline medium. The normal potential of the redox system

$$VO^{2+} + 2\,H_2O \rightleftharpoons VO_3^- + 4\,H^+ + e^-$$

is stated to be 1·0 V.[1] The potential increases with increasing concentration of hydrogen ions; it increases in 1–8 N sulphuric acid medium by 0·3 V. The opposite is the case in alkaline medium; here, the redox potential becomes more negative by 0·22 V on increasing the concentration of sodium hydroxide from 1 to 5 N. Vanadyl (IV) sulphate is therefore a much stronger reducing agent in alkaline solution.

Titrations with vanadyl (IV) sulphate are carried out mostly in sodium or potassium hydroxide solution:

$$VO^{2+} + 4\,OH^- \rightarrow VO_3^- + 2\,H_2O + e^-$$

Titrations with vanadyl (IV) acetate are carried out in glacial acetic acid.

Vanadyl (IV) sulphate behaves with strong reducing agents as an oxidizing agent. According to Jones and Colvin[1a] the standard potential of the system VO^{2+}/V^{3+} is $+ 0.337$. In dilute sulphuric acid solution, the formal redox potential is $+ 0.359 \pm 0.002$ V.[1b]

STANDARD SOLUTION

Standard vanadyl (IV) sulphate solution[2,3,4] is usually prepared by dissolving pure ammonium meta-vanadate (V), or vanadium (V) oxide in an appropriate volume of sulphuric acid; the solution is heated on a water bath to 70°C and sulphur dioxide is passed into the solution. After 2–3 hr reduction, the excess of sulphur dioxide is removed by passing carbon dioxide, the solution is cooled and diluted to the required volume.

The preparation of vanadyl (IV) acetate solution is not so simple. It can be made by boiling vanadium (IV) oxide with acetic acid[5] over a prolonged period or by conversion of vanadyl (IV) sulphate with barium acetate.[6] In both cases, it needs to be recrystallized by evaporation in vacuum and in an inert atmosphere. Therefore, an easier method has been suggested[7] which starts from ammonium meta-vanadate (V) or from vanadium (V) oxide:

A mixture of the oxide, oxalic acid and glacial acetic acid is kept on a water bath at 70°C whilst steadily stirring. The reaction proceeds with formation of carbon dioxide, the mixture thickens and assumes a blue colour. The reaction product is dissolved in glacial acetic acid and diluted to the desired volume.

Acid solutions of vanadyl (IV) sulphate and vanadyl (IV) acetate are stable in air;[8-10] alkaline solutions, on the contrary, are easily oxidized in air. Titrations with vanadyl salts in alkaline solutions are for that reason carried out in an inert atmosphere.

Dean and Herringshaw[10a] studied the oxidation of tetravalent vanadium with atmospheric oxygen in alkaline medium and stated that traces of iron (III) catalyse the oxidation; in the presence of chromium (III) the velocity of the reaction is diminished.

Standardization

The normality of acid standard vanadyl (IV) sulphate or acetate solutions is determined with permanganate or cerium (IV) solution. The more usual standardization with permanganate is carried out in sulphuric acid solution at 75°C.

INDICATOR

Most of the titrations are done potentiometrically or amperometrically. In potentiometric titrations, mostly a platinum indicator electrode is used, and in amperometric titration a rotating platinum electrode.[7]

Willard and Manalo[11] recommended various derivatives of diphenylamine as indicators for the titration of hexacyanoferrate (III) with vanadyl (IV) sulphate; these are: 2-carboxy-2'-methoxy-diphenylamine, diphenylamine sulphonic acid, 2-carboxy-2'-methyl-diphenylamine, 2-carboxy-diphenylamine, 2,2'-dicarboxy diphenylamine, 2-carboxy-2'-bromo-diphenylamine and 2-carboxy-3'-ethoxy diphenylamine.

REVIEW OF DETERMINATIONS

Hexacyanoferrate (III). Hexacyanoferrate (III) can be determined potentiometrically with 0·1 N vanadyl (IV) sulphate solution at 20°C.[12]

In sodium hydroxide solution and in a nitrogen atmosphere, the reaction proceeds according to the equation:

$$[Fe(CN)_6]^{3-} + VO^{2+} + 4\,OH^- \rightleftharpoons VO_3^- + [Fe(CN)_6]^{4-} + 2\,H_2O$$

For the visual titration of hexacyanoferrate (III) in alkaline solution, various diphenylamine derivatives are used as indicators.[11]

This method can be applied to the *indirect determination* of *arsenite, antimonite, chromium, hydrazine sulphate* and *hydrogen peroxide*:

These substances are oxidized at 85–90°C in alkaline solution with a known amount of hexacyanoferrate (III), the excess of which is titrated with vanadyl (IV) sulphate at room temperature.

Chromate. Chromate or dichromate can be determined potentiometrically in alkaline solution, as with hexacyanoferrate (III):[13]

$$CrO_4^{2-} + 3\,VO^{2+} + 7\,OH^- \rightarrow Cr(OH)_3 + 3\,VO_3^- + 2\,H_2O$$

The titration is carried out in 15 per cent sodium hydroxide solution and in a nitrogen atmosphere at room temperature, but also at elevated temperature (up to 70°C). Better results are achieved at elevated temperature. The method is suitable for the determination of chromium and bismuth or of chromium and copper together.

Because of the simple nature of the titrations of hexacyanoferrate (III) and of chromate, conditions have been established which allow both substances to be determined in one sample simultaneously.[14,15]

Copper. Copper (II) ions can be titrated potentiometrically with vanadyl (IV) sulphate in ammoniacal and sodium hydroxide medium[16] at 50°C in an inert atmosphere.

Tomíček and Mandelík[2] worked out a procedure for the potentiometric titration of copper as citrate complex with vanadyl (IV) sulphate in potassium hydroxide or potassium carbonate solution. The reduction proceeds in two steps, first to monovalent and then to metallic copper:

Add 3 ml of 30 per cent citric acid to 5 ml of a 0·1 N copper (II) sulphate solution. After the complex has formed, add potassium hydroxide and potassium carbonate. Titrate at 50–85°C in an inert atmosphere. 120–130 ml of the solution to be titrated should contain between 6–95 mg copper. Nitrate, iron, arsenic, tin, selenium and tellurium salts interfere with the titration. Precipitates eventually formed have to be filtered off.

This method is also used for the determination of copper and chromium as well as of copper and bismuth simultaneously.

Mercury. Divalent complex bound mercury can be reduced to the metal with vanadyl (IV) sulphate in alkaline solution. The determination of mercury[17] is based on the following reaction:

$$K_2HgI_4 + 2\,VOSO_4 + 8\,KOH \rightarrow 2\,KVO_3 + 4\,KI + 2\,K_2SO_4 + 4\,H_2O + Hg$$

Divalent mercury reacts with potassium iodide forming K_2HgI_4 which is then titrated potentiometrically with vanadyl (IV) sulphate in 11–28 per cent sodium hydroxide solution at 80–90°C in a nitrogen atmosphere.

At room temperature, the reaction proceeds only slowly and yields too high results.

Silver. Silver salts can also be titrated potentiometrically with vanadyl (IV) sulphate.[16] The titration is carried out in ammoniacal solution containing sodium hydroxide at 50°C in an inert atmosphere. The method has also been applied to the simultaneous determination of silver and copper.

Bismuth. Bismuth (III) ions in strongly alkaline medium are converted to basic bismuthate (III) which can be determined potentiometrically with vanadyl (IV) sulphate:[2]

$$BiO_2^- + 3\,VO^{2+} + 8\,OH^- \rightarrow Bi + 3\,VO_3^- + 4\,H_2O$$

The determination is carried out at 60–80°C in saturated potassium hydroxide solution and in an inert atmosphere. 120 ml of the solution to be titrated should contain 4–150 mg of bismuth. Nitrate, iron (III), tin (II), selenium (IV), selenium (VI) and tellurium (VI) salts, as well as concentrations of more than 1 per cent of lead (II) salts, interfere with the titration.

The method is suitable for the determination of bismuth and chromium as well as the determination of bismuth and copper simultaneously.[2]

Gold. Trivalent gold can be determined with vanadyl (IV) sulphate in an inert atmosphere according to the following equation:[18,19]

$$AuO_2^- + 3\,VO^{2+} + 8\,OH^- \rightarrow Au + 3\,VO_3^- + 4\,H_2O$$

The determination is carried out potentiometrically in 7·5–30 per cent sodium hydroxide solution at 50–70°C. The recommended hydroxide ion concentration has to be followed precisely, because the potential adjusts itself only slowly at lower alkalinity.

Titanium. Trivalent titanium reacts with vanadyl (IV) sulphate according to the equation:

$$Ti^{3+} + VO^{2+} + 2\,H^+ \rightarrow Ti^{4+} + V^{3+} + H_2O$$

The volumetric determination of titanium (III) is based on this reaction,[19,19a] but the titrant acts here as oxidizing agent. The determination is carried out in 1 N hydrochloric acid solution at room temperature in a carbon dioxide atmosphere. The authors recommend that oxalic acid be added to the solution for a more rapid adjustment of the potential; the same can be achieved by adding tartaric and citric acid.

Lead (IV) acetate, chromic (VI) acid, bromate (V), permanganate. The reducing action of vanadyl (IV) ions has also been applied for titrations in acid medium, mainly in glacial acetic acid. Vanadyl (IV) acetate, in the presence of sodium or ammonium acetate is mostly used instead of vanadyl (IV) sulphate.

Lead (IV) acetate, chromic (VI) acid, bromate (V) and permanganate can thus be titrated potentiometrically in anhydrous glacial acetic acid.[20] Lead (IV) acetate is determined at 20°C, sodium acetate need not be present. Chromic (VI) acid or bromate are titrated at 60–80°C, permanganate at 20°C, but only in the presence of sulphuric acid. The accuracy of the semi-micro determinations is 0·5–1 per cent. Chromic (VI) acid can also be determined amperometrically using a rotating platinum electrode.[7]

Other substances. Amongst other substances, there can be determined with vanadyl (IV) sulphate in acid (generally sulphuric acid) medium:

Cerium (IV) ions[21] (this method has been used for the amperometric determination of cerium in Loparite),[21a] *permanganate*,[22,23] *periodic acid*,[24] *Caro's acid*,[25] *trivalent cobalt and manganese* and *tetravalent manganese* in the form of oxides.[25a]

Vanadium (II) salt can be titrated with vanadyl (IV) sulphate at elevated temperature in the solution of various acids:[26]

$$V^{2+} + VO^{2+} + 2H^+ \rightleftharpoons 2V^{3+} + H_2O$$

Vanadyl (IV) sulphate reacts here as an oxidant, similar as in the determination of titanium (III).

A survey of titrations with vanadyl (IV) sulphate and vanadyl (IV) acetate is given in the literature.[27,28]

REFERENCES

1. WILLARD, H. H. and MANALO, G. D., *Anal. Chemistry*, **19**, 462 (1947).
1a. JONES, G. and COLVIN, J. H., *J. Amer. Chem. Soc.*, **66** 1573 (1944).
1b. SYROKOMSKII, V. S., AVILOV, V. B., *Zavodskaya Lab.*, **15**, 769 (1949).
2. TOMÍČEK, O. and MANDELÍK, J., *Chem. Listy*, **43**, 169 (1949).
3. TOMÍČEK, O., *Chem. Listy*, **32**, 442 (1938).
4. BÍLEK, P., *Collection (Czech. Chem. Comm.)*, **10**, 430 (1938).
5. PREBLUDA, H. J. and PARKS, H., *Klin. Wschr.*, **16**, 1220 (1937).
6. PREBLUDA, H. J., US-Pat. 2135111; *Chem. Abstr.*, **33**, 1104 (1939).
7. NOVOTNÝ, J., *Chem. Listy*, **48**, 1865 (1954).
8. FURMAN, N. H., *J. Amer. Chem. Soc.*, **50**, 1675 (1928).
9. STOUT, L. E. and WHITAKER, G. C., *Ind. Engng. Chem.*, **20**, 210 (1928).
10. VELTICKÝ, B. and MAYER, V., *Chemie*, **4**, 67 (1948).
10a. DEAN, G. A. and HERRINGSHAW, J. F., *Talanta*, **10**, 793 (1963).
11. WILLARD, H. H. and MANALO, G. D., *Anal. Chemistry*, **19**, 167 (1947).
12. FRESNO, C. DEL and VALDÉS, L., *Z. anorg. Chem.*, **183**, 251 (1929).
13. FRESNO, C. DEL and MAIRLOT, E., *An. Real. Soc. Españ. Fisica Quim.*, **30**, 254 (1932).
14. FRESNO, C. DEL and MAIRLOT, E., *An. Real. Soc. Españ. Fisica Quim.*, **31**, 122 (1933).
15. FRESNO, C. DEL and MAIRLOT, E., *Z. anorg. Chem.*, **212**, 331 (1933).
16. FRESNO, C. DEL and MAIRLOT, E., *An. Real. Soc. Españ. Fisica Quim.*, **32**, 280 (1934).
17. FRESNO, C. DEL and LAFUENTE, DE E., *Gazz. Chim. Ital.*, **68**, 619 (1938).
18. FRESNO, C. DEL and MAIRLOT, E., *An. Real. Soc. Españ. Fisica Quim.*, **31**, 531 (1933).

19. FRESNO, C. DEL and MAIRLOT, E., *Z. anorg. Chem.*, **214**, 73 (1933).
19a. MURTY, B. V. S. R. and RAO, G. G., *Talanta*, **8**, 547 (1961).
20. TOMÍČEK, O., STODOLOVÁ, A. and HEŘMAN, M., *Chem. Listy*, **47**, 516 (1953).
21. WILLARD, H. H. and YOUNG, PH., *Ind. Engng. Chem.*, **20**, 972 (1929).
21a. ZOLOTAVIN, V. L. and PONOMAREVA, L. K., *Redkozemel. Elementy (Poluchenie, Analiz, Primenenie)* Izdat. Akad. Nauk. SSSR, Moskow 1959, p. 176.
22. MÜLLER, E. and JUST, H., *Z. anorg. Chem.*, **125**, 155 (1922).
23. TSUBAKI, I. and KIURA, M., *Japan Analyst*, **3**, 137 (1954).
24. HARA, S., *Japan Analyst*, **5**, 163 (1956).
25. BERRY, A. J., *Analyst*, **58**, 464 (1933).
25a. WICKHAM, D. G. and WHIPLE, E. R., *Talanta*, **10**, 314 (1963).
26. MAASS, K., *Z. anal. Chem.*, **97**, 241 (1934).
27. VULTERIN, J., *Chem. Listy*, **53**, 392 (1959).
28. MACDONALD, A. M. G., *Industrial Chemist*, 332 (1956).

CHAPTER 26

ASCORBIC ACID

ASCORBIC ACID (vitamin C) is easily soluble in water (about 1 g in 3 ml of water). Its aqueous solutions exhibit a strong reducing action (they are oxidized readily by atmospheric oxygen). Ascorbic acid has therefore been used widely as a reducing titrant in recent years. By oxidizing agents, it is oxidized to dehydroascorbic acid:

$$\underset{\substack{\| | | \\ O OH OH OH OH}}{C-C=C-\overset{\displaystyle\overset{\displaystyle\,\,O\,\,}{\overbrace{}}}{C}H-CH-CH_2} \rightleftharpoons$$

$$\rightleftharpoons \underset{\substack{\| \| \| \\ O O O OH OH}}{C-C-C-\overset{\displaystyle\overset{\displaystyle\,\,O\,\,}{\overbrace{}}}{C}H-CH-CH_2} + 2\,H^+ + 2\,e^-$$

Erdey and Bodor[1] were the first to measure the redox potential of this system; they state it to be $+\,0\cdot185$ V at 21°C and pH 7 against a normal hydrogen electrode. Rao and Rao[2] observed that the potential changes from $-\,0\cdot012$ V (at pH 8·7) up to $+\,0\cdot326$ V (at pH 1·05).

STANDARD SOLUTION

For the preparation of a 0·1 N standard solution, dissolve 8·806 g of pure ascorbic acid in metal-free distilled water and dilute to 1 l. Add stabilizing substances to the solution before dilution or store under an inert atmosphere.

For titrations in glacial acetic acid,[3] only 0·05 N standard solutions can be prepared because of the low solubility of ascorbic in glacial acetic acid. If 1 mole of anhydrous sodium acetate is added to 1 l. of glacial acetic acid, 0·1 N ascorbic acid solution can also be prepared; this solution is, however, not very stable. If prepared with twice distilled water, 0·1 N ascorbic acid solution is stable for about 24 hr, its titre decreases daily by about 0·3 per cent.[1]

Enzymes of the oxidase type, ultraviolet light and especially traces of heavy metals (copper) accelerate the decomposition catalytically.[2]

The stability of the standard solutions depend on the purity of the substance and of the water. It has been observed[1] that a solution stored in a burette under carbon dioxide decreases its titre during two days and is thereupon stable for more than two weeks.

EDTA and formic acid serve as stabilizers for standard ascorbic acid solutions;[4,5] in their presence it is not necessary to store the solutions in an inert atmosphere. EDTA forms complexes with the traces of heavy metals and so inhibits the catalysed oxidation of ascorbic acid. A solution containing about 0·5 g of EDTA per litre is very stable at 0°C if stored in the dark. The addition of formic acid makes the standard solution sufficiently stable even at room temperature; an approximately 0·1 N standard solution containing 0·1 g of EDTA and 4 g of formic acid per litre, loses at the utmost only 0·1 per cent of its reducing power daily.

The instability of the solutions increases with increasing dilution.[6] If, however, 0·01 N solutions are stabilized by adding 10 per cent of ammonium thiocyanate and kept in the dark, they only lose 1 per cent of their original titre within 30 days.

Oxalic or sulphuric acid stabilize aqueous solutions of ascorbic acid slightly.[2]

Solutions of ascorbic acid in glacial acetic acid are less stable than aqueous solutions. The reducing power of glacial acetic acid solutions decreases daily by about 0·5 per cent. In the presence of sodium acetate, the redox potential becomes noticeably more negative so that the titre of the solution decreases much more rapidly.[3] In most cases the sodium acetate used for lowering the redox potential is therefore added only during the titration.

Standardization

The titre of the standard ascorbic acid solution is usually determined against iodate or iodine solution;[1] Mohr's salt can also be used as standard substance after oxidation with hydrogen peroxide[6] or hexacyanoferrate (III).[7]

Procedure. Add 1 g of potassium iodide to 20 ml of 0·1 N potassium iodate solution; acidify with 5 ml of 2 N hydrochloric acid and titrate with the ascorbic acid solution to the disappearance of the colour caused by the iodine produced. Starch is not used because it decreases the reaction rate.[2]

The normality of standard ascorbic acid solution in glacial acetic acid is determined against a glacial acetic acid solution of bromine or iodimetrically in aqueous solution. The solution of bromine in glacial acetic acid is stored in a dark bottle in a cool place; under these conditions it is very stable. Its concentration is determined in aqueous solutions against As^{3+} or iodimetrically.

INDICATOR

In titrations with ascorbic acid, the equivalence point can either be detected visually, using a suitable redox indicator, or by one of the electrometric methods.

Amongst the redox indicators, Variamine Blue[8] (4-amino-4'-methoxy-diphenylamine) is mostly used in these titrations; it yields satisfactory results over the pH range 0–6. Using this indicator, the following ions can be determined " ascorbimetrically ":
Fe^{3+}, V^{5+}, Ag^+, Hg^{2+}, Au^{3+}, Pt^{4+}, Ce^{4+}, Cr^{6+} and others.

Because Variamine Blue is oxidized by iodine at pH above 2, it is also applicable in iodimetry.

The use of 4-amino-4'-ethoxy-diphenylamine is less suitable than Variamine Blue, because the colour formed in the titration is not so intense and also is less stable.

The colour change of 2,6-dichloro phenol indophenol (used e.g. in the determination of copper (II) ions or of hexacyanoferrate (III)) is well recognizable only at higher concentrations of the indicator in the solution to be titrated; it is therefore necessary to determine by a blank test the additional amount of standard solution consumed for the reduction of the indicator.

Diphenylamine has been used for the titration of iridium (IV) ions.

In some cases, other methods for visual detection of the end-point are also applied, e.g. the colour produced by thiocyanate in the titration of iron (III) ions, or the precipitation of calomel (by reduction of mercury (II) chloride) at the equivalence point in the titration of iodate in the presence of selenious acid as catalyst.

In indirect titrations, the excess of ascorbic acid is usually back-titrated with iodine.

Amongst electrometric methods used in ascorbimetry are:
Potentiometry (for Ag^+, Hg^{2+}, Au^{3+}, Cu^{2+}, $[Fe(CN)_6]^{3-}$, Br_2), amperometry (for Fe^{3+}, V^{5+}, Ce^{4+}, Se^{4+}, Se^{6+}) and " dead-stop " titration (Fe^{3+}).

REVIEW OF DETERMINATIONS

Iron. Pticyn and Kozlov[9] were the first to use ascorbic acid for the determination of iron, they reduced trivalent iron in hydrochloric acid solution with excess of ascorbic acid and back-titrated the unconsumed titrant with iodine solution. The reduction proceeds according to the following equation:

$$2\,Fe^{3+} + C_6H_8O_6 \rightarrow 2\,Fe^{2+} + 2\,H^+ + C_6H_6O_6$$

In 0·1–0·2 N hydrochloric acid solution, trivalent iron can be titrated directly at 60°C with ascorbic acid to the disappearance of the red colour of the thiocyanate iron (III) complex.[1,6,10,11]

The method has proved suitable for the determination of iron in various iron ores,[12] in bauxite, in ceramic and glass materials, in luminescent substances which contain besides the iron, also phosphate and fluoride.

Trivalent iron can likewise be determined visually using Variamine Blue as indicator; near the end, one titrates at 50–60°C, the indicator being only added shortly before the equivalence point because it is partially destroyed by strong oxidizing agents.[8]

In coloured or turbid solutions, the titrations are carried out electrometrically, e.g. by the "dead-stop" method.[1] This method is suitable for the determination of iron in metallic nickel, in ferrochromium and in chromium steels.[12] 1–2 mg iron can be titrated amperometrically with an accuracy of 1 per cent.[13,14] The amperometric detection of the end-point is also used for the determination of iron in clay, fireclay and bauxite.[15]

Chlorite, hypochlorite, chlorine, hypobromite, bromine, hydrogen peroxide, peroxo disulphate, and permanganate. All these substances can be determined indirectly;[16] they are reduced with an excess of iron (II) salt, and the trivalent iron thus produced is titrated with ascorbic acid.

Vanadium. Pentavalent vanadium can be determined visually with ascorbic acid by direct and indirect titration; vanadium (V) is reduced to vanadium (IV). The direct titration proceeds more rapidly whereas the indirect one yields more accurate results.

In the direct titration of pentavalent vanadium in sulphuric acid solution using Variamine Blue as indicator,[8,17] the accuracy of the results depends on the rapidity of the titration. (In a slow titration, the vanadium (V) is also reduced by the dehydroascorbic acid produced.) The first result is therefore used only as an orientation and the actual determination is then carried out rapidly. The accuracy of this method is ± 0·5 per cent.

In the indirect method, an excess of iron (II) sulphate is added to the hydrochloric acid vanadium (V) sample solution. The equivalent amount of iron (III) salt produced by the reaction is titrated with 0·1 N ascorbic acid, using thiocyanate or Variamine Blue as indicator. Because in this case a system with a markedly lower redox potential is titrated than in the direct titration, the determination is more precise, for the dehydroascorbic acid cannot take part in the reaction.[12] The accuracy of the method is ± 0·1 per cent.

Pentavalent vanadium can also be easily titrated amperometrically with ascorbic acid.[18] The titration is carried out in 0·1–1 N sulphuric acid at + 0·9 V with a rotating platinum electrode. As little as 0·04 mg of vanadium can still be determined with an error of ± 0·001 mg. Nickel, manganese, zinc, aluminium, chromium and titanium do not interfere with this determination even in a hundred-fold excess. The titration can be carried out in the presence of tungsten in the ratio of W:V = 1:1. At higher concentrations of tungsten, the results for the vanadium are lower, whereas in titrations of vanadium in the presence of molybdenum in the ratio Mo:V > 20:1 the determination is affected with positive errors. In the presence of iron, the results are still sufficiently precise up to a ratio Fe:V = 10:1.

For the determination of vanadium (V) ions with ascorbic acid in the presence of copper and iron, the potentiometric titration has been suggested,[19] in which the iodine liberated from added potassium iodide is titrated. Iron is masked with fluoride and copper with diphenylamine. At an appropriate combination of masking agents, all three elements can be determined.

Cerium. Cerium (IV) salt can be determined with ascorbic acid, either in buffered solution at pH 2 using Variamine Blue as indicator[20] or amperometrically[18] at + 0·3–0·5 V using a rotating platinum electrode, preferably in 5 N sulphuric acid solution. If iron is also present, concentrated ammonia solution is added dropwise (after the titration of cerium has been finished) until the solution becomes turbid owing to iron (III) hydroxide formed; this is then dissolved with some drops of 0·1 N sulphuric acid, the solution warmed to 40–60°C and the iron (III) ions are determined at 0·0 V.

Copper. The direct titration of copper (II) ions[21] with ascorbic acid is best carried out in ammonium chloride–sodium acetate solution either potentiometrically or visually, using 2,6-dichloro phenol indophenol as indicator.

Mercury. The determination of mercury (II) ions with ascorbic acid (reduction to the metal) is carried out in weakly nitric acid solution free of chloride at 60°C. Variamine Blue is used as indicator.[22] As soon as half of the amount of standard solution has been added, the solution decolorizes; from that, the total consumption of titrant necessary for the complete reduction can be calculated. The decolorization, lasting for some seconds, corresponds to the reduction $2\,Hg^{2+} \rightarrow 2\,Hg^+ - 2\,e^-$. After further addition of standard solution, metallic mercury separates out and the indicator assumes again its original colour, i.e. the colour of its oxidized form. Shortly before the end of the titration, the acidity of the solution is buffered with sodium acetate. A special method of determination has to be used if the solution to be analysed contains chloride ions.

In glacial acetic acid solution containing sodium acetate, mercury (II) ions can be titrated potentiometrically.[3]

An indirect determination of mercury (II) ions has also been recommended, in which the excess of ascorbic acid is back-titrated with standard iodate (V) solution.[23]

Silver. Silver salts are reduced by ascorbic acid to the metal. The titrations are done at 60°C; shortly before the equivalence point, the solution has to be buffered with sodium acetate; Variamine Blue is used as indicator.[8] The potentiometric titration is best carried out using a bright platinum electrode and a glass electrode.[24] The ascorbimetric determination of silver in alloys was described recently.[24a]

Gold. Titrations of gold (III) ions with ascorbic acid (reduction to the metal) can be carried out potentiometrically or amperometrically. When the titration is done with the aid of a gold indicator electrode[25] the solution has to be 0·02–0·13 N hydrochloric or sulphuric acid and 60–70°C warm; when a platinum indicator electrode and a calomel reference electrode are used, gold (III) ions can be titrated with ascorbic acid at pH 1·6–3 at 50°C, if the chloride concentration does not exceed 0·1 N. Mercury (II), copper (II) and iron (III) do not interfere with the determination.[26]

In glacial acetic acid, gold (III) ions can also be titrated directly with ascorbic acid, if sodium acetate is added to the sample solution.[3] The amperometric titration is carried out with the aid of a dropping mercury electrode.[48] In the literature, there is also mentioned the visual titration of gold (III) ions, using Variamine Blue as indicator,[8] but no details are given.

Platinum. Platinum can be determined indirectly. Platinum (IV) salt is reduced with an excess of ascorbic acid in 0·02 N hydrochloric acid solution, the unconsumed amount is back-titrated potentiometrically with 0·1 N standard iron (III) chloride solution.[27]

The possibility is also mentioned in the literature of titrating platinum (IV) ions with ascorbic acid directly and visually, using Variamine Blue as indicator,[8] but no detailed procedure is given.

Iridium. The direct determination of iridium (IV) ions[28] is carried out in 1 N hydrochloric or sulphuric acid, either visually using diphenylamine as indicator, or potentiometrically, using a platinum indicator electrode and a calomel reference electrode. For the amperometric titration of tetravalent iridium, a rotating platinum electrode is used at 0·4–0·5 V (standard calomel electrode) in 0·001–10 N hydrochloric acid solution.[29] At higher acidity, however (from 5 N HCl onward), the reaction proceeds more slowly. The determination can be carried out in the presence of rhodium, platinum, palladium, ruthenium (Ru:Ir $\leq$ 1:50), copper, nickel, selenium, tellurium and iron (in this case add phosphoric acid).

Selenium. Selenium (IV) and selenium (VI) compounds are reduced with ascorbic acid in acid medium to elementary selenium.[30a,b] This has been applied to the amperometric[30] and photometric[31] titration of selenium. The amperometric titration is carried out in solutions of pH 1–2 at 60°C, using a dropping mercury electrode and a calomel electrode at − 0·05 V; this determination is not affected by tellurium and many other elements. The photometric titration (430 mμ) is carried out in the presence of gelatine at a pH lower than 3.

A method has been recommended for the indirect volumetric determination of selenium in technical selenium,[32] in which the excess of ascorbic acid is back-titrated iodimetrically.[32a]

Hexacyanoferrate (III). The determination of hexacyanoferrate (III) with ascorbic acid is carried out in sodium acetate buffered solution. The equivalence point can be detected visually even without an indicator, better, however, with 2,6-dichloro phenol indophenol. In potentiometric titration, a platinum electrode and a calomel reference electrode are applied. This method is suitable for standardizing the normality of the ascorbic acid solution.[7]

Chromate, permanganate,[33] *chlorine, bromine, iodine, hypochlorite, chlorite, hypobromite, bromate, iodate, chloramine.*[34] These substances can oxidize hexacyanoferrate (II) to an equivalent amount of hexacyanoferrate (III), which thereupon is titrated with ascorbic acid. The substances listed can thus be determined indirectly.

Hexacyanoferrate (II). This compound can be determined indirectly: Oxidize hexacyanoferrate (II) with nitrite, remove the excess of nitrite, and titrate the hexacyanoferrate (III) formed with ascorbic acid.[33]

Zinc and lead.[35] Ascorbic acid reduces hexacyanoferrate (III) at a pH higher than 4 already at room temperature. In acid solutions, pH < 4, the reaction proceeds very slowly. If, however, heavy metal ions are added to the hexacyanoferrate (III) solution which can form a precipitate with hexacyanoferrate (II), the redox potential is, even in acid solution, increased so much that hexacyanoferrate (III) reacts fairly rapidly with ascorbic acid. The consumption of ascorbic acid is proportional to the amount of heavy metal ions present, because they bind the hexacyanoferrate (II) ions produced by reduction; in this way, the potential of the system becomes strongly positive. On adding the first excess of ascorbic acid, the concentration of hexacyanoferrate (II) ions is suddenly increased, whereby the system hexacyanoferrate (II) / hexacyanoferrate (III) assumes suddenly a more negative potential. This makes it possible to titrate with ascorbic acid such heavy metal ions, which form with hexacyanoferrate (II) sparingly soluble precipitates of well-defined composition.

This method has been used for the determination of zinc and lead, because both ions produce well-defined, sparingly soluble precipitates with hexacyanoferrate (II) ($K_2Zn_3[Fe(CN)_6]_2$ and $Pb_2[Fe(CN)_6]$).

The sudden potential change at the equivalence point can be observed potentiometrically or using Variamine Blue as indicator.

Tin. For the indirect volumetric determination of tin, a method has been recommended,[36] in which tin (II) ions are oxidized with an excess of iron (III) ions at boiling temperature. The excess of iron (III) ions is titrated with ascorbic acid at 60°C in a solution of pH 1 in the presence of thiocyanate as indicator.

Iodine monochloride, lead (IV) acetate, permanganate, vanadate (V) and chromate. These compounds can be titrated potentiometrically in glacial acetic acid solution with ascorbic acid.[3] The titrations of these systems need, however, still to be studied more extensively, especially with regard to the stoichiometry of the reactions on which they are based.

Chromate. Chromate can be titrated visually with ascorbic acid against Variamine Blue; the blue colour disappears and the solution assumes a green colour owing to the chromium (III) produced.[8]

Chlorate.[12,37,38] Chlorate is titrated with ascorbic acid in 4 N sulphuric acid solution at 60°C in the presence of selenious acid and manganese (II) sulphate. With rigorous stirring, an amount of standard ascorbic solution, about 1 ml less than corresponds to the equivalent, is added; then it is titrated slowly to the end-point, at which the orange colour of selenium appears.

Iodate,[12,39,40] *bromate.*[12,41,42] In the titration of iodate with ascorbic acid in hydrochloric acid solution, the end-point is detected by the disappearance of the colour due to elementary iodine formed by reduction or, in the presence of mercury (II) chloride and selenious acid, by the precipitation of calomel. In the latter case, the reduction of iodate to iodide proceeds without intermediate liberation of iodine (the iodide ions are bound in a complex form by $HgCl_2$). Selenious acid catalyses the precipitation of calomel near the end of the titration.

By this method, it is possible to determine *iodide* and *free iodine*, if the iodide is first oxidized with bromine water to iodate and the excess of bromine water is removed with formaldehyde. This method is suitable for the analysis of tinctures of iodine; a weighed sample is diluted with ethanol and the free iodine is determined with standard ascorbic acid solution; then, the alcohol is evaporated, the solution oxidized with bromine and the total iodide + iodine is determined.

The titration of iodate and iodine with ascorbic acid can also be carried out amperometrically.[43]

Bromate is titrated as in the determination of iodate.

Bromine. Solutions of bromine in glacial acetic acid are applied to the standardization of ascorbic acid solutions, which are used for titrations in non-aqueous media. The end-point is detected potentiometrically.[3]

Chloramine-B. Chloramine-B can be titrated amperometrically in 1 N sulphuric acid solution.[43]

Oxygen. Mohr's salt is oxidized in alkaline solution to trivalent iron by oxygen dissolved in water; after acidifying this solution, the iron (III) salt can be titrated in the usual way with ascorbic acid, using Variamine Blue as indicator. The Winkler method for the determination of oxygen can also be completed by titration with ascorbic acid.[44]

2,6-*Dichloro phenol indophenol*[45] and 2,6-*dinitro phenol indophenol*.[46] These compounds can be titrated amperometrically with ascorbic acid. Stepanov and Sergienko[46a] hint at the possibility of titrating indophenol with ascorbic acid.

Ascorbic acid can also replace thiosulphate in *iodimetric determinations*; the liberated iodine is titrated using Variamine Blue or a chloroform layer as indicator (eventually potentiometrically).[47,47a]

The following substances have been determined by this method: Cl_2, Br_2, ClO_3^-, BrO_3^- IO_3^-, IO_4^-, ClO^-, BrO^-, chloramine, H_2O_2, MnO_2, PbO_2, PbO_2 in red lead, $Cr_2O_7^{2-}$, VO_3^-, MnO_4^-, NO_2^-, oxygen dissolved in water, Fe^{3+}, CN^-, SCN^-, formaldehyde, glycerol, H_2SO_3, Cu^{2+}, I^-, Hg^{2+}, S^{2-}, Sn^{2+}, amidopyrine and the iodine–bromine number.

Recently, ascorbic acid has been used for the titration of trivalent *thallium*,[49,51–52a] *resazurine* and *resorufine*[50] and antibiotics of the *rifamycine* type.[52b]

Thiazine and *thiazone*[53] have been likewise determined with ascorbic acid. Mixtures consisting of *vanadium and iron*,[55] *iron and cerium*,[56] *vanadium, chromium and molybdenum*,[57,57a] as well as the titrimetric determination of *vanadium*,[54] *gold*,[54a] *cobalt*,[54b] *tungsten*[54c] and *cerium*,[54d] using an amperometric end-point indication, have been investigated anew.

Some other communications deal with the high-frequency titration of *silver*[58] and with the indirect determination of *peroxide*,[59] *hydroxylamine*,[60] *silver*, [61] *hydrazine*,[62] *hypochlorite, hypobromite*[63] and *semicarbazide*.[64]

REFERENCES

1. ERDEY, L. and BODOR, E., *Anal. Chemistry*, **24**, 418 (1952).
2. RAO, G. G. and RAO, V. N., *Z. anal. Chem.*, **147**, 338 (1955).
3. ERDEY, L. and RÁDY, G., *Acta Chim. Acad. Sci. Hung.*, **15**, 81 (1958).
4. ERDEY, L. and BODOR, E., *Magyar Kém. Folyóirat*, **58**, 295 (1952).
5. ERDEY, L. and BODOR, E., *Z. anal. Chem.*, **136**, 109 (1952).
6. FLASCHKA, H. and ZAVAGYL, H., *Z. anal. Chem.*, **132**, 170 (1951).
7. ERDEY, L. and SVEHLA, G., *Z. anal. Chem.*, **150**, 407 (1956).
8. ERDEY, L. and BODOR, E., *Z. anal. Chem.*, **137**, 410 (1953).
9. PTITSYN, B. V. and KOZLOV, N. A., *Zhur. Anal. Khim.*, **4**, 35 (1949).
10. ERDEY, L., *Magyar Kém. Folyóirat*, **56**, 262 (1950).
11. ERDEY, L. and BODOR, E., *Magyar Kém. Folyóirat*, **56**, 277 (1950).
12. ERDEY, L., *Zhur. Anal. Khim.*, **8**, 356 (1953).
13. ERDEY, L. and KARSAY, A., *Acta Chim. Acad. Sci. Hung.*, **9**, 43 (1956).
14. ZHDANOV, A. K., KHADEYEV, V. A. and KAO, A. L., *Uzbek. Khim. Zhur.*, N1, 27 (1958); *Ref. Zhur. Khim.*, N20, 119 (1958).
15. USATENKO, J. I. and BEKLESHKHOVA, G. R., *Zavodskaya Lab.*, **20**, 266 (1954).
16. ERDEY, L., BUZÁS, I. and VIGH, K., *Periodica polytechn.*, **3**, 1 (1959); *Chem. Abstr.*, **53**, 16804 (1959).
17. ERDEY, L., BODOR, E. and BUZÁS, L., *Acta Chim. Acad. Sci. Hung.*, **7**, 277, 287 (1955).

18. GALLAI, Z. A., TIPCOVA, V. G. and PESHKOVA, V. M., *Zhur. Anal. Khim.*, **12**, 469 (1957).
19. YOSHIMURA, CH. and FUJITANI, T., *J. Chem. Soc. Japan, Pure Chem. Sect.*, **76**, 304 (1955); *Chem. Zbl.*, **130**, 6245 (1959).
20. JACH, Z., PACOVSKÝ, J. and SVACH, M., *Z. anal. Chem.*, **154**, 185 (1957).
21. ERDEY, L. and SIPOSS, G., *Z. anal. Chem.*, **157**, 166 (1957).
22. ERDEY, L. and BUZÁS, I., *Acta Chim. Acad. Sci. Hung.*, **8**, 263 (1955).
23. RAO, G. G. and RAO, N. V., *Z. anal. Chem.*, **150**, 29 (1956).
24. ERDEY, L. and BUZÁS, H., *Acta Chim. Acad. Sci. Hung.*, **4**, 195 (1954).
24a. ERDEY, L., RÁDY, G. and GIMESY, O., *Acta Chim. Hung.*, **32**, 151 (1962).
25. PSHENICYN, N. K. and GINZBURG, C. I., *Izvest. Sektora Platiny i Drugikh Blagorodnykh Metallov*, N30, 171 (1955).
26. ERDEY, L. and RÁDY, G., *Talanta*, **1**, 159 (1958).
27. MAKSIMJUK, E. A., *Izvest. Sektora Platiny i Drugikh Blagorodnykh Metallov*, N30, 180 (1955); *Ref. Zhur. Khim.*, N22, 52190 (1955).
28. PSHENICYN, N. K. and PROKOF'EVA, I. V., *Izvest. Sektora Platiny i Drugikh Blagorodnykh Metallov*, N30, 176 (1955); *Ref. Zhur. Khim.*, N1, 1151 (1956).
29. PSHENICYN, N. K. and EZERSKAYA, N. A., *Zhur. Anal. Khim.*, **14**, 81 (1959).
30. SIMON, V. and GRIM, V., *Chem. Listy*, **48**, 1774 (1954).
30a. BATALIN, A. K., *Zhur. Anal. Khim.*, **15**, 507 (1960).
30b. BATALIN, A. K., *Trudy Orenburgsk. Selskokhoz. Inst.*, **10**, 339 (1960); *Chem. Abstr.*, **55**, 17347 (1961).
31. YOSHIMURA, CH., *J. Chem. Soc. Japan, Pure Chem. Sect.*, **78**, 5 (1957); *Ref. Zhur. Khim.*, N16, 53449 (1958).
32. ZAYKOVSKII, F. V., *Sovremennye Metody Analiza Metallov, Metallurgizdat*, p. 142 (1955); *Ref. Zhur. Khim.*, N29, 362 (1956).
32a. SHUSHIN, M. V. and BURNEVIN, D. Z., *Glasnik Khemiskog Drushtva Beograd*, **23-24**, 297 (1958/59); *Ref. Zhur. Khim.*, **17 D85**, 125 (1961).
33. ERDEY, L. and SVEHLA, G., *Z. anal. Chem.*, **163**, 6 (1958).
34. ERDEY, L. and SVEHLA, G., *Z. anal. Chem.*, **167**, 164 (1959).
35. ERDEY, L. and PÓLOS, L., *Z. anal. Chem.*, **153**, 401 (1956).
36. STAN SUCIN, M. and ZELINGHER, R., *An. Univ. " C. J. Parhon ", Ser. stiint. natur.*, N10, 39 (1956); *Ref. Zhur. Khim.*, N6, 19564 (1957).
37. ERDEY, L. and BODOR, E., *Magyar Kém. Folyóirat*, **57**, 78 (1951).
38. ERDEY, L. and BODOR, E., *Z. anal. Chem.*, **133**, 265 (1951).
39. ERDEY, L., BODOR, E. and BUZÁS, H., *Z. anal. Chem.*, **134**, 22 (1951).
40. ERDEY, L., BODOR, E. and PAP, I. B., *Magyar Kém. Folyóirat*, **57**, 234 (1951).
41. ERDEY, L., BODOR, E. and BUZÁS, H., *Z. anal. Chem.*, **134**, 412 (1951).
42. ERDEY, L., BODOR, E. and PAP, I. B., *Magyar Kém. Folyóirat*, **58**, 129 (1952).
43. GALLAI, Z. A., TIPCOVA, V. G. and PESHKOVA, V. M., *Vestnik Moskov. Univ., Ser., Mat., Mekhan., Astron., Fiz., Khim.*, **13**, 209 (1958); *Chem. Abstr.*, **53**, 4009 (1959).
44. ERDEY, L. and SZABADVÁRY, F., *Acta Chim. Acad. Sci. Hung.*, **4**, 325 (1954).
45. OGAWA, T., *J. Electrochem. Soc. Japan*, **26**, 105 (1958); *Ref. Zhur. Khim.*, N18, 64238 (1959).
46. OGAWA, T., *J. Electrochem. Soc. Japan*, **26**, 323 (1958); *Anal. Abstr.* **6**, 1835 (1959).
46a. STEPANOV, B. I., SERGIENKO, V. A., *Trudy Komisii Anal. Khim. Akad. Nauk. SSSR, Otdel. Khim. Nauk.*, **5**, 274 (1954).

47. ERDEY, L., BODOR, E. and PÁPAY, M., *Acta Chim. Acad. Hung.*, **5**, 235 (1955).
47a. SINGH, B., SAHOTA, S. S. and MANKOTIA, M. S., *Res. Bull. Panjab Univ.*, **11**, 221 (1960); *Chem. Abstr.*, **56**, 2870 (1962).
48. MILENKO, V. and SUŠIĆ, M., *Bull. Inst. Nucl. Sci.* " *Boris Kidrich* ", **7**, 25 (1957).
49. DRAGULESCU, C. and HERESCU, S., *Bull. Stiint. Tehn. Inst. Politehn. Timisoara*, **3**, 237 (1958); *Chem. Zbl.*, **131**, 8970 (1960).
50. RŮŽIČKA, E., *Z. anal. Chem.*, **168**, 414 (1959).
51. ERDEY, L., VIGH, K. and BUZÁS, I., *Acta Chim. Hung.*, **26**, 85 (1961).
52. ERDEY, L., VIGH, K. and BUZÁS, I., *Acta Chim. Hung.*, **26**, 93 (1961).
52a. BHATNAGAR, R., BHATNAGAR, M. L. and MATHUR, N. K., *J. Sci. Ind. Res. (India)*, **21B**, 44 (1962); *Chem. Abstr.*, **57**, 39 (1962).
52b. GALLO, G. G., CHIESA, L. and SENSI, P., *Farmaco (Pavia), Edie. Sci.*, **17**, 668 (1962); *Chem. Zbl.*, **135** N1540 (1964).
53. RŮŽIČKA, E. and KOTOUČEK, M., *Z. anal. Chem.*, **180**, 429 (1961).
54. SINGH, D., *J. Sci. Res. Banaras Hindu Univ.*, **10**, 46 (1959); *Chem. Abstr.*, **55**, 7164 (1961).
54a. VARMA, A., *J. Sci. Res. Banaras Hindu Univ.*, **12**, 31 (1961); *Chem. Abstr.*, **57**, 7893 (1962).
54b. SINGH, D., PRASAD, B. B. and VARMA, A., *J. Sci. Res. Banaras Hindu Univ.*, **12**, 267 (1961/62); *Ref. Zhur. Khim.*, **15G**, 113 (1963).
54c. SINGH, D., VARMA, A., *J. Sci. Res. Banaras Hindu Univ.*, **12**, 326 (1961/62); *Ref. Zhur. Khim.*, **15G**, 108 (1963).
54d. ZHDANOV, A. K. and RYACANOVA, I. M., *Uzb. Khim. Zhur.*, **N4**, 18 (1962); *Ref. Zhur. Khim.*, **14G**, 82 (1963).
55. SINGH, D. and VARMA, A., *J. Sci. Res. Banaras Hindu Univ.*, **11**, 125 (1960/61).
56. SINGH, D., VARMA, A. and AGARWALA, V. S., *Z. anal. Chem.*, **183**, 172 (1961).
57. SINGH, D. and VARMA, A., *J. Sci. Res. Banaras Hindu Univ.*, **11**, 202 (1960/61); *Chem. Abstr.*, **56**, 9400 (1962).
57a. SINGH, V. B. and PRASAD, B. B., *J. Sci. Res. Banaras Hindu Univ.*, **12**, 116 (1961); *Chem. Abstr.*, **57**, 7897 (1962).
58. ERDEY, L. and VÁNDORFFY, M. T., *Acta Chim. Acad. Sci. Hung.*, **35**, 381 (1963); *Ref. Zhur. Khim.*, **1G**, 67 (1964).
59. ERDEY, L., SVEHLA, G. and KOLTAI, L., *Analyt. Chim. Acta*, **27**, 164 (1962).
60. ERDEY, L., KOLTAI, L. and SVEHLA, G., *Analyt. Chim. Acta*, **27**, 363 (1962).
61. ERDEY, L., KOLTAI, L. and SVEHLA, G., *Analyt. Chim. Acta*, **27**, 498 (1962).
62. ERDEY, L., SVEHLA, G. and KOLTAI, L., *Analyt. Chim. Acta*, **28**, 398 (1963).
63. ERDEY, L. and VIGH, K., *Talanta*, **10**, 439 (1963).
64. KASA, I. and ERDEY, L., *Acta Chim. Hung.*, **39**, 21 (1963).

CHAPTER 27

HYDROQUINONE AND SIMILAR REDUCING AGENTS

IN REDUCING power hydroquinone is about equivalent to iron (II). Its redox potential is: + 0·68 V in acid solution. Hydroquinone is oxidized to quinone by oxidants:

$$C_6H_4(OH)_2 \rightleftharpoons C_6H_4O_2 + 2\,H^+ + 2\,e^-$$

Compared with other reducing agents, hydroquinone is very stable in weakly acid solution. It has, nevertheless, only very recently been examined systematically as a titrant, apart from some sporadic earlier investigations. Many inorganic systems can be determined on the micro-range by visual or electrometric titration with hydroquinone. The titrations are carried out in air. It is an advantage in practice that iron (III) salts do not affect volumetric determinations in which hydroquinone is used as titrant.

STANDARD SOLUTION

For the preparation of a 0·1 N hydroquinone solution, dissolve 5·50 g of the substance in 1 l. of hot water.[1] The solution becomes more stable if HCl or H_2SO_4 is added, to give a 1 per cent concentration.

Standardization

Hydroquinone solution is standardized against standard dichromate solution in 20 per cent H_2SO_4, using diphenylamine as indicator, or potentiometrically.[1,2] It is recommended that the solution be standardized one day after its preparation. The velocity of the reaction can be increased by titrating at 40–50°C; even then the last drops of the reagent must be added in intervals of 30–60 sec, whilst stirring the solution constantly. A recent publication deals with the determination of hydroquinone solutions with regard to their use as titrant;[69] the authors recommend to standardize the solution with cerium (IV), hexacyanoferrate (III) (in the presence of Zn^{2+}) or with dichromate.[69,70]

It has been shown from various studies that even weakly acid 10^{-5} N hydroquinone solutions are stable for several months. Occasionally a brown colour is formed but this is without influence on the titre. Neutral hydroquinone solutions are less stable; in alkaline solution a brown precipitate separates,[3-7] which is probably similar to humic acids in structure.[39]

INDICATOR

Titrations with hydroquinone can be carried out visually, using diphenylamine, ferroin or o-dianisidine and some other substances as indicators; or potentiometrically (with a platinum indicator electrode having a large surface, or with a gold electrode), or amperometrically (with a rotating or vibrating platinum electrode).

REVIEW OF DETERMINATIONS

Halogens and halogen compounds. The determination of halogens and their compounds are listed in Table 4.

TABLE 4. TITRATION OF HALOGENS AND THEIR COMPOUNDS WITH HYDROQUINONE

Substance to be determined	Optimal medium	Indicator	Remarks
I_2	pH 6–8 (NaHCO$_3$)	starch solution[1] potentiometry[1]	1–10 μg can be determined
Br_2	1–3% HCl, 20% H$_2$SO$_4$	o-dianisidine[8] potentiometry[1]	
Cl_2, ClO^-, active chlorine	1–15% HCl 20% H$_2$SO$_4$	potentiometry[1] o-dianisidine[8]	as little as 0·1 μg chlorine can still be determined.
BrO_3^-	1–3% HCl 20% H$_2$SO$_4$	potentiometry[1]	
IO_3^-	20% H$_2$SO$_4$	potentiometry[1]	

Hexacyanoferrate (III). Hexacyanoferrate (III) can be titrated in about 50 per cent sulphuric acid solution against ferroin or potentiometrically.[1] The method is applied to the indirect determination of those heavy metals[50] (e.g. Zn, Ag) which form a precipitate with hexacyanoferrate (II), and (see p. 177) to the indirect determination of chlorine, bromine and hypochlorite.[50]

Thallium. The potentiometric titration of thallium (III) salts proceeds best in 5–20 per cent H$_2$SO$_4$. The determination is not affected by many of the common cations.[9]

Lead. In investigations of the oxidation of organic compounds, in the first instance of α-hydroxy-acids, an excess of lead acetate solution is used. The unconsumed amount can be back-titrated potentiometrically with hydroquinone,[46,47] or against ferroin.[48] This method serves mainly for the determination of the velocity of the reaction in oxidations carried out with lead (IV) acetate.

Iridium. Iridium (IV) salts can be titrated visually as well as potentiometrically, best in 5 per cent H_2SO_4 or 10 per cent HCl. Dichlorobenzidine[10] o-dianisidine[11] or diphenylamine[12] have been recommended as indicators. In these titrations, platinum, palladium and rhodium salts do not interfere.[13] If the concentration of rhodium is higher, iridium has to be first separated with thioacetamide.[14] As little as 10 μg iridium can still be titrated potentiometrically;[15] 10^{-5} M iridium (IV) solutions can be titrated amperometrically.[49] Because of its accuracy, this method is suitable for the standardization of hydroquinone solutions.[16]

Chromium. Dichromate can be determined in 10–20 per cent H_2SO_4 or in 5 per cent HNO_3 ($Cr^{VI} \rightarrow Cr^{III}$) at 40 to 50°C against diphenylamine or potentiometrically (rather slow titration). By means of potentiometric titration, as dilute as 10^{-4} N dichromate solutions can still be determined. Iron (III) ions and a number of other cations do not interfere with the method, even when present in high concentrations. At a ratio of Fe: Cr > 1000:1, it is, however, advisable to add fluoride or phosphoric acid. The method is used in steel analysis,[1] and to check the chromium content in washings from chromium melts in the presence of hydroxide and carbonate.[68] The end-point of the titration can also be detected amperometrically, using a rotating platinum electrode.[17]

Vanadium. The titration of vanadate (V) with hydroquinone ($VO_3^- \rightarrow VO^{2+}$) is essentially the same as the determination of dichromate. This method has been tested in the analysis of steels[1] and of graphite slate with a content of vanadium of about 1 per cent.[18] Traces of vanadium have been determined in petroleum coke by amperometric titration.[19] Dichromate interferes, but can be selectively reduced beforehand.

Cerium. The determination of cerium (IV) salts is one of the most sensitive titrations with hydroquinone. It can either be carried out visually using ferroin[1] as indicator, or potentiometrically[1,20,21] or amperometrically[17,72] (even in 10^{-5} N solutions) in 2–15 per cent sulphuric, nitric or acetic acid solution. Because this method is affected only by dichromate, vanadate (V) and permanganate, it is suitable for the determination of cerium in monazite sand, in " incandescent bodies " (Auer gas lamp), in metals containing cerium, in various alloys, in waste-products from raw materials,[52,22] as well as for the analysis of cerium-containing pharmaceutical products[23] and other compounds. Cerium is oxidized to the tetravalent state with peroxidisulphate in the presence of silver ions as catalyst. The reaction of cerium (IV) salts with hydroquinone has also been investigated in glacial acetic acid and in acetonitrile.[53]

Cerium, vanadium and chromium simultaneously. Two distinct potential changes can be distinguished in the potentiometric titration of the mixture of cerium (IV) and vanadate (V) with hydroquinone, corresponding to the stepwise reduction of the two ions;[24] cerium (IV) and chromate (VI) can be determined in the same way successively. The titration of a mixture of dichromate and vanadate (V) is less reliable. In the determination of all three substances together the reduction of cerium (IV) salt proceeds first; the next potential changes correspond to the reduction of dichromate and finally to the reduction of VO_3^- to VO^{2+}; the reproducibility of this determination depends on the mutual proportion of the three compounds.

Gold. Gold (III) salts are reduced with hydroquinone in very weakly acid, neutral or alkaline solution to the metal. The reaction can be followed visually, potentiometrically or amperometrically. The potentiometric titration is best carried out in a solution of about pH 6, containing gold as chloride, at room temperature or at elevated temperature (50–60°C).[1] The presence of nitrate does not interfere; it can therefore be titrated immediately after dissolving the metal in *aqua regia* and diluting the solution. A large excess of common metals as well as smaller amounts of platinum, rhodium and iridium do not interfere, palladium, on the contrary, does.[1] In titrations of gold with hydroquinone,[25-27] *o*-dianisidine,[28-30,54-57] as well as 3-methyl benzidine, 3,3'-diethyl benzidine,[58] benzidine, toluidine[57] are used as indicators; in potentiometric determinations, gold electrodes[31] are applied; amperometric methods are carried out with rotating platinum electrodes.[32] These methods have been used in the determination of gold in solutions from the cyanide process,[30] in noble metal alloys,[31] in ores which contain selenium and tellurium,[32] in gold ores (ČSSR),[33] in alloys with copper and silver,[34] in pharmaceutical products containing gold in organic or inorganic combination[35] and in the urine of patients who have been medicated with these preparations. As little as 1 μg gold has been determined potentiometrically in this way.[36] The possibility of titrating gold in the presence of platinum and palladium has been studied recently.[62a,63]

Ruthenium. Ezerskaja and Pshenicyn[49] described the amperometric microtitration of ruthenium with hydroquinone (rotating platinum standard calomel electrode) in 6 N hydrochloric acid solution.

Copper. Copper (II) salts can be reduced quantitatively with hydroquinone to copper (I) salts in solutions containing acetate, sufficient thiocyanate and ammonium salts; an inert atmosphere is not necessary.[37] This titration, not affected by various metals, has been used in the rapid analysis of copper in aluminium bronzes,[37] in pharmaceutical products[38] and in the indirect determination of sugar after its reaction with Fehling's solution.[38]

Silver. Silver salts can be reduced in alkali hydroxide solutions, containing a sufficient amount of ammonium salts and acetate, to metallic silver;[39] this has been used in the potentiometric determination of silver salts.

Manganese. Manganese (III) salts can be very reliably titrated with hydroquinone. This method is applied on the one hand to the standardization of manganese (III) solutions (complexed with diphosphate),[51,69] on the other hand to the determination of manganese after it has been oxidized to the trivalent state with bromate in phosphoric acid solution.[40] The method is suitable for the rapid analysis of various materials.

The principle of titrating higher valency states of manganese with hydroquinone has been applied to the indirect, rapid and simple determination of oxygen in industrial waste waters:

Manganese (III) hydroxide, formed by oxidation with oxygen, is titrated potentiometrically or visually (ferroin as indicator).[64]

A similar method (titration of higher valency states of manganese in pyrophosphate solution, using diphenylamine as indicator) is suitable for routine analysis of oxygen and oxidants in industrial waste waters.[67]

In the *indirect determination*, the excess of hydroquinone is titrated as usual with standard dichromate or cerium (IV) solution.

Higher oxides of lead and manganese.[41] An excess of hydroquinone reduces lead (IV), manganese (IV) or manganese (III) oxide in acid solution to lead (II) and manganese (II) salt. The unconsumed amount of reagent is then back-titrated with cerium (IV) solution against ferroin, or with dichromate solution against diphenylamine. Iron (III) and copper (II) salts do not interfere with the determination.

Peroxidisulphate.[42] Peroxidisulphate is quantitatively reduced to sulphate in acid medium with excess of hydroquinone under the catalytic action of silver ions. The unconsumed hydroquinone is back-titrated with standard cerium (IV) salt solution using ferroin as indicator. Amongst the anions, chloride, bromide, nitrate, chlorate do not interfere with the determination; peroxides, however, do affect the titration.

Cerium.[43] Cerium (IV) salts react with manganese (II) ions in weakly acid solution according to the following equation:

$$2\,Ce^{4+} + Mn^{2+} + 2H_2O \rightarrow 2\,Ce^{3+} + MnO_2 + 4\,H^+$$

The hydrated manganese (IV) oxide thus formed can be titrated with hydroquinone after it has been separated. Dichromate or vanadate (V) do not react in this way.

Azides. The exchange reaction of azides with silver chromate is used in their indirect determination:

$$2\,NaN_3 + Ag_2CrO_4 \rightarrow 2\,AgN_3 + Na_2CrO_4$$

After the silver azide thus formed has been separated, the liberated chromate is titrated with hydroquinone.[44]

Halide and sulphate.[44] Both ions can, on principle, be determined by the same method, which has been applied for azide. In the determination of sulphate, barium chromate is used as exchange partner.

Zn, Ag and In can also be determined indirectly: an excess of hexacyanoferrate (III) (unknown amount) is added and titrated with hydroquinone solution; hexacyanoferrate (II) is thus formed and precipitates with the metal ion to be determined. The end of the reaction is indicated by a distinct change of potential.[50]

Apart from hydroquinone, other compounds of the series of divalent (catechol) and trivalent (pyrogallol and phloroglucinol) phenols, of *o-*, *m-* and *p*-phenylenediamines, of *o-*, *m-* and *p*-aminophenols and others have also been investigated as titrants.[45] Only compounds substituted in the *p*-position have proved suitable as titrants, such as p-*aminophenol*, p-*phenylenediamine, quinhydrone* and the very stable *metol*. All these compounds are applied in a way similar to hydroquinone.[15,45,59,60] The only difference is noticeable in the titration of gold (III) salts with *p*-phenylenediamine at pH 6; here two potential changes appear, the first one corresponding to the stoichiometric reaction ($Au^{III} \rightarrow Au^I$); this is accompanied by the formation of a yellowish-brown precipitate which dissolves again when the titration is continued ($Au^I \rightarrow Au$).

All the compounds mentioned above, however, offer no advantage as titrants over hydroquinone.

Recently, the experience gained in titrations with hydroquinone has been discussed extensively in several publications.[61,62,65,66,71]

REFERENCES

1. SIMON, V. and ZÝKA, J., *Chem. Listy*, 1646 (1955).
2. KOLTHOFF, I. M., *Rec. Trav. Chim. Pays-Bas*, **45**, 745 (1926).
3. MRÁZ, V., SIMON, V. and ZÝKA, J., *Chem. Listy*, **52**, 1083 (1958).
4. EULER, H. VON and BRUNIUS, E., *Z. phys. Chem.*, **139**, 615 (1928).
5. DUBRISAY, R. and SAINT-MAXEN, A., *Compt. Rend.*, **189**, 694 (1929).
6. ELLER, W. and KOCH, W., *Ber. dt. chem. Ges.*, **53**, 1469 (1920).
7. ETTEL, V. and POSPÍŠIL, J., *Chem. Listy*, **51**, 505, 1153 (1957).
8. MILAZZO, G. and PAOLINI, L., *Mikrochem. ver. Mikrochim. Acta*, **36/37**, 255 (1951).
9. KREJZOVÁ, E., SIMON, V. and ZÝKA, J., *Chem. Listy*, **52**, 936 (1958).
10. POLLARD, W. B., *Bull. Inst. Mining and Metallurgy*, **9**, 497 (1948).
11. MILAZZO, G. and PAOLINI, L., *Mh. Chem.*, **81**, 155 (1955).
12. PSHENICYN, N. K. and PROKOF'EVA, I. V., *Izvest. Sektora Platiny i Drugikh Blagorodnykh Metallov*, **28**, 229 (1954); *Chem. Abstr.*, **49**, 3723 (1955).
13. PSHENICYN, N. K. and GINZBURG, S. I., *Izvest. Sektora Platiny i Drugikh Blagorodnykh Metallov*, **25**, 192 (1950); *Chem. Abstr.*, **50**, 9934 (1956).
14. JACOBSON, E., *Analyst*, **84**, 106 (1959).
15. BERKA, A. and ZÝKA, J., *Chem. Listy*, **50**, 829 (1956).
16. RYABCHIKOV, D. I., *Zhur. Anal. Khim.*, **1**, 47 (1946); *Izvest. Sektora Platiny i Drugikh Blagorodnykh Metallov*, **22**, 35 (1948).
17. NOVÁK, V., SIMON, A. and ZÝKA, J., Unpublished Studies.
18. SIMON, V. and ZÝKA, J., *Rudy* (in print).

19. ŘEZÁČ, Z. and DVOŘÁK, J., *Chem. Průmysl*, **8**, 499 (1958).
20. FURMAN, N. H. and WALLACE, JR., J. H., *J. Amer. Chem. Soc.*, **52**, 1443 (1930).
21. BOGDANOV, S. G. and SUKHOBOKOVA, N. S., *Zhur. Anal. Khim.*, **6**, 344 (1951).
22. MRÁZ, L., SIMON, V. and ZÝKA, J., *Chem. Listy*, **52**, 1354 (1958).
23. KREJZOVÁ, E., SIMON, V. and ZÝKA, J., *Českoslov. Farm.*, **6**, 438 (1957).
24. MRÁZ, L., SIMON, V. and ZÝKA, J., *Chem. Listy*, **52**, 1089 (1958).
25. POLLARD, W. B., **62**, 597 (1937).
26. JAMIESON, A. R. and WILSON, R. S., *Analyst*, **63**, 702 (1938).
27. BELCHER, R. and STEPHAN, W. I., *Analyst*, **76**, 45 (1951).
28. ZVYAGINCEV, O. E., SHABARIN, S. K., VOROB'EVA, V. A. and BOKHAREVA, A. P., *Trudy Vses. Konferencii po Analit. Khimii, Akad. Nauk SSSR*, **1**, 375 (1939); *Chem. Abstr.*, **63**, 126 (1943).
29. MILAZZO, G., *Analyt. Chim. Acta*, **3**, 126 (1949).
30. ZHARKOVA, Z. P. and ZHATCHEVA, E. J., *Nauch. Isled. Trud. Centr. Nauch. Isled. Inst. Vspomag. Izd. i Zapas. Detal. i Tekst. Oborud.*, **44** (1956); *Anal. Abstr.*, **5**, 36 (1958).
31. RYABCHIKOV, D. I. and KNYAZEVA, G. V., *Doklady Akad. Nauk SSSR*, **25**, 601 (1939); *Chem. Abstr.*, **34**, 4011 (1940).
32. REISHAKHRIT, L. S. and SUKHOBOKOVA, N. S., *Zhur. Anal. Khim.*, **12**, 146 (1957).
33. TRDLIČKA, Z., ROSENKRANC, O., KUPKA, F. and VALCHA, Z., *Čas. Moravsk. Musea*, **42**, 17 (1957).
34. CZAPLIŃSKY, A. and TROKOWICZ, J., *Chem. analit. (Warszawa)*, **4**, 463 (1959); *Chem. Abstr.*, **53**, 12953 (1959).
35. DOLEŽAL, J., HÖFER, J. and ZÝKA, J., *Českoslov. Farm.*, **8**, 557 (1959).
36. DOLEŽAL, J., HÖFER, J. and ZÝKA, J., *Českoslov. Farm.*, **8**, 578 (1959).
37. PAVLÍKOVA, M. and ZÝKA, J., *Z. anal. Chem.*, **172**, 321 (1960).
38. BERKA, A. and ZÝKA, J., *Českoslov. Farm.*, **8**, 576 (1959).
39. BERKA, A., TICHÝ, M. and ZÝKA, J., *Z. anal. Chem.*, **182**, 335 (1961).
40. DOLEŽAL, J., KLUH, I. and ZÝKA, J., *Z. anal. Chem.*, **177**, 14 (1960).
41. KREJZOVÁ, E., SIMON, V. and ZÝKA, J., *Chem. Listy*, **52**, 976 (1958).
42. ROTTOVÁ, O. and ZÝKA, J., *Collection (Czech. Chem. Comm.)*, **24**, 2809 (1959).
43. MRÁZ, L., SIMON, V. and ZÝKA, J., *Chem. Listy*, **51**, 1828 (1958).
44. KREJZOVÁ, E., SIMON, V. and ZÝKA, J., *Chem. Listy*, **51**, 1764 (1957).
45. SIMON, V. and ZÝKA, J., *Chem. Listy*, **50**, 360 (1956).
46. BERKA, A. and ZÝKA, J., *Chem. Listy*, **52**, 930 (1958).
47. TOMÍČEK, O. and VALCHA, J., *Chem. Listy*, **44**, 283 (1950).
48. BERKA, A., *Českoslov. Farm.*, **8**, 561 (1959).
49. PSHENICYN, N. K. and EZERSKAYA, N. A., *Zhur. Anal. Khim.*, **14**, 81 (1959).
50. BLAŽEK, J., DOLEŽAL, J. and ZÝKA, J., *Z. anal. Chem.*, **180**, 241 (1961).
51. KŘÍŽOVÁ, J., Diploma Thesis, Charles University, Prague, 1960.
52. ARUSHEYEVA, A. A. and BEZUGLYI, D. V., *Zhur. Anal. Khim.*, **16**, 683 (1961).
53. RAO PRABHAKAR, G. and MURTHY VASUDAVA, A. R., *Z. anal. Chem.*, **182**, 358 (1961).
54. MILAZZO, G., *Rend. Inst. Super. Sanità*, **11**, 801 (1948).
55. ZVYAGINCEV, O. E., *Zolotaya Promyshlennost*, **3**, 36 (1939).
56. POLLARD, W. B., *Trans. Inst. Mining and Metallurgy*, **41**, 434, 454 (1931/32).
57. PESHKOV, I. A., *Nauch. Trudy Tulsk. Gornovo Inst.*, **229** (1958); *Chem. Abstr.*, **56**, 7988 (1962).

58. BELCHER, R. and NUTTEN, J., *J. Chem. Soc.*, 550 (1951).
59. POLLARD, W. B., *Trans. Inst. Mining and Metallurgy*, 32, 242, 249 (1922/23).
60. REISHAKHRIT, L. S. and SUKHOBOKOVA, N. S., *Uchennye Zapiski Leningrad. Gosud. Univ.*, 19, 150 (1960); *Chem. Abstr.*, 55, 9159 (1961).
61. ZÝKA, J., *Zavodskaya Lab.*, 27, 1075 (1961).
62. ZÝKA, J., *Wiss. Zeitschrift der Humboldt-Universität, Berlin, Math.-naturwiss. Reihe*, X, 3/4, 415 (1961).
62a. PSHENICYN, N. K., GINZBURG, S. I. and PROKOF'EVA, I. V., *Zhur. Anal. Khim.*, 17, 343 (1962).
63. ISHIBASHI, M., SHIGEMATSU, T. and SHIBATA, S., *Japan Analyst*, 7, 646 (1958).
64. BERKA, A. and HOFMANN, P., *Analyt. Chim. Acta*, 28, 196 (1963).
65. BERKA, A., VULTERIN, J. and ZÝKA, J., *Chemist-Analyst*, 51, 88, 92 (1962).
66. ZÝKA, J. and BERKA, A., Proceedings " Feigl Anniversary Symposium " Birmingham 1962, p. 383.
67. BERKA, A. and HOFMANN, P., *Chem. Průmysl*, 13, 287 (1963).
68. BERAN, P. and BERKA, A., *Chem. Průmysl*, 13, 20 (1963).
69. BRINKMAN, U. A. TH. and SNELDERS, H. A. M., *Talanta*, 11, 47 (1964).
70. BRINKMAN, U. A. TH. and SNELDERS, H. A. M., *Talanta*, 11, 47 (1964).
71. ZÝKA, J., *Rudy*, 12, 177 (1964).
72. ZHDANOV, A. K. and RYAZANOVA, T. M., *Uzbekh. Khim. Zhur.*, 18 (1962); *Anal. Abstr.*, 11, 179 (1964).

CHAPTER 28

HYDRAZINE SULPHATE

HYDRAZINE and its compounds, well known as active reducing agents in preparative inorganic and organic chemistry, have been used so far in chemical analysis mainly for the reduction of some cations to the metal, e.g. in gravimetric determination. In these reductions, hydrazine is usually oxidized to nitrogen:

$$N_2H_4 \rightarrow N_2 + 4H^+ + 4e^-$$

Under certain conditions, small amounts of azoimides or ammonia can also be formed.

The reaction given above, valid in acid medium, can also be formulated for alkaline medium:

$$N_2H_4 + 4OH^- \rightarrow N_2 + 4H_2O + 4e^-$$

Latimer[1] states the redox potential of the first reaction to be $E_0 = 0.23$ V and for the second reaction $E_0 = -1.16$ V. Both values are exceptionally low. When titrations are done in alkaline solution, higher values are always found.[2] The formal redox potential in 1 N hydrochloric acid is $+ 0.65$ V, in 5 per cent sodium hydroxide solution $- 0.55$ V and in sodium hydrogen carbonate solution $+ 0.35$ V.

The literature on hydrazine and its salts has been reviewed in a monograph, giving the references up to 1950.[2a]

Only in recent years have solutions of hydrazine salts been investigated as volumetric reagents,[2] and then mainly in potentiometric determinations. A solution of the acid, monobasic hydrazine sulphate $N_2H_4 \cdot H_2SO_4$ is used. This compound is easily available in a pure and well-defined form which is often used as a standard substance in acid–base titrations. Standard hydrazine sulphate solutions have been applied to the reductimetric determinations of a number of inorganic compounds.[38,44] With strong oxidizing agents, such as permanganate or cerium (IV) salts, however, hydrazine sulphate does not react in a defined ratio.

STANDARD SOLUTION

For the preparation of 0·1 M standard solution, gradually dissolve 13·013 g of pure $N_2H_4H_2SO_4$ in water and make up to 1 l. Less pure material has to be recrystallized from water and dried at 150°C. Hydrazine sulphate solutions are stable even at boiling temperature; this is still true for 10^{-4} M solutions. In alkaline medium, hydrazine sulphate solutions decompose.

INDICATOR

The majority of volumetric determinations with hydrazine sulphate are carried out potentiometrically or by " dead-stop " titration[2] (e.g. the determination of bromine). Iodine can be titrated visually using starch solution,[3] or hypochlorite and hypobromite using lucigenin as chemi-luminescent indicator.[10] In the titration of bromate or bromine, the end-point can be detected by the addition of iodide and starch solution.[7,8]

REVIEW OF DETERMINATIONS

Halogen compounds. The titration of chlorine[9] and bromine[2] proceeds in mineral or acetic acid solution quantitatively, whereas the determination of iodine has to be carried out in hydrogen carbonate,[3] or weakly alkaline buffered solution.[4] This titration has proved satisfactory in many indirect determinations in which the added excess of iodine (especially in the determination of organic substances),[31] or the iodine liberated from potassium iodide[32] is titrated. The titration of bromine with hydrazine[5] may be used instead of iodimetric analysis in all those cases in which the quantitative determination of liberated bromine is the final measurement, e.g. in the determination of substances through the action of bromate and bromide.[6,7,8] This titration is also used in the determination of active chlorine and hypochlorite[9] or in the indirect determination of permanganate[3] and selenate.[43] In the titration of hypobromite and hypochlorite, lucigenin in alkaline solution has been recommended as indicator.[10]

Bromate ($BrO_3^- \rightarrow Br^-$)[5], *iodate* or *periodate*[11] (the reduction takes place either to I^+ or to I_2, dependent on the acidity) can be titrated potentiometrically, even in the presence of many other anions; the determination has to be carried out in acid solution. In a pertinent publication,[29] the reduction of periodate to iodate ($IO_4^- \rightarrow IO_3^-$) in acid solution, containing traces of ruthenium salts as catalyst, has also been mentioned.

Periodate can be titrated potentiometrically in a solution buffered with sodium tetraborate or hydrogen carbonate; it is reduced to iodate.[29a]

Bromate can be titrated in the presence of iodide using starch solution as indicator.[12,13] Titration with hydrazine has also been recommended for the standardization of iodine monochloride solutions.[14]

Hexacyanoferrate (III). Hexacyanoferrate (III) can be determined very accurately in alkali hydroxide solution at elevated[3] or room temperature, if zinc salt has been added to the solution, or osmium (VIII) oxide is used as catalyst.[15,16]

Dichromate and vanadate (V). The direct potentiometric titration of dichromate and vanadate[17,18] (reduction to Cr^{3+} and VO^{2+}) yields unprecise results. If, however, hydrazine solution is added in excess and the unconsumed part is back-titrated, the results are satisfactory.[19,20] A direct potentiometric titration of vanadate (0·04–0·001 N VO_3^- in 0·5–4 N H_2SO_4) in the presence of OsO_4 as catalyst was described recently.[40]

Gold. Gold (III) salts are reduced with hydrazine sulphate solution to the metal in weakly acid medium at elevated temperature; as little as $5 \cdot 10^{-5}$ M gold (III) solution can thus be titrated, even in the presence of many other ions.[21] This method can therefore be used on the micro-scale. Iridium (IV) and palladium (II) salts in higher concentrations affect the determination, but they cannot be titrated themselves with hydrazine.

Silver. Silver ions can be titrated potentiometrically with hydrazine sulphate (reduction to metallic silver) in solutions which contain 5–35 per cent of potassium hydroxide and such an amount of ammonium ions that the ammine complex is formed; this determination can also be used in the presence of zinc (II), nickel (II), lead (II), aluminium, cadmium salts and others.[22] A similar titration has been published recently.[39]

Copper. Copper (II) salts are determined potentiometrically by their reduction to copper (I) in a solution which contains acetate and, moreover, sufficient chloride and thiocyanate ions.[23] The titration can likewise be done in 2–4 N ammonia solution, but boiling temperature is necessary.[23a]

Mercury. The reduction of the complex mercury (II) iodide (or bromide) with hydrazine proceeds in alkaline medium according to the equation:

$$2 K_2HgI_4 + N_2H_4 + 4 KOH \rightarrow 2 Hg + N_2 + 8 KI + 4 H_2O$$

This reaction can be carried out potentiometrically.[35,36] For the titration of the tetraiodo-complex, a 23–33 per cent solution of potassium hydroxide is most suitable, whereas the reduction of the tetrabromo-complex proceeds better in a less alkaline medium (0·75–5 per cent KOH).[30]

Thallium. The potentiometric titration of thallium (III) compounds with hydrazine sulphate yields very accurate results.[35,36]

Out of the great number of *indirect* determinations carried out with hydrazine sulphate, that of *selenium (IV)* salts (reduction to elementary selenium) may be mentioned; here the unconsumed reagent is back-titrated with bromate solution.[24]

Selenate. In the presence of KBr + HCl, selenate is reduced to selenite and the liberated bromine is titrated with hydrazine;[43] the presence of selenites does not interfere.

Higher *manganese* and *lead oxides* can also be determined, based on their reaction with hydrazine.[33,41]

Molybdate. Potentiometric titration of molybdate (even if slow) can be done in HCl, H_2SO_4 or H_3PO_4 solution. The best results are obtained in 3–15 N H_2SO_4 at 95–97°C; the first potential jump corresponds to $Mo^{VI} \rightarrow Mo^V$, the second one to the formation of lower valency complexes.[42]

Hydrazine reacts with nitrite in acid solution rapidly and quantitatively according to the equation:

$$N_2H_4 + 2\,HNO_2 \rightarrow N_2 + 2\,HNO + 2\,H_2O$$
$$\swarrow\searrow$$
$$H_2O\quad N_2O$$

This reaction can be applied to the accurate potentiometric determination of hydrazine (or hydrazine salts respectively)[25] or to the determination of *nitrite*. In the latter case, one has to titrate with the nitrite solution to be determined. This method is very well suited for the determination of nitrite concentration in pharmaceutical products.[26] Less suitable is the indirect determination in which an excess of hydrazine is added to the nitrite solution and the unconsumed part is then titrated.[27,28]

Recently, Vulterin studied some hydrazine derivatives as titrants. He found that the isonicotinic acid hydrazide which is stable in solution, proved as satisfactory as hydrazine sulphate. In some cases, the reducing power of isonicotinic acid hydrazide is even more distinct.[34,37,38]

REFERENCES

1. LATIMER, W. M., *Oxidation Potentials. The Oxidation States of the Elements and their Potentials in Aqueous Solutions*, 2nd Ed., Prentice Hall, New York, 1953.
2. ZÝKA, J. and VULTERIN, J., *Collection (Czech. Chem. Comm.)*, **20**, 804 (1955).
2a. AUDRIETH, L. F. and OGG, B. A., *The Chemistry of Hydrazine*, New York, 1951.
3. VULTERIN, J. and ZÝKA, J., *Chem. Listy*, **48**, 1762 (1954).
4. DESHMUKH, J. S. and BAPAT, M. G., *Z. anal. Chem.*, **156**, 269 (1957).
5. VULTERIN, J. and ZÝKA, J., *Chem. Listy*, **48**, 1745 (1954).
6. BERKA, A. and ZÝKA, J., *Českoslov. Farm.*, **8**, 17 (1959).
7. SZEKERES, L. and MOLNÁR, G. L., *Magyar Kém. Lapja*, **13**, 302 (1958).
8. BALÁSFALVY-ZERGÉNYI, M., GRÖBLER, G., FAKETE-PAP, E. KELLNER, A., MOLNÁR, L. G., NAGY, M., SUGÁR-NOVÁK, E. and SZEKERES, L., *Magyar Kém. Folyóirat*, **64**, 249 (1958).
9. VULTERIN, J. and ZÝKA, J., *Chem. Listy*, **48**, 1768 (1954).
10. ERDEY, L. and BUZÁS, J., *Acta Chim. Acad. Sci. Hung.*, **6**, 127 (1955).
11. ZÝKA, J., *Chem. Listy*, **48**, 1754 (1954).
12. SZEKERES, L., *Magyar Kém. Folyóirat*, **63**, 274 (1957).

13. SZEKERES, L., SUGÁR, E. and PAP, E., *Z. anal. Chem.*, **161**, 38 (1958).
14. ČÍHALÍK, J. and TEREBOVÁ, K., *Chem. Listy*, **50**, 1134 (1956).
15. SANT, S. B., *Z. anal. Chem.*, **168**, 112 (1959).
16. SANT, S. B., *Analyt. Chim. Acta*, **19**, 205 (1958).
17. VULTERIN, J. and ZÝKA, J., unpublished studies.
18. HOLST, G., *Svensk. Kem. Tidskr.*, **43**, 2 (1931); *Chem. Abstr.*, **25**, 4196 (1931).
19. DESHMUKH, G. S. and BAPAT, M. G., *Analyt. Chim. Acta*, **14**, 225 (1956).
20. MITRANESCU, M., *Acad. rep. popul. Romîne*, **4**, 79 (1958); *Chem. Abstr.*, **52**, 2659 (1958).
21. ZÝKA, J., *Chem. Listy*, **48**, 1768 (1954).
22. VULTERIN, J. and ZÝKA, J., *Collection (Czech. Chem. Comm.)*, **25**, 206 (1960).
23. BERKA, A. and ZÝKA, J., *Českoslov. Farm.*, **8**, 578 (1959).
23a. MITRANESCU, M., *Acad. Rep. Populare Romine, Baza cercetari stiint. Timisoara, Studii cercetari stiint.*, Ser. stiinte chim., **5**, 83 (1958); *Chem. Abstr.*, **55**, 234 (1961).
24. SUSEELA, B., *Z. anal. Chem.*, **147**, 13 (1955).
25. VULTERIN, J. and ZÝKA, J., *Chem. Listy*, **50**, 364 (1956).
26. MARXOVA, I. and ZÝKA, J., *Českoslov. Farm.*, **5**, 218 (1956).
27. VAYSMAN, G. A., *Aptechnoe Delo*, **3**, 9 (1954).
28. HENERMANN, R. F., *J. Assoc. off. Agric. Chemists*, **38**, 651 (1955).
29. GLEU, K. and KATTHÄN, W., *Chem. Ber.*, **86**, 1077 (1953).
29a. BERKA, A. and ZÝKA, J., *Českoslov. Farm.*, **8**, 136 (1959).
30. VULTERIN, J., *Collection (Czech. Chem. Comm.)*, **26**, 317 (1961).
31. PANWAR, K. S., RAO, S. P. and GAUR, J. N., *Analyt. Chim. Acta*, **25**, 218 (1961).
32. PANWAR, K. S., MATHUR, N. K. H. and RAO, S. P., *Analyt. Chim. Acta*, **24**, 541 (1961).
33. YOVANOVICH, M. S., POPOVICH, D. V. and VUKASINOVICH, J. I., *Glasnik Chemiskog Drushtva Beograd*, **23/24**, 507 (1958/59).
34. VULTERIN, J., Doctoral Thesis, Charles University, Prague 1962.
35. BERKA, A. and BUSEV, A. I., *Analyt. Chim. Acta*, **27**, 493 (1962).
36. DRAGULESCU, G. and MITRANESCU, M., *Acad. Rep. Populare Romine, Baza cercetari stiint. Timisoara, Studii cercetari stiinte Chim.*, **8**, 195 (1961).
37. VULTERIN, J. and ZÝKA, J., *Talanta*, **10**, 891 (1963).
38. BERKA, A., VULTERIN, J. and ZÝKA, J., *Chemist-Analyst*, **52**, 56, 60 (1963).
39. FACSKO, G. and MINGES, R., *Bull. Stiint. Technic. Inst. Politechnic Timisoara*, **6**, 63 (1961); *Chem. Abstr.*, **57**, 11854 (1962).
40. DRAGULESCU, C. and MITRANESCU, M., *Acad. Rep. Populare Romine, Baza cercetari stiint. Timisyara, Studii cercetari stiinte Chim.*, **9**, 21 (1962).
41. MANOLJOVIC, C. M. and GUDALO, B., *Bull. Soc. Chim. (Beograd)*, **22–23**, 109 (1960–61).
42. ANKUDIMOVA, E. V., *Trudy Novocherkass. Politek. Inst.*, **143**, 11 (1963); *Ref. Zhur. Khim.*, **19G**, 2 (1963).
43. BLANKA, B., HUDEC, P., MOŠNA, P. and TOUŽÍN, J., *Collection (Czech. Chem. Comm.)*, **28**, 3434 (1963).
44. ZÝKA, J., *Rudy*, **12**, 1172 (1964).

CHAPTER 29

SODIUM NITRITE

STANDARD sodium nitrite solutions are useful in the determination of organic substances. The major number of the determinations with sodium nitrite is based on easily proceeding diazotization or nitrosation reactions of organic substances. The reactions usually proceed according to the following equations:

$$RNH_2 + NaNO_2 + 2 HCl \rightarrow (RN\equiv N)Cl + NaCl + 2 H_2O$$
$$RNHR' + NaNO_2 + HCl \rightarrow RN(NO)R' + NaCl + H_2O$$

Some compounds, however, react with sodium nitrite to form other reaction products, such as nitrogen, dinitrogen monoxide, nitrate and so on.

STANDARD SOLUTION

Standard sodium nitrite solution is prepared by dissolving the pure salt in an appropriate amount of distilled water. Aqueous solutions of alkali and alkaline earth nitrites are only sufficiently stable if they are not too dilute. In very dilute solutions, nitrite is easily oxidized to nitrate. Lunge[1] observed that solutions which contain 1 mg of nitrite in 1 l. of water are oxidized quantitatively after some weeks. Concentrated solutions (0·1 M and stronger) are sufficiently stable for a period of four weeks.[2]

Standardization

For the standardization, sulphanilic acid is commonly used as primary standard. This is readily accessible in a sufficiently pure state and is about 18 times more rapidly diazotized without catalyst than aniline.[3]

Because sulphanilic acid contains variable amounts of water of crystallization (1–2 molecules), it must be transformed to the anhydrous form at 105°C.

The standardization is carried out as follows:[4]

Dissolve 3–3·5 g of sulphanilic acid in a hot solution of 10 ml of ammonia and 150 ml of water; then add 20 ml of concentrated hydrochloric acid, cool down to 7–8°C and titrate with the 0·1 N sodium nitrite solution.

p-Amino benzoic acid, p-amino ethyl benzoate,[5] hydrazine sulphate[2] or permanganate[6] solutions can likewise be used as standard substances.

INDICATOR

In titrations with sodium nitrite, the equivalence point is determined either visually or electrometrically.

Potassium iodide–starch paper is commonly used as a sufficiently sensitive indicator.[7] It often happens that the end-point of the diazotization reaction must be determined in the presence of iron (III) ions. This is the case if the aromatic amine has been prepared by reduction of the original nitro compound with iron, e.g. in the reduction of p-nitrophenol, p-nitrotoluene and others. Here, the potassium iodide–starch paper fails completely, because iron (III) salt reacts in the same way as nitrous acid. The so-called " sulphone-reagent " is then used, which reacts specifically only in the presence of nitrous acid. It is, basically, 4,4'-diamino diphenyl methane-2,2'-sulphone or its N,N'-tetramethyl derivative.[8] Besides these compounds, acridine yellow, metanil yellow, safranine-T and acridine orange have also been suggested as indicators.[9,10] All these compounds are so-called external indicators (" spot indicators ").

Electrometric determination of the end-point is to be preferred. The advantages of potentiometric titrations[11–15] consist mainly in the fact that highly coloured solutions can also be titrated, even in non-aqueous media and in the presence of iron (III) ions.[16] In some cases, cooling of the solution before the titration can be neglected. It is possible to get good results even at relatively higher temperatures (60°C)[12,12a] " Dead-stop "[17,18] or amperometric titration[19,20] may also be applied. The electrometric determinations are more reliable and accurate than the visual ones; when using suitably designed apparatus or certain electrodes,[21] these methods can also be applied in the micro-range.[22]

REVIEW OF DETERMINATIONS

Primary aromatic amines. These compounds are titrated with sodium nitrite in the presence of hydrochloric or sulphuric acid, according to the equation mentioned above. For the titrimetric determination, it is important that the diazotization proceeds with sufficient rapidity[23–25a] to avoid a loss of nitrous acid, which would then lead to unprecise results. Therefore, the solution to be diazotized is cooled with ice during the titration.

The procedure for the actual determination depends on the solubility of the sample to be analysed. If the amine is soluble in dilute hydrochloric acid, and if the titration is not influenced by the too early precipitation of diazo salt of only limited solubility, then a fiftieth of a mole is usually weighed in.

After the addition of 50 ml of 1 N hydrochloric acid, the solution is diluted with water to 300 ml, cooled to 5–10°C with ice and titrated with 1 N sodium nitrite solution.

If the sample is not sufficiently homogenous or if it is in the form of a paste (as often happens in the analysis of dyestuff intermediate products), then a fifth of a mole of the sample is weighed in.

Compounds which contain the group —SO₃H, —COOH (amino sulphonic acids, amino acids) and certain others, need to be dissolved first in dilute ammonia solution or by adding sodium carbonate, and are then acidified. Substances which precipitate again on acidification are successfully titrated with so-called hydrotrope reagents, e.g. with a saturated sodium naphthalene sulphonate solution,[26] or with a saturated solution of the condensation product of 1- and 2-sodium naphthalene sulphonate with formaldehyde. But in the first instance, the sodium salt of *m*-xylene sulphonic acid[27] is used for this purpose; this acts extremely hydrotrope on the amino compound as well as on the diazo compounds formed and, moreover, possesses also stabilizing properties. If such hydrotrope reagents are applied, most of the compounds can be titrated at 25°C, because the velocity of the reaction is about trebled by these substances.

The following compounds can be analysed: 1-amino anthraquinone, 4-amino azobenzene, 1-naphthyl amine-4-sulphonic acid, 2-amino-5-naphthol-7-sulphonic acid and other substances.[28] If no hydrotrope reagents are used, the determination can be done in such a way that only about 95 per cent of the sodium nitrite necessary for the diazotization is added to the alkaline solution of the amine; then, the sample is titrated in hydrochloric acid solution dropwise to the equivalence point. In the literature, there is also described a semi-direct method,[4] in which about half of the necessary standard sodium nitrite solution is added and then titrated slowly to the end-point, or also an indirect method[29-33] in which the excess of sodium nitrite is back-titrated with standard solutions, such as sulphanilic acid, *p*-nitraniline, aniline hydrochloride, 2-naphthol and others.

In industry, the results of diazotization titrations are frequently expressed by the so-called *nitrite number*. This nitrite number refers to the number of grams of sodium nitrite which corresponds to 1000 g of sample.

In the determination of some aromatic amines, a different procedure has to be followed. If, for example, the amino group is acetylated it must first be liberated by saponifying before it can be diazotized. In this way, *p*-nitro acetanilide[34] or naphthol AS (β-hydroxy naphtholic acid anilide)[35] are determined. Some diazotization titrations must be catalysed by sodium bromide; others depend on the concentration of the substance to be analysed.[36]

Amongst the compounds with a primary aromatic group, which can be determined according to one of the above-mentioned compounds, there are also some important pharmaceutical preparations[37-41] such as procaine, benzocaine, *p*-amino benzoic acid, *p*-amino salicylic acid, *p*-amino hippuric acid, isonicotinic acid hydrazide and other compounds. *p*-Amino benzoic acid and its derivatives can be titrated visually in the presence of potassium bromide, using tropeoline 00 as indicator.[42] Various aryl amines[42a] can be titrated visually with sodium nitrite against diphenylamine as indicator.[42a] Sulphonamides[43-47c] and phenacetine[48] can be determined potentiometrically or visually.

Aromatic diamines. Most aromatic diamines can be easily diazotized, but only with some of them does the reaction proceed quantitatively. Atanasiu and Velculescu worked out a potentiometric titration of benzidine with sodium nitrite and applied this method to the determination of sulphate.[12] Butt and Stagg[49] occupied themselves likewise with the potentiometric determination of benzidine. 1,4-Diamino anthraquinone,[53] *p*-phenylene diamine,[50-52] 4-amino diphenyl amine[54] and other substances[55,56,28] can also be titrated with sodium nitrite. If the direct titration is not feasible, sodium nitrite is added in excess and the unconsumed reagent is determined with aniline hydrochloride or sulphanilic acid.[57]

Nitro and nitroso compounds. Some aromatic nitro and nitroso compounds can be titrated indirectly by diazotization. On principle, the method is based on the reduction of the nitro or nitroso groups with zinc in hydrochloric or acetic acid medium; then the amino groups thus formed are diazotized. The method is applied to the determination[58] of nitrobenzene, *m*-nitrobenzene sulphonic acid, *o*-nitroanisol, 4,4'-dinitrostilbene-2,2'-disulphonic acid (important in dye works),[59] of the pharmaceutically important diethyl-*p*-nitrophenyl phosphate,[60] of chloramphenicol,[61] *p*-nitroso phenol and others. Some azo dyes are analogously determined.[62]

Secondary and tertiary aromatic amines. Sodium nitrite reacts in hydrochloric acid with secondary aromatic amines to form nitrosamines:

$$RNHR' + NaNO_2 + HCl \rightarrow RN(NO)R' + NaCl + H_2O$$

R' can be an alkyl as well as an aryl radical. The conditions for the nitrosation are the same as for the diazotization (for the nitrosation potassium bromide can likewise be used as catalyst). The equivalence point is usually determined with potassium iodide–starch paper. Potentiometric indication of the end-point does not give the results to be expected in some cases.[15]

The determination of diphenylamine[63] is an example for a nitrosation-titration.

Tertiary amines can likewise be transformed to nitroso compounds with sodium nitrite in hydrochloric acid solution. *N*-dimethyl aniline may be mentioned as an example which is nitrosated according to the equation:[57]

$$C_6H_5N(CH_3)_2 + NaNO_2 + HCl \rightarrow (NO)C_6H_4N(CH_3)_2 + NaCl + H_2O$$

For details on the conditions for the titration the reader is referred to the literature.[28,37,64,65]

Phenols and cyclic ketones. Some phenols react quantitatively with sodium nitrite to form nitroso compounds. This can be applied for example to the determination of resorcinol which can be nitrosated unlike catechol and hydroquinone, according to the equation:[66-68]

$$\text{C}_6\text{H}_4(\text{OH})_2 + 2\text{NaNO}_2 + 2\text{HCl} \rightarrow \text{C}_6\text{H}_2(\text{OH})_2(\text{NO})_2 + 2\text{NaCl} + 2\text{H}_2\text{O}$$

The titration is carried out like the diazotization titration.[15] Resorcinol can also be determined in pharmaceutical products, even in the presence of other substances.[38]

Some cyclic ketones behave similarly to phenols, so e.g. the important 3-methyl-1-phenyl-5-pyrazolone which is nitrosated according to the equation:

$$\underset{\substack{|\\ \text{N}\\ \diagdown\\ \text{C}_6\text{H}_5}}{\text{O}=\text{C}-\text{NH}} \overset{\text{H}-\text{C}=\text{C}-\text{CH}_3}{\diagup} + \text{NaNO}_2 + \text{HCl} \rightarrow \underset{\substack{|\\ \text{N}\\ \diagdown\\ \text{C}_6\text{H}_5}}{\text{O}=\text{C}-\text{NH}} \overset{\text{ON}-\text{C}=\text{C}-\text{CH}_3}{\diagup} + \text{NaCl} + \text{H}_2\text{O}$$

This reaction, which even in aqueous solution proceeds with satisfactory velocity, is the basis for the titrimetric determination.[66]

The determination of 1-naphthol[68a] is also based on a nitrosation-reaction. The actual titration with sodium nitrite is carried out potentiometrically in a solution containing sodium-*m*-xylene sulphonate, using a platinum indicator electrode and a calomel reference electrode.

Hydrazine and hydrazine derivatives. Hydrazine sulphate reacts with sodium nitrite rapidly and quantitatively to form nitrogen and dinitrogen monoxide. The potentiometric determination of hydrazine salts is based on this reaction;[2] it proceeds in 7–10 per cent hydrochloric, 15–30 per cent perchloric, 10–12 per cent sulphuric and 50 per cent phosphoric acid solution. As little as 0·1–100 mg of hydrazine salt can thus be determined with good results, even in the presence of a hundred-fold excess of nitrates or ammonium salts. Urea and hydroxylamine interfere with this determination.

Phenylhydrazine can be titrated potentiometrically with standard sodium nitrite solution in 10 per cent hydrochloric acid medium;[2] by this reaction, phenylnitrosohydrazine is formed which very easily turns to the respective azide, or to the amine and dinitrogen monoxide. 2,4-Dinitrophenylhydrazine can be titrated similarly; this compound is used for the determination of aldehydes.

Semicarbazide and thiosemicarbazide. In the potentiometric titration of semicarbazide, the following reaction takes place:

$$NH_2-NH-CO-NH_2 + HNO_2 \rightarrow N_2O + 2NH_3 + CO_2$$

Optimal results are achieved in 10–15 per cent hydrochloric acid.[2] Thiosemicarbazide reacts similarly with sodium nitrite, its potentiometric titration is carried out in 1–5 per cent hydrochloric or sulphuric acid, in 20–30 per cent phosphoric acid or also in 1–15 per cent perchloric acid solution.[2,69]

Hydrazides of aliphatic acids. Hydrazides of aliphatic acids, such as acetic, propionic, butyric, phenylacetic and phenoxyacetic acid can be determined with sodium nitrite solution.[69a] After dissolving the sample in benzene, the titration is carried out in 0·03–0·05 M hydrochloric acid in the presence of potassium bromide. According to the authors, the relative error is 1 per cent.

In the literature, there are also described determinations of hydrazino phthalazine,[70] carbonic acid hydrazides and further compounds.[71,71a]

Other determinations. In addition can be determined directly in hydrochloric acid medium with standard sodium nitrite solution:

Aromatic sulphinic acids,[72] monosubstituted 1,3,4-oxadiazole,[72a] N-arylhydroxylamines[72b] and (in sulphuric acid medium) azides,[73] amido sulphonic acid,[74] and isonicotinoyl hydrazone.[74a]

Amongst inorganic substances, hexacyanoferrate (III),[75] hydrogen peroxide[76] and permanganate[77,77a] are titrated with sodium nitrite in the presence of zinc sulphate.

Willard and Young[78] recommend sodium nitrite as a reductimetric reagent for the determination of cerium (IV) ions. Cerium (III) ions can also be determined after they have been oxidized with peroxidisulphate. The potentiometric titration is carried out in sulphuric or nitric acid medium at 40–50°C; it can be used for the indirect determination of chromium:

Chromium (III) ions are oxidized with cerium (IV) sulphate; the excess of the latter is titrated with sodium nitrite. This method has been applied to the determination of small amounts of chromium (III) in the presence of large amounts of dichromate.[79]

An indirect determination of pentavalent vanadium has also been described; this is based on its reduction with an excess of sodium nitrite and back-titration of the unconsumed reagent with permanganate.[80]

Many of the determinations mentioned here have been re-examined with electrometric indication methods.[81-83]

Recently, Mathur and co-workers[84] used alkyl nitrite as titrant for the determination of primary aromatic amines and of compounds with active methyl groups. These titrations have been carried out potentiometrically at room temperature, using a platinum indicator electrode. Matrka gave a review of titrations with sodium nitrite.[28]

REFERENCES

1. LUNGE, G., *Z. angew. Chem.*, **15**, 1 (1902).
2. VULTERIN, J. and ZÝKA, J., *Chem. Listy*, **50**, 364 (1956).
3. ELOFSON, R. L., EDSBERG, R. and MECHERLY, P. M., *J. Elektrochem. Soc.*, **97**, 166 (1950).
4. Fiat Microfilm PBL 17692, Suppl. S.2; IG Analysenvorschrift No. 18.
5. JUREČEK, M., *Organická analysa*, 2nd Ed., p. 394, Nakladatelství ČSAV, Prague 1957.
6. *Československý lékopis*, 2nd Ed., p. 493, Státní zdravotnické nakladatelství, Prague 1954.
7. LASTOVSKII, R. P., *Technické rozbory ve výrobě meziproduktů a barviv*, p. 114, Průmyslové vydavatelství, Prague 1952.
8. FIERZ-DAVID, M. E., BLANGEY, L., *Grundlegende Operationen der Farbenchemie*, 4th Ed., p. 131, Springer, Vienna 1938.
9. FROST, H. F., *Analyst*, **68**, 51 (1943).
10. AGREN, A., *Svensk. farm. Tidskr.*, **55**, 229 (1951).
11. MÜLLER, E. and DACHSELT, E., *Z. Elektrochem.*, **31**, 662 (1925).
12. ATANASIU, I. A. and VELCULESCU, A. I., *Ber. dt. chem. Ges.*, **65**, 1080 (1932).
12a. MATRKA, M., PODSTATA, J. and SÁGNER, Z., *Chem. Průmysl*, **12**, 549 (1962).
13. BELEŇSKIJ, L. and SOKOLOV, I., *Promyšlenost organičeskoj chimii*, **1**, 618 (1936).
14. SINGH, B. and AHMED, G., *Ind. chem. Soc.*, **15**, 416 (1938).
15. MATRKA, M., *Chemie*, **8**, 13 (1952).
16. MATRKA, M. and STEINER, K., *Chem. Průmysl*, **6**, 471 (1956).
17. MATRKA, M., *Org. chem. a technol.*, **2**, 199 (1952).
18. SCHOLTEN, H. G. and STONE, K. G., *Anal. Chemistry*, **24**, 749 (1952).
19. MATRKA, M., *Org. chem. a technol.*, **4**, 45 (1954).
20. RYBÁŘ, D. and SKŘIVAN, V., *Českoslov. Farm.*, **5**, 147 (1956).
21. LORD, S. and ROGERS, L., *Anal. Chemistry*, **26**, 284 (1953).
22. LITVENENKO, L. M. and GREKOV, A. P., *Zhur. Anal. Khim.*, **10**, 164 (1955).
23. SMID, H. and MUHR, G., *Ber. dt. chem. Ges.*, **70**, 421 (1937).
24. ABEL, E., SMID, H. and WEISS, J., *Z. phys. Chem.*, **147A**, 49 (1930).
25. UENO, S. and SUZUKI, T., *J. Soc. Chem. Ind. Japan, Spl.*, **36**, 615B (1933).
25a. TROKOVITZ, D., *Chem. Anal. (Warszawa)*, **8**, 107 (1963).
26. BIOS, 1493, 59 (1946).
27. HAASE, J., *Chem. Listy*, **45**, 222 (1951).
28. MATRKA, M., *Chemie*, **10**, 635 (1958).
29. SCHULTZ, A. and VAUBEL, W., *Z. Farben-Ind.*, **1**, 37, 149, 339 (1902).
30. LUNGE, G., *Chemiker-Ztg.*, **28**, 501 (1904).
31. UENO, S. and SEKIGUCHI, H., *J. Soc. Chem. Ind. Japan*, **37**, 235 (1934).
32. JONES, D. O. and LEE, H. R., *Ind. Engng. Chem.*, **16**, 948 (1924).
33. FIERZ-DAVID, H. F. and BLANGEY, L., *Grundlegende Operationen der Farbenchemie*, 5. Aufl., p. 365, Springer, Vienna 1943.
34. Fiat Microfilm PBL 17692, p. 2531; IG Analysenvorschriften No. 77.

35. LASTOVSKII, R. P., *Technické rozbory ve výrobě meziproduktů a barviv*, p. 132, Průmyslové vydavatelství, Prague 1952.
36. IG Analysenvorschriften No. 125.
37. HRDÝ, O., JINDRA, A., JUNG, Z., ŠLOUF, A. and ZÝKA, J., *Kvantitativní Analysa léčiv*, p. 127, 423, Státní zdravotnické nakladatelství, Prague 1957.
38. DUŠINSKÝ, G. and GRUNTOVÁ, Z., *Potenciometrické titrace vo farmácii a príbuzných oboroch*, p. 241, Vydavatelstvo SAV, Bratislava 1956.
39. SCOTT, P. G. W., *J. Pharm. Pharmacol.*, **4**, 686 (1952).
40. KLEFLIN, Z. and ŠUMANOVIČ, K., *Croat. chem. Acta*, **30**, 181 (1958).
41. BLAŽEK, J. and STEJSKAL, Z., *Českoslov. Farm.*, **7**, 23 (1958).
42. VASILIEV, R., WERMESCHER, B., COSMIN, A., MANGU, M. and BURNEA, I., *Lucrarile prezentate conf. natl. farm.*, p. 147, Bukarest 1958.
42a. TUN TAO, *Yao Hsueh Hsueh Pao*, **5**, 97 (1957).
43. CALAMARI, J. A., HUBATA, R. and ROTH, P. B., *Ind. Engng. Chem., Anal. Ed.*, **14**, 534 (1942).
44. LA ROCCA, J. P. and WATERS, K. L., *J. Amer. Pharm. Assoc., Sci. Ed.*, **39**, 521 (1950).
45. KUCHARSKÝ, J., ERBEN, J. and MIKULE, V., *Českoslov. Farm.*, **3**, 284 (1954).
46. VASILIEV, R., COSMIN, A., WERMESCHER, B., MANGU, M. and BURNEA, I., *Farmacia (Bukarest)*, **6**, 327 (1958).
47. *Československý lékopis*, 2nd Ed., p. 759, Státní zdravotnické nakladatelství, Prague 1954.
47a. MATHUR, N. K., RAO, S. P. and NARAIN DAU, *Analyt. Chim. Acta*, **24**, 474 (1961).
47b. DZOTTSOTI, S. Kh., *Azerbaidshan. Khim. Zhur.*, **4**, 95 (1960).
47c. UENO, K. and TACHIKAWA, T., *Japan Analyst*, **11**, 554 (1962).
48. DUŠINSKÝ, G., *Chem. Zvesti*, **6**, 201 (1952).
49. BUTT, L. T. and STAGG, H. E., *Analyt. Chim. Acta*, **19**, 208 (1958).
50. ZORINOVA, M. D., *Zavodskaya Lab.*, **13**, 1180 (1947).
51. ALLAN, Z. J. and MUŽÍK, F., *Chem. Listy*, **48**, 52 (1954).
52. MATRKA, M., *Org. chem. a technol.*, **5**, 495 (1955).
53. MATRKA, M., *Org. chem. a technol.*, **5**, 426 (1955).
54. MATRKA, M., *Org. chem. a technol.*, **5**, 478 (1955).
55. MATRKA, M. and SÁGNER, Z., *Chem. Průmysl*, **9**, 288 (1959).
56. VOROŽCOV, N. N., *Základy synthesy polotvarů a barviv*, 2nd part, p. 308, Státní nakladatelství technické literatury, Prague 1953.
57. PHILLIPS, J. and LOWY, A., *Ind. Engng. Chem., Anal. Ed.*, **9**, 381 (1937).
58. LASTOVSKII, R. P., *Technické rozbory ve výrobě meziproduktů a barviv*, p. 167, Průmyslové vydavatelství, Prague 1952.
59. INUKAI, K. and MAKI, I., *Reports of the Government Industrial Research Institute Nagaya*, **1**, 101 (1952).
60. BLAŽEK, J., *Českoslov. Farm.*, **7**, 455 (1958).
61. VASILIEV, R., WERMESCHER, B., COSMIN, A., MANGU, M. and BURNEA, I., *Lucrarile prezentate conf. natl. Farm.*, p. 147, Bukarest 1958.
62. MATRKA, M., *Sborník přednášek z. I. celostátního koloristického sjezdu v Pardubicích*, p. 132, VÚOS, Pardubice 1955.
63. LASTOVSKII, R. P., *Technické rozbory ve výrobě meziproduktů a barviv*, p. 137, Průmyslové vydavatelství, Prague 1952.
64. NELYUBINA, A., *Anilinokrasochnaya Promyshlennost*, **4**, 120 (1934).
65. REVERDIN, F. and DE LA HARPE, C., *Ber. dt. chem. Ges.*, **22**, 1004 (1889).

66. VOROŽCOV, N. N., *Základy synthesy polotovarů a barviv*, 1. part, pp. 117 and 119. Státní nakladatelství technické literatury, Prague 1953.
67. ROSTOVTSEVA, J. and SHEMYAKIN, M., *Anilinokrasochnaya Promyshlennost*, **3**, 457 (1933).
68. VENC, S. and SUZUKI, T., *Z. anal. Chem.*, **117**, 428 (1939).
68a. MATRKA, M., *Collection (Czech. Chem. Comm.)*, **25**, 964 (1960).
69. MISS, A. and IANKU, S., *Rev. chim. (Bukarest)*, **5**, 399 (1954).
69a. GREKOV, A. P. and MARAKHOVA, M. S., *Zhur. Anal. Khim.*, **16**, 643 (1961).
70. RUGGIERI, R., *Farmaco (Pavia) Ed. prat.*, **11**, 571 (1956).
71. LITVIENENKO, L. M., ARLOZOROV, D. G. and KOROLEVA, V. I., *Ukrain. Khim. Zhur.*, **22**, 527 (1956).
71a. MATRKA, M. and NAVRÁTIL, F., *Chem. Průmysl*, **10**, 361 (1960).
72. PONZINI, S., *Farm. sci. e tec. (Pavia)*, **2**, 198 (1947).
72a. GREKOV, A. P. and SHVAYKA, O. P., *Zhur. Anal. Khim.*, **15**, 731 (1960).
72b. SHVAYKA, O. P. and PROTSHENKO, E. G., *Zhur. Anal. Khim.*, **18**, 410 (1963).
73. REITH, H. F., Doctoral Thesis, Utrecht 1929.
74. BOWLER, W. W. and ARNOLD, E. A., *Anal. Chemistry*, **19**, 336 (1947).
74a. NINO, N., *Farmacia (Sofia)*, **12**, 27 (1962); *Chem. Abstr.*, **57**, 6026 (1962).
75. SANT, B. R., *Analyt. Chim. Acta*, **19**, 523 (1958).
76. REICHERT, J. S., MCNEIGHT, S. A. and RUDEL, H. W., *Ind. Engng. Chem., Anal. Ed.*, **11**, 194 (1939).
77. LUNGE, G., *Ber. dt. chem. Ges.*, **10**, 1073 (1877).
77a. BELLEMARE, P. A. and SABOURIN, R. G., *Titrimetric Methods*, Proc. Symp. Cornwall, Ont., Can. 33 (1961); *Chem. Abstr.*, **57**, 4020 (1962).
78. WILLARD, H. H. and YOUNG, P., *J. Amer. Chem. Soc.*, **50**, 1379 (1928); **51**, 139 (1929).
79. MONNIER, D. and ZWAHLEN, P., *Helv. Chim. Acta*, **39**, 1859 (1956).
80. PAIS, J. and PATAKI, L., *Magyar Kém. Folyóirat*, **62**, 289 (1956).
81. ENOKI, T. and MORISAKA, K., *Yakugaku Zasshi*, **78**, 432 (1958).
82. SAXENA, R. S. and BHATNAGAR, C. S., *Naturwiss.*, **44** (22), 583 (1957).
83. SITARAMIAH, G. and SHARMA, R. S., *Current Sci.*, **24**, 334 (1955).
84. MATHUR, N. K., ROA, S. P. and DAU NARAIN, *Analyt. Chim. Acta*, **23**, 312 (1960).

CHAPTER 30

SOME OTHER OXIDIZING AND REDUCING TITRANTS

OWING to the limited size of this book, it is not possible to deal in separate chapters with all reagents which have been used recently in titrimetric analysis. These reagents should, however, at least be mentioned in a short review for the sake of completeness.

Oxidimetric reagents. Recently, increasing attention has been paid to the comparatively stable solution of *potassium manganate* (VI).[1] Potassium manganate (VI) solution is suitable for the determination of arsenic (III),[2,3] antimony (III),[4] chromium (III)[5] and tellurium (IV)[6] compounds. Thallium (I) compounds,[65] hydrogen peroxide, manganese (II) salts[66] and formic acid[67] can likewise be determined with potassium manganate. The main advantage of manganate (VI) as a titrant is its stability even at elevated temperatures.[68] Den Boef, Polak and Fronk[69,79,80] have summarized the possibilities of its use, especially for oxidations of organic compounds.

Compounds which contain chlorine in higher valency states are likewise used as titrants. Thus, a solution of *potassium chlorate* has been applied to the potentiometric titration of some organic aromatic compounds[7] as well as of vanadium (II)[8] and tin (II) salts.[9]

In some publications[11-14a] the determination of various inorganic systems with standard *chlorate (III)* solutions has been described.[84,85] The solution of this reagent in glacial acetic acid is especially well suited for the determination of iodide in the presence of a great amount of chloride.[14]

N-Chlorobenzamide was found to be similar to chloramine-T.[89]

Standard *potassium peroxidisulphate* solutions are suitable for the titration of many inorganic substances which have been previously oxidized with iodine monochloride,[16] or for the titration of hydrazine and related compounds.[15,83]

Arsenic (III), antimony (III), tin (II), iron (II) salts and others can be titrated with standard *perbenzoic acid* solution in chloroform.[17] Numerous publications deal with the use of *copper (II) salt* solutions in titrimetric analysis. These solutions are mostly applied to the determination of reducing sugars (with Fehling's solution or with the complex of copper (II) salt and trihydroxiglutaric acid)[63] following a modified " dead-stop " end-point titration,[18-21] using e.g. two copper electrodes.[21] A method has also been described in which methylene blue is used as indicator.[62] Copper (II) solutions are also suitable for the potentiometric determination of hydroxylamine and hydrazine derivatives; this can be utilized for the indirect determination of carbonyl groups.[22] Substances which contain SH-groups, such as cystein, can be oxidized by amperometric titration with copper (II) sulphate solution;[64,23-26] some strongly reducing inorganic systems, such as the lower valency states of tungsten and molybdenum compounds[27] can likewise be titrated with copper (II) sulphate.

Some authors recommend the use of *gold (III) chloride*[28] for the amperometric oxidation of tocopherol to the respective quinone[29,30] and of antibiotics of the rifamycine type.[88]

Alkaline *mercury (II) iodide* solution has been suggested for the direct and indirect titration of some reducing sugars, of some photographic developers and of formaldehyde;[31] *diantimonate (V)* has been proposed for the titration of tervalent molybdenum.[32]

This review of oxidimetric reagents can be completed by reference to 2,6-*dichloro phenol indophenol* which can be used for the electrometric determination of ascorbic acid[33-36,81,86] in natural materials[36] and to *indigo* which can serve for the visual titration of dithionite.[37]

Reductimetric reagents. Reference is made to a number of publications in which *phenyl arsine oxide* is used for the determination of free halogens (chlorine and bromine); this method has proved very useful, especially in the control of water.[38-45]

Arsenate (V),[46,49] gold (III),[70,71,47] cerium (IV),[92] manganese (III)[82] and copper (II) salts,[48] selenate (IV) and tellurate (IV),[49,50] iodate, hypochlorite[50] and chlorine[87] can be titrated in acid solution potentiometrically, conductimetrically and amperometrically with *potassium iodide*. The reduction of Os (VIII) → Os (VI) proceeds in alkaline solution.[51]

Thiosulphate which is commonly used for the determination of iodine, is suitable for the potentiometric titration of iron (III) (catalysed with Cu^{2+}), thallium (III), copper (II) salts, iodate, bromate and quinone.[52]

Gold (III) salts can be titrated with *sulphite* solution.[72] *Cerium (III) salt* is used for the titrimetric determination of hexacyanoferrate (II).[73]

The titration of the iron (III) thiocyanate complex with a solution of *ferrocene* in ethanol[74] is a very sensitive and selective determination; the chloride of 2,4-dinitro-*N*-phenyl pyridine is used as indicator.

Arsenic (III) chloride and *catechol* are suitable for the reductimetric determination of some inorganic systems by potentiometric titration in glacial acetic acid.[53]

Solutions of various *organic dyes* have also been suggested as reductimetric reagents. The following determinations have been done on the micro scale.

Iron (II) salts and hydrogen peroxide,[54,55] hypochlorite and chlorine in water[55a,56,75] and chlorate[57] or vanadate (V)[76] with an excess of *methyl orange* solution.

Tin (II) salts and cobalt carbonyl can be titrated with *methylene blue*,[58,77] and chlorine and bromine with some derivatives of *phenoxazine*.[59,60]

Tin (II), titanium (III) and chromium (II) salts have been determined with 7-*oxiphenthiazone*-2.[61,78]

Solutions of *hydroxylammonium sulphate* and *chloride* as volumetric reagents were studied recently.[90]

For a survey of various oxidation–reduction titrants in the determination of organic compounds, the reader is referred to the monograph of Ashworth.[91]

REFERENCES

1. DEN BOEF, G., *Z. anal. Chem.*, **166**, 321 (1959).
2. DEN BOEF, G., DEN BOEF-NUGTEREN, J. and VAN LAAR, B., *Z. anal. Chem.*, **166**, 422 (1959).
3. ISSA, I. M., KHALIFA, H. and ALLAM, M. G. E., *Z. anal. Chem.*, **172**, 21 (1960).
4. DEN BOEF, G. and DAALDER, A., *Z. anal. Chem.*, **167**, 430 (1959).
5. DEN BOEF, G. and DAALDER, A., *Z. anal. Chem.*, **172**, 360 (1960).

6. ISSA, L. M., ALLAM, M. G. E. and AMER, M. M. A., *Z. anal. Chem.*, **172**, 82 (1960).
7. SINGH, B. and SINGH, S., *J. Indian Chem. Soc.*, **16**, 346 (1939).
8. ERDEY, L. and MÁZOR, L., *Acta Chim. Acad. Sci. Hung.*, **3**, 469 (1953).
9. KULWARSKAJA, R. M., *Z. anal. Chem.*, **89**, 199 (1932).
10. PAUL, R. CH. and SINGH, A., *J. Indian Chem. Soc.*, **35**, 294 (1958).
11. JACKSON, D. T. and PARSONS, J. L., *Ind. Engng. Chem., Anal. Ed.*, **9**, 250 (1937).
12. BROWN, E. G., *Analyt. Chim. Acta*, **7**, 494 (1952).
13. YUTEMA, L. F. and FLEMING, T., *Ind. Engng. Chem., Anal. Ed.*, **11**, 375 (1939).
14. ŠKRAMOVSKÝ, S., TAUER, Z. and NOVOTNÝ, J., *Chem. Listy*, **48**, 1335 (1954).
14a. MINCZEWSKI, J. and GLABICZ, N., *Talanta*, **5**, 179 (1960).
15. SINGH, B., SAHOTA, S. S. and SINGH, I., *Z. anal. Chem.*, **162**, 256 (1958).
16. SINGH, B. and ALUYA, N. S., *J. Indian Chem. Soc.*, **35**, 508 (1958).
17. SINGH, B., SAHOTA, S. S. and SINGH, A., *Z. anal. Chem.*, **169**, 106 (1959).
18. COALSTAD, S. E., *J. Soc. Chem. Ind.*, **65**, 230 (1946).
19. NIEDERL, J. B. and MÜLLER, R. H., *J. Amer. Chem. Soc.*, **51**, 1356 (1929).
20. DAGETT, W. L., CAMPBELL, A. W. and WHITMAN, J. L., *J. Amer. Chem. Soc.*, **45**, 1043 (1923).
21. CHUBB, L. and HARTLEY, A. W., *Analyst*, **83**, 311 (1958).
22. BUDĚŠÍNSKÝ, B., *Chem. Listy*, **52**, 2292 (1958).
23. KOLTHOFF, I. M. and STRICKS, W., *Anal. Chem.*, **23**, 763 (1951).
24. KOLTHOFF, I. M. and STRICKS, W., *J. Amer. Chem. Soc.*, **73**, 1728 (1951).
25. KOLTHOFF, I. M. and WILLEFORD, JR., B. R., *J. Amer. Chem. Soc.*, **79**, 2656 (1957).
26. KOLTHOFF, I. M. and WILLEFORD JR., B. R., *J. Amer. Chem. Soc.*, **80**, 5673 (1958).
27. VLASÁK, F., *Collection (Czech. Chem. Comm.)*, **10**, 278 (1938).
28. KARRER, P. and KELLER, H., *Helv. Chim. Acta*, **21**, 116 (1938); **22**, 617 (1939).
29. SMITH, L. I., KOLTHOFF, I. M. and SPILLANE, L. J., *J. Amer. Chem. Soc.*, **64**, 646 (1942).
30. BEAVER, J. J. and KUNITZ, H., *J. biol. Chem.*, **152**, 363 (1944).
31. ŠKRAMOVSKÝ, S., *Sborník I. celost. prac. konf. anal. chemie*, p. 180, Nakl. ČSAV, Prague 1953.
32. DOLEŽAL, J., MOLDAN, B. and ZÝKA, J., *Collection (Czech. Chem. Comm.)*, **24**, 3769 (1959).
33. ZUMAN, P. and PROCHÁZKA, Z., *Chem. Listy*, **47**, 357 (1953).
34. BOGDANOVA, V. A., *Gigiena i Sanitaria*, **13**, 31 (1948).
35. LIEBMANN, H. and AYRES, A. D., *Analyst*, **70**, 421 (1945).
36. CURTIS, R. C., *Analyst*, **83**, 54 (1958).
37. Unpublished studies.
38. HANDERSON, W. L., *Sewage and Ind. Wastes*, **24**, 1467 (1952); *Chem. Abstr.*, **47**, 2407 (1953).
39. JOHANNESSON, J. K., *Analyst*, **83**, 155 (1958).
40. MARKS, H. C., *J. New Engl. Water Works Assoc.*, **66**, 1 (1952); *Chem. Abstr.*, **46**, 5230 (1952).
41. MARKS, H. C. and co-workers, *Sewage Works*, **21**, 23 (1949).
42. MARKS, H. C. and CHAMBERLIN, S., *Anal. Chemistry*, **24**, 1885 (1952).
43. MARKS, H. C., WILLIAMS, D. B. and GLASGOW, G. A., *J. Amer. Water Assoc.*, **43**, 201 (1951).

44. MARKS, H. C. and JOINER, R. R., *Anal. Chemistry*, **20**, 1197 (1948).
45. MARKS, H. C., JOINER, R. R. and STRANDSKOV, Z. B., *Water Sewage*, **95**, 175 (1948).
46. DUŠINSKÝ, G. and GRUNTOVÁ, Z., *Potentiometrické titrácie vo farmácii a príbuzných odboroch*, p. 258, Nakl. SAV, Bratislava 1956.
47. SOMEYA, K., *Science Reports Tohoku Imp. Univ.*, **19**, 124 (1930).
48. SONGINA, O. A., *Trudy Komissii Anal. Khim. Akad. Nauk SSSR, Otdel. Khim. Nauk*, **4**, 116 (1952).
49. SONGINA, O. A. and VOYLOSHNIKOVA, A. P., *Zavodskaya Lab.*, **24**, 1331 (1958).
50. TOMÍČEK, O., *Potentiometrické titráce*, p. 132, Nakl. JČMF, Prague 1941.
51. RYABCHIKOV, D. I., *Zhur. Anal. Khim.*, **1**, 47 (1946).
52. ČŮTA, F., *Collection (Czech. Chem. Comm.)*, **6**, 383 (1934); **7**, 33 (1935).
53. TOMÍČEK, O., STODOLOVÁ, A., and HEŘMAN, M., *Chem. Listy*, **47**, 516 (1953).
54. ALMÁSSY, G. and DESZÖ, I., *Magyar Kém. Folyóirat*, **61**, 300 (1955).
55. ALMÁSSY, G. and DESZÖ, I., *Acta Chim. Acad. Sci. Hung.*, **8**, 59 (1957).
55a. STANKOVIČ, L., *Chem. Zvesti*, **14**, 275 (1960).
56. MOLT, E. L., *Chem. Wbl.*, **52**, 265 (1956).
57. CHARLOT, G., *Analyt. Chim. Acta*, **1**, 314 (1947).
58. LEUTWEIN, F., *Z. anal. Chem.*, **120**, 233 (1940).
59. RUŽIČKA, E., *Collection (Czech. Chem. Comm.)*, **24**, 2062 (1959).
60. RUŽIČKA, E., *Chem. Listy*, **52**, 1716 (1958).
61. RUŽIČKA, E., *Collection (Czech. Chem. Comm.)*, **25**, 1691 (1960).
62. TAKAHASHI, M., *Bull. Soc. Chem. Japan*, **33**, 178 (1960).
63. ABLOV, A. V. and BATYR, D. G., *Zhur. Anal. Khim.*, **15**, 734 (1960).
64. SANT, S. B. and SANT, B. R., *Anal. Chemistry*, **31**, 1879 (1959).
65. ISSA, I. M. and ALLAM, M. G. E., *Z. anal. Chem.*, **175**, 421 (1960).
66. POLAK, H. L., *Z. anal. Chem.*, **176**, 34 (1960).
67. ISSA, I. M. and ALLAM, M. G. E., *Z. anal. Chem.*, **175**, 103 (1960).
68. POLAK, H. L. and DEN BOEF, G., *Z. anal. Chem.*, **175**, 265 (1960).
69. DEN BOEF, G. and POLAK, H. L., *Talanta*, **9**, 271 (1962).
70. MURRAY, K. A. and KRIEGE, M., *South African Ind. Chemist*, **9**, 110 (1955).
71. SOMEYA, K., *Z. anorg. Chem.*, **187**, 354 (1930).
72. LENHER, L., *J. Amer. Chem. Soc.*, **35**, 733 (1913).
73. FURMAN, N. H. and FENTON, JR., A. J., *Anal. Chemistry*, **32**, 745 (1960).
74. WOLF, L., FRANZ, H. and HENNIG, H., *Z. Chem.*, **1**, 27 (1960); **2**, 220 (1961).
75. STANKOVIČ, L., *Chem. Zvesti*, **14**, 275 (1960).
76. ZIGALKINA, T. S. and CHERKASSOV, A. J., *Zhur. Anal. Khim.*, **16**, 505 (1961).
77. IWANAGA, R., *Bull. Chem. Soc. Japan*, **35**, 247 (1962).
78. RUŽIČKA, E. and KOTOUČEK, M., *Z. anal. Chem.*, **183**, 351 (1961).
79. POLAK, H. L., FRONK, H. T. and DEN BOEF, G., *Z. anal. Chem.*, **189**, 411 (1962).
80. POLAK, H. L., FRONK, H. T. and DEN BOEF, G., *Z. anal. Chem.*, **189**, 377 (1962).
81. SPAETH, E. E., BAPTIST, V. H. and ROBERTS, M., *Anal. Chemistry*, **34**, 1342 (1962).
82. DONOSO, G., DOLEŽAL, J. and ZÝKA, J., *Analyt. Chim. Acta*, **29**, 70 (1963).
83. JOVANOVIĆ, M. S. and KALINIĆ, M. P., *Bull. Soc. Chim. Beograd*, **27**, 289 (1962).
84. SPACU, P., GHEORGHIU, C. and PARALESCU, J., *Z. anal. Chem.*, **195**, 321 (1963).

85. SPACU, P., BREZEANU, P. M., GHEORGHIU, C. and CRISTUREAN, E., *Rev. Chim. A.S.I.I.*, **12**, 723 (1961).
86. HUBER, C. O. and STAPELFELDT, H. E., *Anal. Chem.*, **36**, 315 (1964).
87. CEAUSESCU, D., *Stud. Cercet. Chim. Cluj*, **8**, 281 (1957); *Anal. Abstr.* No. 161 (1959).
88. GALLO, G. G., CHIESA, L. and SENSI, P., *Farmaco (Pavia), Ediz. Sci.*, **17**, 668 (1962).
89. SINGH, B., VERMA, B. C. and SAFRA, R. L., *Z. anal. Chem.*, **196**, 323 (1963).
90. BUJÁKOVÁ, A. and ZÝKA, J., Unpublished results.
91. ASHWORTH, M. R. F., *Titrimetric Organic Analysis*, I, Direct Methods Chemical Analysis. Interscience, New York, 1964.
92. ZHDANOV, A. K. and RYAZANOVA, T. M., *Uzbekh. Khim. Zhur.*, 18 (1962); *Anal. Abstr.*, **11**, 917 (1964).

AUTHOR INDEX

Page references in italic are to names occurring in the text.

ABDALLA, A. 54
ABDEL-WAHAB, M. F. 54
ABDUL AZIM, A. A. 8, 9
ABEL, E. 191
ABLOV, A. V. 198
ABUSKER, K. M. 117
ACKERMANN, L. 35
ADAMS, R. M. *18*, 27
AFANASJEV, B. N. *44*, *45*, 46
AGERWALA, V. S. 171
AGASYAN, P. A. 26
AGREN, A. 191
AHMED, G. 191
AIROLDI, R. 46
AJMAL, M. 13
ALEXEYEVA, N. N. 117
ALIMARIN, I. P. 17, 75
ALLAM, M. G. E. 9, 196, 197, 198
ALLAN, Z. J. 192
ALLES, J. A. 27
ALLINI, P. 65
ALMÁSSY, G. 198
ALUYA, N. S. 197
AMER, M. M. A. 197
AMES, S. R. 75
ANAND, V. D. 26
ANKUDIMOVA, E. A. 91, 184
ANTROPOV, V. I. 90
AQUALDO, A. 46
ARAVAMUDEN, G. 91
ARBOROSOV, D. G. 193
ARENDS, W. 50, 51
ARMIN, A. M. 144
ARNOLD, E. A. 193
ARRIBAS, S. J. 135
ARUSHEYEVA, A. A. 178
ARUTYUNYAN, A. A. 136
ASENSI MORA, G. 92
ASHWORTH, M. R. 199
ASMANOV, A. *106*, 116
ASO 54
ATANASIU, I. A. *188*, 191

ATTKINS, G. 35
AUDRIETH, L. F. 183
AUERBACH, C. 118
AULICH, M. 153
AVILOV, V. B. 159
AVRUNINA, A. M. 130
AWAD, S. A. 8, 9
AYRES, A. D. 197
AZIM, S. M. A. 46

BABUCHKIN, S. A. 135
BAGSHAVE, B. 25
BALÁSFALVY-ZERGÉNYI, M. 183
BALKRISHNAN, E. 34, 35
BANERJEE, P. CH. 153,154
BANHAM, M. F. 119
BANICK, JR., W. M. 90
BANNISTER, G. L. 129
BANYAI, E. 27
BAPAT, M. G. 26, 27, 28, 34, 35, 129, 183, 184
BAPTIST, W. H. 198
BARAKAT, M. F. 117
BARAKAT, M. Z. 54
BARCZA, L. 97
BARD, A. J. *108*, 117, 126
BARON, M. S. 64
BARTHA, L. G. 126
BATALIN, A. K. 170
BATYR, D. G. 198
BAUMBACH, H. L. 116
BAWN, C. E. G. 100
BEAVER, J. J. 197
BECCARI, E. 27
BECK, G. 17
BECK, M. T. 103
BEESON, C. 116
BEKLESKHOVA, G. R. 169
BELCHER, R. 9, 12, *30*, 34, 35, *49*, 50, 74, 75, *115*, 119, 125, 129, 135, 136, *138*, 139, 146, 178, 179

AUTHOR INDEX

BELENKIJ, L. 191
BELLEMARE, P. A. 193
BELLIDO, I. S. 25, 136
BENNEWITZ, R. 125
BERAN, P. 179
BERGSHOEFF, G. 154
BERKA, A. 8, 17, 28, 45, 46, 51, 54, 75, 81, 82, 93, 177, 178, 179, 183, 184
BERRY, A. J. 160
BEST, H. 12
BEZUGLYI, D. 178
BHATNAGAR, C. S. 193
BHATNAGAR, M. L. 118, 154, 171
BHATNAGAR, R. 154, 171
BHATTY, M. K. 35, *115*, 119
BÍLEK, P. 159
BINDER, F. 146
BINOUN, L. 50
BIOS 191
BISHOP, E. *38*, 40, 45, 46, 47, 90, 130
BITSKEI, J. *34*, 35, 129
BLAKELY, J. D. 129
BLANGEY, R. 191
BLANKA, B. 184
BLAŽEK, J. 178, 192
BLOM, J. 27
BOCK, R. 129
BODOR, E. 126, *161*, 169, 170, 171
BOEF, G. DEN *194*, 196, 198
BOEF-NUGTEREN, J. 196
BOGDANOV, S. G. 178
BOGDANOVA, V. A. 197
BOISSON, S. 75
BOKHAREVA, A. P. 178
BOLLENBACH, H. 25
BOMMER, V. H. 100
BONDARENKO, N. V. 64
BONNER, V. D. 35
BÖTTCHER, F. 153
BOTTEI, R. S. 119
BÖTTGER, K. 45
BÖTTGER, W. 45
BOWLER, W. W. 193
BOZSAI, I. 26
BRACHMANN, W. 45
BRADBURY, F. R. *133*, 135, 136
BRENNECKE, E. *97*, 117
BREZEANU, P. M. 199
BRICKER, C. E. 100
BRINKMAN, U. A. 179
BRINTZINGER, H. 116, 117, 118

BRITTON, H. T. 27
BROWNING, P. E. 28
BROWNS, E. G. 197
BRUCHHAUSEN, F. 45
BRUCKENSTEIN, S. 103
BRUNIUS, E. 177
BRUNNER, E. 100
BUBEN, F. 65
BUBL, E. 75
BÜCHNER, E. 81
BÜCHNER, K. 118
BUDĚŠÍNSKÝ, B. 27, 51, 197
BUEHRER, TH. 117
BUJÁKOVÁ, A. 199
BUKANOVA, A. E. 139
BUKHAROV, P. S. 27
BURGER, K. 51, 65
BURMEISTER, E. 116
BURNEA, I. 192
BURNEVIN, D. 170
BURRIEL-MARTI, F. 135, 136
BUSEV, A. I. *113*, 117, 118, 119, *140*, *141*, 142, 184
BUTLER, J. P. 26
BUTT, L. T. *188*, 192
BUZÁS, H. 129, 170
BUZÁS, I. *33*, 35, *101*, 103, 169, 170, 171, 183
BUZÁS, L. 129, 169

CALAMARI, J. A. 192
CALLAN, T. 97
CAMPBELL, A. W. 197
CAMPE, A. 36
CAPIZZI, F. M. 97
CARLI, B. 46
CARLSON, C. E. 130
CATTELAIN, J. 35
CEAUSESCU, D. 199
CELLINI, R. F. 97
CERANA, A. 35
CHAMBERLIN, S. 197
CHANG, YE, SIA. 26
CHARLOT, G. *23*, 26, 198
CHATTERGEE, K. C. 136
CHENG, J. A. 17
CHERKASHINA, T. V. *110*, 117, 118
CHERKASOV, A. J. 198
CHERKASOV, V. M. 92
CHERKOVNITSKAYA, I. A. 91
CHERNIKHOV, J. A. 118

AUTHOR INDEX

CHIESA, L. 171, 199
CHIRNSIDE, R. C. 25
CHLEBOVSKÝ, T. 96
CHLOPIN, N. J. 130
CHUB, L. 197
ČÍHALÍK, J. 55, 64, 184
CLARK, J. D. 46
CLASS, J. R. 129
CLAEYS, S. 36
CLULEY, J. H. 25
CLUTTERBUCK, P. W. 75
COALSTAD, S. E. 197
COLLENBERG, O. 144
COLVIN, J. H. 153, 155, 159
CONANT, J. B. 27, 153
CONN, I. 97
COOKE, W. D. 109, 117
CORBETT, J. A. 129
CORIN, M. N. 125
COSMIN, A. 192
COURTOIS, J. 74, 75, 76, 81
CRAWFORD, A. B. 90
CRIEGEE, R. 72, 75, 76, 81
CRISTUREAN, E. 199
CRITCHFIELD, F. E. 50
CROWELL, W. R. 116
CSÁNYI, I. 129
CSIK, I. 47
CUNNINGHAM, T. R. 26
CURTIS, R. C. 197
ČŮTA, F. 198
CUTHILL, R. 35
CUTTER, H. B. 153
CZAPLIŃSKI, A. 178

DAALDER, A. 196
DACHSELT, E. 191
DAGETT, W. 197
DAS GUPTA, R. N. 129
DAESS, A. M. 9, 144, 146
DAUKSHAS, K. 92, 126
DAVISON, G. A. 25
DEAHL, T. J. 100
DEAN, G. 156, 159
DESAI, M. W. 97
DESHMUKH, G. S. 22, 26, 34, 35, 45, 46, 183, 184
DESPHANDE, G. M. 97
DESZÖ, I. 198
DÉVORÉ, P. 26
DEWASNES, P. 26

DIAZ-FLORES, C. A. 92, 97
DICKENS, P. 25
DIEHL, H. 26, 142
DIEHL, K. 146
DIETMANN, H. 129
DIETZEL, R. 45
DIMROTH, O. 76, 81, 104, 116
DOBRESCU, F. 126
DOLCETTA, M. 45
DOLEŽAL, J. 12, 26, 34, 75, 81, 96, 138, 142, 178, 197, 198
DOMANGE, L. 108, 117
DONOSO, G. 198
DÖRING, TH. 117
DRAGOI, I. 47
DRĂGULESCU, C. 47, 171, 184
DRAHOŇOVSKÝ, J. 138
DRUMMOND, A. Y. 8, 9, 12
DUBRISSAY, R. 177
DUKE, F. 8
DUŠEK, O. 12
DUŠINSKÝ, G. 27, 50, 51, 192, 198
DUVAL, C. 46
DVOŘÁK, J. 178
DVOŘÁK, V. 81
DWORZAK, R. 119
DZOTTSOTI, S. 192

EASTON, W. 45
EDSBERG, R. 191
EDWARDS, E. G. 133, 135, 136
EHSAN, ALI 35
EICHELBERGER, R. L. 118
EID, S. C. 8
EKIMYAN, M. G. 13, 135, 136
ELDERIDGE, E. F. 146
ELLER, W. 177
ELLIS, C. M. 149, 154
ELOFSON, R. L. 191
ENOKI, T. 193
ERBEN, J. 192
ERDEY, L. 33, 35, 46, 75, 81, 97, 101, 103, 126, 129, 161, 169, 170, 171, 183, 197
EROMINA, Z. 92
ESHVAR, M. C. 34, 35, 45, 46
ESKEVICH, V. F. 91
ETARD, A. 12
ETTEL, V. 177
EULER, H. VON 177
EVERED, D. F. 54

AUTHOR INDEX

EWING, D. P. 146
EZERSKAYA, N. A. 170, *175*, 178

FACSKO, G. 103, 184
FARA, M. 82
FARAH, M. Y. 142
FARKAS, L. 35
FATHALLA, A. H. 9
FEIL, E. 125
FEKETE, L. 91
FEKETE-PAP, E. 183
FENTON, JR., A. J. 198
FIALKOV, J. 64
FICHTER, H. 100
FIDLER, J. *140*, *141*, 142
FIERZ-DAVID, H. E. 191
FIGAROVÁ, M. 35
FILIPI, J. 97
FILIPOVIČ, P. 34
FILONOVA, V. 46
FINKELSCHTEIN, D. N. 135
FLANIGAN, D A. 26
FLASCHKA, H. 8, 136, *138*, 139, 169
FLATT, R. *110*, *113*, 116, 117
FLEMING, T. 197
FLEURY, P. 74, 75, *76*, 81
FLORENCE, T. M. 97
FOGEL'SON, H. J. 126
FORBES, G. S. 116
FORHENCZ, M. 35
FOERSTER, F. 142, 153
FRANZ, H. 198
FREHDEN, O. 97
FREI, V. 82
FREIBERGER, F. *20*, 25
FRESENIUS, R. 125
FRESNO, C. DEL 25, 46, 159, 160
FRIEBERGER, F. 75
FRIEDMAN, A. H. 35
FRIEDRICH, B. 119
FRISONA, G. 82
FRISTER, F. 81, *104*, 116
FROLKINA, V. A. 118
FRONK, H. T. *194*, 198
FROST, H. F. 191
FRUMINA, N. S. 26
FUDGE, A. J. 119
FUJITANI, T. 170
FUREY, J. J. 26
FURMAN, N. H. *18*, 27, 64, 119, 129, 142, 159, 178, 198

GALLAI, Z. A. 96, 117, 118, 170
GALL, H. 9
GALLUS-OLENDER, J. 129
GALLO, G. G. 171, 199
GAPCHENKO, M. V. 142, 153
GARETT, G. 8
GAUCHMANN, M. S. 129
GAUDEFROY, G. 90
GAUR, J. N. 92, 93, 184
GAVANESCU, D. 47
GENGRINOVICH, A. I. *55*, 64, 125
GENTELE, I. G. 27
GERO, A. 65
GEYER, R. *110*, 117, 126, *143*, 144
GHEORGHIU, C. 198, 199
GIBBONS, D. 139, 146
GILBERT, J. M. 51
GIMESY, O. 170
GINSBERG, S. I. 139
GINZBURG, C. I. 170, 177, 179
GIUFFRE, L. 97
GLABICZ, N. 197
GLASE, J. R. 129
GLASGOW, G. A. 197
GLEU, K. 25, *128*, 129, 184
GOLBRAYKH, Z. E. 138
GOLDSTONE, N. I. 35
GOLSE, J. 35
GOODSON, A. 116
GORDON, H. T. 35
GORIN, G. 27
GORYACHEVA, G. S. *115*, 119
GORYUSHINA, V. G. *110*, 117, 118
GÖRNE, J. 125
GOTO, H. 46
GOTTFRIED, J. 118
GOTTLIEB, S. *83*, *85*, 91
GOWDA, H. S. 91, 92
GRANDSCHAMP, M. 75
GREATHOUSE, L. H. 75
GREINER, G. 129
GREKOV, A. P. 191, 193
GRIM, V. 170
GRINBERG, A. A. 138
GRÖBLER, G. 183
GROVER, K. C. *34*, 35
GRUBE, G. *104*, 116
GRUNTOVÁ, Z. *50*, 51, 192, 198
GUBELBANK, S. M. 92
GUDALO, B. 184
GUPTA, M. P. 75
GUREVICH, V. G. 92

AUTHOR INDEX

Gusev, S. I. *150*, 154
Gustavson, R. 146
Guthe, A. 144

Haase, G. 126
Haase, J. 191
Hadidy, A. E. 9
Hagedorn, H. C. 27
Hahn, F. L. 25
Hála, E. 27
Hall, A. 26
Hall, D. 100
Hall, L. 26
Hall, W. T. 130
Haller, J. 129
Hamdy, M. 9
Handerson, W. L. 197
Hanzlík, E. 45
Hara, S. 160
Harpe, C. de la 192
Hartley, A. M. 26
Hartley, A. W. 197
Hartmann, H. 9
Hashmi 35
Hatfield, M. R. *105*, 117
Hayashi, S. 26
Hazel, F. 117
Head, F. S. 75
Heising, G. B. 64
Helbig, W. 119
Henermann, R. F. 184
Hennig, H. 198
Henze, G. *143*, 144
Herescu, S. 171
Herigel, E. 129
Heřman, M. 160, 198
Herringshow, J. F. *156*, 159
Heubner, C. F. 75
Heusinger, W. 50
Heyrovský, A. 50
Higginson, W. C. E. 100
Hildebrand, H. 17
Hillson, H. D. 129
Hinton, H. D. 46
Hirsjärvi, V. P. 136
Hiscox, D. J. 24, 27
Hobson, J. D. 25
Höfer, J. 178
Hofmann, P. 179
Hofman-Bang, N. 100
Höhne, R. 97

Hokoyama, E. 97
Holeček, V. 96
Höleman H. 117, 126
Holder, G. 130
Holluta, J. 8
Holst, G. 184
Höltje, R. *110*, 117, 126
Holzman, G. 46
Horrobin, S. 97
Hostetter, J. C. 125
Hrdý, O. 192
Hubata, R. 192
Hudec, P. 184
Hudlický, K. 54
Hume, D. N. 25
Huwy, C. S. 125

Ianku, S. 193
Ibadov, A. J. 64
Iftikar Ali, S. 142
Iglesies Castano, J. M. 129
Ikegami, H. 12, 13
Iljin, W. 35
Illman, H. *105*, 117
Imhof, J. G. 8, 9
Inczédy, I. 35, 103
Ingbermann, A. K. 51
Inukai, K. 192
Ionesco-Matiu, A. 26, 27
Ionescu, A. 27
Iranzo, I. R. 96
Ishibashi, M. 12, 179
Issa, I. M. 8, 9, 144, 146, 196, 197, 198
Issa, R. M. 8, 9
Iwamatsu, H. 97
Iwanaga, R. 198

Jablczynski, K. 116
Jach, Z. 170
Jackson, D. T. 197
Jacob, W. 75
Jacobs, M. B. 35
Jacobson, E. 177
Jahns, S. 100
Jakovenko, G. D. 129
Jamieson, A. R. 178
Janata, J. 81
Jančík, F. 65
Jander, G. 74, 135

JACQUELAIN, C. R. 77, 81
JAŠEK, M. 34, 35
JELLINEK, K. 29, 34, 97, 103
JENNINGS, V. J. 38, 40, 45, 46, 47
JENSEN, B. N. 27
JENŠOVSKÝ, L. 17
JERMAN, L. 96
JERSCHKEWICZ, H. G. 117
JIMENO, S. A. 129
JOB, A. 100, 103
JOHANNESSON, J. K. 197
JOHNSON, E. H. 65
JOINER, R. R. 198
JONES, D. V. 191
JONES, G. 155, 153, 159
JONES, J. R. 90, 93
JOSHI, M. K. 22, 26, 27
JOVTSCHEFF, A. 54
JUCKER, H. 114, 119
JUDEVITSCH, E. A. 64
JUHLIN, O. 50
JUNG, Z. 192
JUNGERMANN, E. 46
JUREČEK, M. 191
JUST, H. 160

KACZKOWSKI, J. 27
KADIČ, K. 9
KADYROV, J. K. 64
KAGAN, F. E. 64, 65
KALENCHUK, G. E. 118
KALIDAS, CH. 92
KALINIČ, M. P. 198
KALMYKOVA, N. V. 126
KALNÝ, J. 26
KALZITA, J. 46
KAMMORI, O. 91
KANKANYAN, A. G. 47
KAO, A. L. 169
KAO, S. S. 117, 118
KÁPLÁR, L. 46
KAPUR, S. R. 46
KARANTASSIS, T. 126
KARLYSHEVA, K. F. 97
KARRER, P. 197
KARSAY, A. 169
KARSTEN, P. 154
KASA, I. 171
KASHYAP, G. P. 64, 65
KATTHÄN, W. 128, 129, 184

KAUFMANN, H. P. 48, 50, 51
KEILY, H. J. 25
KELLER, H. 197
KELLNER, A. 129, 183
KETOVA, L. A. 150, 154
KEYWORTH, D. A. 17
KHADEEV, V. A. 45, 64, 129, 169
KHAJASI 54
KHAKIMOVA, V. K. 26
KHALIFA, H. 9, 196
KHATUM, S. 35
KHU-CZI-FAN 97
KHUNDHAR, M. H. 35
KIBA, T. 119
KIBOKU, M. 22, 25, 26, 35
KIES, B. I. 154
KILLHEFFER, V. J. 46
KING, S. 82
KIRTCHIK, H. 26
KISS, S. A. 129
KITAGAWA, H. 13
KITAGAWA, T. 125
KITAHARA, S. 12, 13
KITHASHIMA, S. 100
KIURA, M. 160
KLEFLIN, Z. 192
KLIMENKO, J. V. 91, 93
KLUH, I. 12, 178
KNUDSON, C. M. 146
KNYAZEVA, G. V. 178
KNYAZEVA, R. N. 74
KOCH, W. 177
KOCHORYAN, A. T. 13
KOLTAI, L. 171
KOLTHOFF, I. M. 9, 25, 29, 30, 31, 34, 35, 49, 50, 65, 74, 75, 96, 103, 125, 129, 130, 177, 197
KOLYGA, S. 118
KOMÁREK, K. 25
KOMAROVA, L. A. 91
KOMAROVSKI, A. S. 46
KONDO, G. 27
KONSTANTINOV 95, 97
KOPANICA, M. 26
KÖRBL, J. 27, 65
KORENMAN, I. M. 46
KORNEVA, L. E. 64
KOROLEVA, V. I. 193
KÖSZEGI, D. 31, 35
KOTOUČEK, M. 171, 198
KOVÁCS, K. M. 125
KOVÁTS, J. 17

AUTHOR INDEX

Kozlov, N. A. *163*, 169
Kraft, I. 81
Kratochvil, B. 26, 142
Krause, H. 119
Krebs, P. 103
Krejzová, E. 177, 178
Kresteff, W. 29, 34
Kriege, M. 198
Křížová, J. 12, 178
Krüll, F. 126
Kryuchkova, G. N. 135
Kubina, H. 64
Kubišta, Z. 25
Kubrakova, A. I. 64
Kucharský, J. 192
Kühn, W. 34, 97, 103
Kukhment, M. L. 125
Kulkarni, V. P. 35
Kulvarskaja, R. M. 197
Kunitz, H. 197
Kupka, F. 178
Kurtenacker, H. 90, 92
Kyuno, M. E. 100

Laar, B. van 196
Ládányi, L. 65
Ladha, G. S. 125
Lafuente, E. de 159
Laitinen, H. A. 34
Land, H. 9
Lang, R. *83, 85*, 90, 91, 92, 130
Lange, J. 74, 75
Lantos, J. 125
Lastovskii, R. P. 191, 192
Latimer, W. M. 17, *55*, 64, 138, 183
Laur, A. *31*, 35, 130
Lee, H. R. 91
Lee, T. 46
Lehmann, G. 9
Lemke, O. 46
Lenher, I. 198
Leonard, G. W. 25
Leutwein, F. 198
Levesley, P. 9, 12
Levy, B. *49*, 50
Levy, F. 45
Lewin, M. 35
Lhota, Z. 27
Lhotka, J. F. 12
Liebmann, H. 197
Li Gyn *113*, 117, 118, 142

Lingane, J. J. 26, *111*, 116, 117, 118, 125, 129
Listek, S. S. 129
Littler, J. S. *90*, 93
Litvinenko, L. M. 191, 193
Loeffler, L. J. 100
Longstaff, J. V. 50
López, J. A. 97
Lord, S. 191
Loub, J. 82
Lowy, A. 192
Lucena-Conde, F. 25, 135, 136
Luchman, E. 25
Lucka, B. 51
Lunge, G. 191, 193
Luther, R. 146
Lux, H. *105*, 117

Maas, K. 153, 160
Maassen, G. 25
Macara, T. 46
Macdonald, A. M. G. 93, 154, 160
Mahan, W. A. 129
Mairlot, E. 159, 160
Majumdar, R. 118
Maki, I. 192
Maksimyuk, E. A. 97, 138, 170
Malaprade, L. 66, *71*, 74, 75
Malatesta, L. 17
Malik, W. U. 13, 117, 142
Malkina, L. A. 118
Manalo, G. D. 25, 27, 90, *156*, 159
Manchot, W. 48, *49*, 50
Mandelík, J. *157*, 159
Mangu, M. 192
Mankotia, M. S. 171
Mansurkhanova, I. 64
Manzoor, Elahi 35
Marakhova, M. S. 193
Markgraf, H. 100
Marks, H. C. 129, 197, 198
Martin, H. 96, 97
Martinchenko, I. *113*, 118
Marxová, I. 184
Mathur, N. K. 154, 171, 184, *191*, 192, 193
Matrka, M. 17, 50, 97, *151*, 153, 154, 191, 192, 193
Matsuo, E. 136
Matsuyama, G. 9, 34, 50, 75, 125
Mayer, V. 159

MÁZOR, L. 75, 97, 197
MCCOY, H. N. 146
MCMILLAN, A. 45
MCNABB, W. M. 117
MCNEIGHT, S. A. 193
MDIVANI 126
MECHERLY, P. M. 191
MEEHAN, E. J. 103
MEHROTRA, R. C. 35, 117, 118, 119, 154
MEISEL, T. 81
MEITES, L. 153
MELAMED, S. I. 74
MELIKSETYAN, A. P. 135
MELOJAN, P. G. 47
MENYHÁRTH, P. 9
MERCK, E. 46, 105
MERRIT, L. L. 75
METZL, A. 100
MEULEN, J. H. VAN DER 50
MEYER, G. 125
MEYER, J. 153
MICHIE, A. E. 146
MIKHAIL, S. Z. 142
MIKKELSON, V. J. 27
MIKULE, V. 192
MILAZZO, G. 177, 178
MILENKO, V. 171
MILLER, C. O. 129
MINATO, H. 103
MINCZEWSKI, J. 113, 118, 197
MINGES, R. 184
MISS, A. 193
MITRĂNESCU, M. 184
MITTAL, R. K. 154
MIURA, K. 26
MIYAKE, S. 126
MOLDAN, B. 96, 142, 197
MOLNÁR, I. G. 129, 183
MOLOTKOVA, A. S. 9
MOLT, E. L. 198
MONNIER, D. 193
MORALES, A. 82
MORACHEVSKII, J. V. 91
MORGULIS, S. 35
MORI, M. 100
MORISAKA, K. 193
MORITA, S. 91
MORREN, L. 34
MOŠNA, P. 184
MOUSA, A. A. 116
MOUSSA. G. M. 54

MRÁZ, V. 177, 178
MUHR, G. 191
MÜLLER, E. 27, 74, 75, 125, 126, 129, 130, 138, 160, 191
MÜLLER, R. 46
MÜLLER, R. H. 197
MURAKAMI, T. 27, 126
MURAKI, I. 111, 113, 116, 118
MURRAY, K. A. 198
MURRAY, W. N 142
MURTAZAEV, A. M. 64
MURTHI, R. V. V. S. 92
MURTHY, A. R. V. 46, 47
MURTHY, T. K. S. 97
MURTHY VASUDAVA, A. R. 178
MURTY, B. V. S. R. 91, 93, 160
MURTY, S. V. S. S. 46
MUSHA, S. 45, 125
MUSTAFIN, I. S. 26
MUŽÍK, F. 192
MYERS, L. S. 35

NABARS, G. M. 35
NAGY, G. 91
NAGY, M. 183
NAHAN, R. K. 9
NARAIN DAU 192, 193
NARUSHKEVICHIUS, L. 92
NASCISMENTO, R. 97
NATARAJAN, R. 97
NAVRÁTIL, F. 17, 50, 97, 193
NAZARENKO, I. I. 118
NAZARENKO, V. A. 91
NELJUBINA, A. 192
NĚMEC, I. 81
NEMODRUG, A. A. 97, 154
NÉRSESOVA, S. V. 138
NESSLE, G. J. 97
NEUMANN, B. 125
NICOLET, B. N. 75
NIEDERL, J. B. 197
NIEDRACH, L. 118
NIEMANN, G. 46
NIERIKER, R. 97, 117
NIKOLAYEVA, E. R. 114, 118
NINO, N. 193
NOLL, A. 45
NORKUS, P. K. 36, 129
NORTHROP, J. P. 50
NOSENKOVA, N. G. 92
NOVÁK, P. 45

NOVÁK, V. 177
NOVOTNÝ, J. 159, 197
NOYES, A. A. 100, 125
NUSSBAUM, R. 35
NUTTEN, J. 179

OBERHAUSER, F. 48, 49, 50
OGAWA, K. 45, 170
OGG, B. A. 183
OKÁČ, A. 126
OLENOVICH, N. L. 85, 91
OLSSON, O. 144
ONO, K. 97
ORESHKO, V. F. 97, 154
ORLOWSKI, M. 50, 51
OSCHATZ, F. 117
OVANESIAN, A. 47
OVSEPYAN, E. N. 135, 136

PACOVSKÝ, J. 170
PAIS, J. 193
PALLAUD, R. 119
PALMER, H. E. 25, 28
PANWAR, K. S. 93, 184
PAOLONI, L. 177
PAP, E. 184
PAP, I. B. 170
PÁPAY, M. 171
PARALESCU, J. 198
PARKS, H. 159
PARSONS, J. L. 197
PATAKI, L. 193
PAUL, R. CH. 96, 197
PAVELKA, O. 8
PAVLÍKOVÁ, M. 178
PECSOK, R. L. 116, 118
PERLIN, A. S. 82
PERMULTES-HEYMANN, B. 50
PESHKOV, I. A. 178
PESHKOVA, V. M. 93, 117, 170
PETRASHEN, V. I. 91
PETROPOULOS, A. G. 108, 117
PETROVA, V. A. 92
PHILLIPS, J. 192
PHILLIPS, L. 27
PICCARDI, G. 65
PICCINI, A. 153
PINXTEREN, J. A. C. VAN 27
PODE, J. S. F. 9
PODSTATA, J. 191

PODOLSKAYA, V. I. 118
POGREBINSKAYA, M. I. 135
POLAK, H. L. 194, 198
POLLARD, W. B. 177, 178, 179
PÓLOS, L. 170
POLUBNAYA, E. T. 27
PONOMAREVA, L. K. 118, 160
PONZINI, S. 193
POPESCO, A. 26
POPOVICH, D. V. 184
POSDNIAKOWA, A. 50
POSPÍŠIL, J. 177
POSTIS, J. DE 105, 117
POETHKE, W. 45
PRASAD, B. B. 171
PREBLUDA, R. J. 159
PRESTON, J. M. 129
PŘIBIL, R. 26, 97, 98, 109, 117, 144
PŘIPLATOVÁ, E. 26
PROCHÁZKOVÁ, Z. 197
PROČKE, O. 8
PROFFIT, P. M. C. 25
PROKOF'EVA, I. V. 139, 170, 177, 179
PROKOPCHIK, J. 129
PROSHENKOVA, N. N. 92
PROTSHENKO, E. G. 193
PRUNNER, G. 27
PSHENICYN, N. K. 139, 170, 175, 177, 178, 179
PTITSYN, B. V. 138, 163, 169
PUGH, M. 136
PUNGOR, E. 65
PUZDRENKOVA, I. V. 17, 75

RADHAKRISHNAMURTY, CH. 91
RÁDY, G. 81, 169, 170
RAMANJANEYULU, J. V. S. 91
RANK, B. 81
RAO, B. K. 90, 91, 92, 97
RAO, B. K. S. 125
RAO, B. V. 92
RAO, G. G. 86, 90, 91, 92, 93, 97, 98, 160, 169, 170
RAO, J. G. 91
RAO, K. B. 92
RAO, M. S. V. 45
RAO, N. V. 170
RAO, P. G. 178
RAO, P. V. 91
RAO, S. P. 92, 184, 192
RAO, S. R. 46

RAO, U. V. 91
RAO, V. B. 92
RAO, V. M. 92
RAO, V. N. 169
RAO, V. R. S. 47
RAPAPORT, L. I. 65
RAPPAPORT, F. 74
RATHI, H. S. 92
RAZNATOVSKA, V. F. 65
REDMOND, J. C. 26
RECHKINA, L. G. 45
REEVES, R. A. 81
REHMAN, R. A. 46
REICHEL, J. 126
REICHERT, J. S. 193
REIFER, I. 74
REILLEY, CH. *18*, 27
REISHAKHRIT, L. S. 178, 179
REITH, H. F. 193
REUTER, F. 75
REVERDIN, F. 192
ŘEZÁČ, Z. 9, 35, 178
RICHTER, E. 35
RICHTER, H. W. 116
RIENÄCKER, G. 116, 117
RILEY, R. F. 27, 35
RIUS, A. 92, 96, 97
RIZK, H. A. M. 117
RJANICHEVA, M. I. 91
ROA, S. O. 193
ROBERTS, H. S. 125
ROBERTS, M. 198
ROBIN, J. 26
ROCCA, J. P. LA 192
RODIS, F. 116, 117
ROGERS, L. 191
ROSCOE, H. E. 154
ROSENKRANC, O. 178
ROSENINSKY, D. R. 100
RÖSSLER, S. 75
ROST, B. 117, 118
ROSTED, C. O. 27
ROSTOVTSEVA, J. 193
ROTH, P. B. 192
ROTHEMAN, M. 35
ROTTOVÁ, O. 178
RUDEL, H. W. 193
RUFF, O. 9
RUGGIERI, R. 193
RUPP, E. 45
RUSSEL, A. S. 154
RUSSO, C. 97

RUTTER, T. F. 153
RUŽIČKA, E. 171, 198
RŮŽIČKA, J. 64
RYABCHIKOV, D. J. 118, 138, 139, 177, 178, 198
RYAZANOVA, T. M. 103, 171, 179, 199
RYBA, O. 27
RYBÁŘ, D. 191

SABOURIN, R. G. 193
SACKUR, O. 8
SADR, M. M. 54
SADUSK, JR., J. F. 116
SAFRA, R. L. 199
SÁGNER, Z. 50, 97, *151*, 153, 154, 191
SAINT-MAXEN, A. 177
SAITO, K. *11*, 12, *54*, 91
SALOVIUS, B. 136
SAMEK, B. 45
SANDELL, E. B. 129
ŠANDL, Z. 26, 27
SANDVED, K. 144
SANT, B. R. 27, 28, 193, 198
SANT, S. B. 25, 28, 184, 198
SARAN, M. S. 65
SARMA, L. S. 92
SASTRI, M. N. *86*, 91, 92, 93
SATO, N. *11*, 12
SAXENA, R. S. 193
SCAGLIARINI, S. 26
SCHACH, C. I. 65
SCHÄFER, H. 35
SCHELL, C. 100
SCHIEFERDECKER, W. 117, 118
SCHIEMANN, G. 45
SCHLECHT, I. *104*, 116
SCHLEICHER, A. 129
SCHLESINGER, H. J. 8
SCHLOFFER, F. *109*, 117
SCHLÜTTING, W. 96
SCHNEER, A. 9
SCHOLEFIELD, F. 129
SCHOLTEN, H. G. 191
SCHORMÜLLER, J. 97
SCHULEK, E. 9, 51, 65
SCHULTZ, A. 191
SCHUPP, O. 117
SCHVARZBURD, M. M. 65
SCHWEIBOLD, J. 74
SCHWEIZER, R. 81

AUTHOR INDEX 211

SCHWICKER, A. 35, 74
SCOTT, P. G. 192
SEASE, J. W. 46
SEETHARANARAYA, S. 97, 98
SEKERKA, I. 34, 35
SEKIGUCHI, H. 191
SENSI, G. 97, 171, 199
SERGIENKO, V. A. *169*, 170
SEVEARINGEN, F. H. 26
SHABARIN, S. K. 178
SHAKER, M. 54
SHAMY, H. K. EL- *110*, 116, 117, 118, 142, 146
SHARADA, K. 47
SHARMA, B. 27, 129
SHARMA, R. S. 193
SHATKO, P. P. *107*, 116, 118
SHEHAB, S. K. 54
SHEINTSIS, O. G. 153
SHEKA, J. A. 97
SHEMYAKIN, M. 193
SHERIF, I. M. EL- 8, 146
SHIBATA, M. 100
SHIBATA, N. 13
SHIGEMATSU, T. 12, 179
SHIMKO, A. *113*, 118
SHIMOMURA, S. 27
SHINN, L. 75
SHLYAKMAN, M. J. 118
SHUSHIN, M. V. 170
SHVAYKA, O. P. 193
SIEBLER, G. 45
SIEGER, H. 9
SIEMS, H. B. 8
SIERRA, F. 92
SILAEVA, E. V. *86*, 91
ŠIMEK, M. 126
SIMON, J. 27, 50, 51
ŠIMON, J. 26
SIMON, V. 34, 75, 170, 177, 178
SIMON-FIALA, J. 126
SINGER, E. 118
SINGER, K. 50
SINGH, A. 9, 13, 46, 75, 197
SINGH, B. 9, *33*, 35, 45, 46, 47, 51, 64, 65, 75, 90, 91, 92, 171, 191, 197, 199
SINGH, D. 171
SING, G. 9, 46
SING, I. 197
SINGH, M. 46, 51
SINGH, R. 90, 92
SINGH, R. P. 47
SINGH, S. 35, *38*, 91, 197
SINGH, V. B. 171
SINN, V. 34, 129
SION, H. 36
SIPOSS, G. 170
SIRANSI 54
SITARAMIAH, G. 193
SKOOG, D. A. 90, 92
ŠKRAMOVSKÝ, S. 193, 197
SKŘIVAN, V. 191
SLAVÍK, J. *113*, 116
ŠLOUF, A. 192
SMETANA, B. 153
SMID, H. 191
SMITH, G. F. *74*, 75, 90
SMITH, J. R. 46
SMITH, L. I. 197
SNELDERS, H. A. M. 179
SODOMKA, J. 51
SOKOLOV, I. 126, 191
SOLYMOSI, F. *21*, 25, 26, 27, 28, 47, 129, 130
SOMEYA, K. 25, 118, 198
SOMIDEVAMMA, G. 91, 92
SOMMER, F. *110*, *113*, 116, 117
SONGINA, O. A. 8, 198
SONI, S. K. 47
SOOD, K. CH. 45, 46
SOSNOWSKII, B. A. 126
SPACU, P. 47, 198, 199
SPAETH, E. E. 198
SPENGLER, G. A. 17
SPILLANE, I. J. 197
SPURNÝ, K. 96
STAGG, H. E. *188*, 192
STAHN, R. 116
STAMM, H. 8, 9
STANKOVIČ, I. 198
STAN SUCIN, M. 170
STAPELFELDT, H. E. 199
STEDEFEDER, J. 96
STEFANOWSKII, V. F. 199, 129
STEIN, W. 125, 126
STEINER, K. 191
STEJSKAL, Z. 192
STENGER, V. A. 9, *29*, *31*, 34, 50, 75, 125, 129
ŠTĚPANOV, B. I. *169*, 170
STEPHAN, W. J. 178
STEPIN, V. V. *86*, 91, 92
STEWART, J. J. 142

STODOLOVÁ, A.   160, 198
STONE, H. W.   116, 118
STONE, K. G.   17, 191
STOUT, I. E.   159
STRANDSKOV, Z. B.   198
STRICKS, W.   34, 197
STRUBL, R.   96
SUÁREZ ACOSTA, R.   118
SUCHARDA, B.   45
SUCHOMELOVÁ, L.   81, 138
SUGÁR, E.   *120*, 125, 184
SUGÁR-NOVÁK, E.   183
SUKHOBOKOVA, N. S.   178, 179
ŠUMANOVIČ, K.   192
SUPRUN, P. P.   65
SURANOVA, Z. P.   *85*, 91, 92
SURYANARAYANA, M.   91, 97
SUSEELA, B.   26, 184
SUŠIĆ, M.   171
SUZUKI, T.   191, 193
SVACH, M.   170
SVEHLA, G.   169, 170, 171
ŠVESTKA, L.   *109*, 117
SWENN, S.   100
SWETSER, P. B.   100
SWIFT, E. H.   46
SYMONS, M. C. R.   8
SYROKOMSKII, V. S.   *66*, 74, *83*, *87*, 90, 91, 92, 93, *109*, 117, 159
SZABADVÁRY, F.   170
SZABÓ, CH.   129
SZABÓ, Z.   125
SZABÓ, Z. G.   *120*, 125, 126, 129
SZARVAS, P.   125
SZEBELLEDY, L.   129
SZEKERES, L.   *127*, *128*, 129, 183, 184

TABORSKA, H.   51
TACHIKAWA, T.   192
TAEGENER, W.   8
TAI, S. K.   117, 118
TAKAHASHI, M.   198
TAKAHASHI, T.   126
TANINO, K.   12, 13
TANDON, J. P.   117, 118, 119, 154
TÄNZLER, K.   138
TARAYAN, V. M.   13, 135, 136
TARTAR, H. V.   125
TATSUZAVA, M.   35
TATWAWADI, S. V.   26, 27, 28, 135
TAUER, Z.   197

TÄUFEL, K.   45
TEODORESCU, G.   47
TERADA, K.   119
TEREBOVÁ, K.   64, 184
TERENTYEV, A. P.   *115*, 119
THORNTON, W. M.   116
THUN, H.   36
THYAGARAJAN, B. S.   28
TICHÝ, M.   178
TIPCOVA, V. G.   170
TODD, N.   35
TOMÍČEK, O.   8, *20*, *22*, 25, 26, 27, *33*, 34, 35, 45, 50, 64, 77, 81, 96, 159, 160, 178, 198
TOMLINSON, H. M.   119
TOMSICEK, W. J.   25
TONGBERG, C. O.   27
TOUHEY, W. O.   26
TOURKY, A. R.   116, 142, 144, 146
TOUSSAINT, L.   129
TOUŽÍN, J.   184
TOWNEND, J.   12
TOYBAEV, B. K.   8
TRAUBE, W.   116
TRDLIČKA, Z.   178
TREADWELL, W. D.   97, 117, 129
TRIBALAT, S.   118
TROFIMOVA, S. G.   136
TROKOWICZ, J.   178, 191
TRUSOV, J. P.   136
TRZEBIATOWSKI, W.   126
TSCHEPELEWETZKY, M.   50
TSENG, J. A.   75
TSUBAKI, I.   12, 13, 90, 91, 160
TUN TAO   192
TURKIEWICZ, E.   126
TYLOVÁ, M.   27

UBBELOHDE, A. R. J. P.   *10*, 12
UDALCOVA, N. I.   91
UENO, K.   192
UENO, S.   191
UOSUKAINEN, M.   136
UPOR, E. L.   91
URALSKAYA, A. V.   45
URBANSKI, S.   91
USATENKO, J. I.   169
UZEL, R.   144

VALCHA, J.   50, 64, 77, 81, 178

AUTHOR INDEX

Valcha, Z. 178
Valdés, L. 25, 159
Vándorffy, M. T. 171
Vaníčková, E. 51
Varga, A. 26
Vargolici, V. 27
Varma, A. 92, 171
Vasil'ev, D. 126
Vasiliev, R. 192
Vasina, N. T. 118
Vaubel, W. 191
Vavrejnová, D. 64
Vaysman, G. A. 184
Velculescu, A. I. *188*, 191
Veltický, B. 159
Venc, S. 193
Venkatamma, N. C. 91
Verdi-Zade, A. A. 92
Verloop, E. 27
Verma, B. C. 13, 65, 199
Viczian, B. 129
Vidic, E. 9
Vigh, K. 169, 171
Vinkovetskaya, S. J. 91
Vlasák, F. 97, 197
Vogel, A. I. *149*, 154
Voyloshnikova, A. P. 198
Voráček, J. 64
Vorlíček, J. 35
Vorobeva, V. A. 178
Vorožcov, N. N. 192, 193
Vortman, G. 146
Votiss, M. 17
Všetečka, L. 46
Vukasinovich, J. I. 184
Vulterin, J. 26, 27, 28, 75, 93, 103, 154, 160, 179, 183, 184, 191
Vydra, F. 97, 98

Wahab, M. F. 54
Wakkad, S. E. S. El– 117
Wallace, jr., J. H. 178
Waddil, H. C. 27
Wardlaw, W. 142
Warynski, M. 126
Wasilewska, D. 27
Watanabe, S. 91
Waters, K. L. 192
Waters, W. A. 8, 9, 12, 75, 90, 93
Wegelin, E. 74
Weinmann, H. 74

Weiss, G. 97
Weiss, J. 125, 191
Weiss, L. 9
Wermescher, B. 192
Werner, W. 81
Wesly, W. 129
West, D. M. 90, 92
West, T. S. 12, 135, 136, 139, 146
Whiple, E. R. 160
Whitaker, G. C. 159
White, A. G. 100
Whitmann, J. L. 197
Whitmoyer, R. B. 27
Wickham, D. G. 160
Wiercinski, J. 118
Wilie, E. 35
Willard, H. H. 25, 27, 64, 75, 90, 100, 146, *156*, 160, *190*, 193
Willeford, jr., B. R. 197
Williams, D. B. 197
Wilson, A. D. 91
Wilson, R. S. 178
Winefordner, J. D. 25
Winogradoff, L. 97
Witry-Schwachtgen, G. 154
Wittmann, G. 22, 26
Wodkiewicz, I. 118
Wolf, F. 45
Wolf, L. 198
Wölfel, K. *34*, 35
Wolfmann, H. 100
Wood, A. J. 119
Wormell, R. L. 142
Woermer, D. E. 34
Wulff, I. 100
Würdig, G. 97

Xanthakos, T. S. 100

Yamazaki, T. S. 100
Yardley, J. T. 25
Yokosuka, S. 26
Yoshimura, Ch. 35, 170
Yoshimura, T. 93
Yost, D. M. 35
Young, J. H. 129
Young, P. 160, *190*, 193
Young, R. S. 26
Young, S. W. 126
Yovanovich, M. S. 184, 198

YU-HENG-SUI 97
YUTEMA, L. F. 197

ZAIMIS, P. 9, 11, 118, 130
ZÁLAY, E. 126
ZANKO, A. M. 118, 119, 130
ZARINSKII, V. A. 118
ZAVAGYL, H. 169
ZAVAROV, G. V. 129
ZAYAN, S. EL-DIN 118, 146
ZAYKOVSKII, F. V. 170
ZELAWSKI, W. 27
ZELINGHER, R. 170
ZERGÉNYI-BALÁS, G. 129
ZHARKOVA, Z. P. 178
ZHATCHEVA, V. J. 178
ZHDANOV, A. K. 45, 64, 103, 129, 169, 171, 179, 199
ZIGALKINA, T. S. 198
ZINTL, E. *109*, 116, 117, 118, 119, 130
ZOLOTAVIN, V. L. 160
ZOLOTUKHIN, V. K. 9
ZORINOVA, M. D. 192
ZUMAN, P. 27, 197
ZVYAGINZEV, O. E. 178
ZWAHLEN, P. 193
ZWENIGORODSKAYA, V. M. 91
ZWEŘINA, J. 130
ZÝKA, J. 12, 26, 27, 46, 51, 54, 75, 81, 82, 93, 96, 103, 138, 142, 177, 178, 179, 183, 184, 191, 192, 197, 198, 199

SUBJECT INDEX

Acetaldehyde, determination with:
 chloramine-T 43
 potassium permanganate 7
 vanadium sulphate 153
Acetals, determination with potassium hexacyanoferrate (III) 25
Acetanilide, determination with copper (III) compounds 17
Acetoacetic acid, determination with copper (III) compounds 17
Acetone, determination with:
 hypohalites 34
 potassium permanganate 7
 vanadium sulphate 153
Acetoxime, determination with chromium salts 116
Acetylene compounds, determination with chromium salts 115
Acyl derivatives of α-amino alcohols, oxidation with periodic acid and its salts 73
Adrenalin, determination with potassium hexacyanoferrate (III) 24
Active methylene groups, determination with:
 alkyl nitrite 191
 potassium hexacyanoferrate (III) 23
Alcohols, determination with:
 periodic acid 73
 vanadium (V) compounds 90
 vanadium sulphate 153
Aldehyde, determination with:
 chloramine-T 43, 44
 periodic acid 73
Aliphatic acid, determination with sodium nitrite 190
Alkyl nitrite, determination of:
 active methyl groups 191
 primary aromatic amines 191
Allyl alcohol, determination with N-bromosuccinimide 54
Allyl amine, determination with N-bromosuccinimide 54

Allylisobutyl-barbituric acid, determination with N-bromosuccinimide 54
Allylisopropyl-barbituric acid, determination with N-bromosuccinimide 54
Allyl isothiocyanate, determination with periodic acid 71
Aluminium, determination with chloramine-T 44
Amidopyrine, determination with:
 iodine monochloride 63
 potassium permanganate in alkaline solution 7
 vanadium (V) compounds 90
Amido sulphonic acid, determination with sodium nitrite 190
Amines, determination with:
 manganese (III) compounds 12
 sodium nitrite 186, 188
 vanadium sulphate 153
Amines, aromatic primary, secondary, tertiary, determination with sodium nitrite 186, 188
Amino acids, determination with:
 copper (III) compounds 16
 hypohalites 34
 periodic acid 73
α-Amino acids, determination with:
 periodic acid 73
 vanadium (V) compounds 90
Amino alcohols, determination with periodic acid 72, 73
Aminoazobenzene, determination with chromium salts 115
p-Aminophenol, determination with hydroquinone 177
Aminoguanidonium chloride, determination with:
 iodine monochloride 63
 vanadium (V) compounds 89

SUBJECT INDEX

Ammonium salts, determination with:
  bromine 49
  hydrogen peroxide 103
  hypohalites 31
Anaesthesine, determination with iodine monochloride 63
Analgin, determination with:
  iodine monochloride 63
  vanadium (V) compounds 90
1,4-Anhydro erythritol, determination with lead (IV) acetate 80
1,4-Anhydro mannitol, determination with lead (IV) acetate 80
Aniline, determination with:
  copper (III) compounds 17
  iodine monochloride 63
Aniline blue, determination with chromium (II) salts 116
Anthranilic acid, determination with iodine monochloride 63
Anthraquinone, determination with:
  chromium (II) salts 115
  vanadium sulphate 152
Anthraquinone-2-sulphonic acid, determination with chromium (II) salts 115
Anthraquinone-2,7-disulphonic acid, determination with chromium (II) salts 115
Antibiotics (rifamycine type), determination with gold chloride 195
Antifebrine, determination with iodine monochloride 63
Antimonite, determination with vanadium (IV) sulphate 157
Antimony, determination with:
  bromine 49
  chloramine-T 39, 43, 44
  chromium salts 107, 112
  copper (III) compounds 15
  hypohalites 31
  iodine monochloride 57, 63
  iron salts 96
  manganese (III) compounds 11
  perbenzoic acid 195
  periodic acid 68, 69, 70
  potassium hexacyanoferrate (III) 19
  potassium manganate 194
  potassium permanganate 3
  tin (II) chloride 123
  vanadium (V) compounds 88
  vanadium (II) sulphate 150

Antimony–copper, determination with chromium (II) salts 112
Antimony–copper–tin, determination with chromium (II) salts 112
Antimony–tin, determination with chromium (II) salts 112
Antipyrine, determination with iodine monochloride 63
Aromatic amines, determination with:
  alkyl nitrite 191
  sodium nitrite 186, 188
Aromatic sulphonic acid, determination with hypohalites (hypochlorite) 34
Aromatic compounds, organic, determination with potassium chlorate 194
Arsenate (V), determination with potassium iodide 195
Arsenic, determination with:
  bromine 49
  chloramine-T 39, 43
  chromium salts 114
  copper (III) compounds 11
  hydrogen peroxide 103
  hypohalites 31
  iodine monochloride 57, 63
  iron salts 96
  manganese (III) compounds 11
  mercury compounds 135
  perbenzoic acid 195
  periodic acid 68, 69, 70
  potassium hexacyanoferrate (III) 19
  potassium manganate 194
  potassium permanganate 3
  vanadium (V) compounds 84
Arsenic (III) chloride, determination of inorganic systems 196
Arsenic (III) salts, determination with N-bromosuccinimide 54
Arsenite:
  determination of:
    bromine 128
    caro acid 128
    cerium 128
    chlorine 128
    chlorites 128
    dichromate 128
    hexacyanoferrate (III) 128
    hypohalites 128
    iodine 128
    manganese 128

SUBJECT INDEX

periodate 128
permanganate 128
determination with:
  N-bromosuccinimide 54
  vanadium (IV) sulphate 157
indicator 127
references 129
review of determinations 128
standard solution 127
N-Arylhydroxylamines, determination with sodium nitrite 190
Ascorbic acid:
  determination of:
    bromate 167, 168
    bromine 164, 167, 168
    cerium 165, 169
    chloramine 167
    chloramine B 168
    chlorate 168
    chlorine 164, 167
    chlorite 164, 167
    cobalt 169
    copper 165
    chromate 167, 168
    2,6-dichloro phenol indophenol 169
    2,6-dinitro phenol indophenol 169
    free iodine 168
    gold 166, 169
    hexacyanoferrate (III) 167
    hydrazine 169
    hydrogen peroxide 164
    hydroxylamine 169
    hypobromite 164, 167, 169
    hypochlorite 164, 167, 169
    iodate 167, 168
    iodine 167, 168
    iodine monochloride 168
    iodide 168
    iridium 166
    iron 163
    iron-cerium 169
    lead 167
    lead (IV) acetate 168
    mercury 165
    oxygen 168
    permanganate 164, 167, 168
    peroxide 169
    peroxodisulphate 164
    platinum 166
    resazurine 169
    resorufine 169
    rifamycine 169
    selenium 166
    semicarbazide 169
    silver 165, 169
    thallium 169
    thiazine 169
    thiazone 169
    tin 167
    tungsten 169
    vanadate 168
    vanadium 164, 169
    vanadium-chromium-molybdenum 169
    vanadium-iron 169
    zinc 167
  determination with:
    bromine 50
    N-bromosuccinimide 54
    chloramine-T 42, 43
    2,6-dichloro phenol indophenol 195
    lead (IV) acetate 80
    manganese (III) compounds 12
    periodic acid 69, 70
    potassium hexacyanoferrate (III) 25
  indicator 163
  iodimetric determinations 169
  procedure 162
  references 169
  review of determinations 163
  standardization 162
  standard solution 161
Azides, determination with:
  hydroquinone 176
  sodium nitrite 190
Azobenzene, determination with vanadium (II) sulphate 151
Azo compounds, determination with chromium salts 115
Azo dyes, determination with:
  chromium (II) salts 115
  vanadium (II) sulphate 151
Azoxybenzene, determination with chromium salts 115
Azoxycompounds, determination with chromium salts 115

Barbituric acid, derivatives of, determination with iodine monochloride 63

SUBJECT INDEX

Barium peroxide, determination with potassium permanganate 6
Benzalazine, determination with vanadium (V) compounds 89
Benzaldehyde, determination with:
  chloramine 43
  hypohalites 34
  potassium permanganate 7
  vanadium sulphate 153
Benzalsemicarbazone, determination with vanadium (V) compounds 89
Benzene sulphinic acid, determination with hypohalites 34
Benzophenone, determination with vanadium (II) sulphate 153
Benzylalcohol, determination with potassium permanganate 7
Benzylmercaptan, determination with lead (IV) acetate 80
Bismarck brown, determination with chromium (II) salts 116
Bismuth, determination with:
  chromium salts 109, 112
  vanadium (II) sulphate 150
  vanadium (IV) sulphate 158
Bismuthon, determination with iodine monochloride 63
Bromate, determination with:
  ascorbic acid 167, 168
  chloramine-T 43
  hydrazine sulphate 181
  mercury compounds 133
  molybdenum (III), (V) compounds 141
  thiosulphate 196
  tin (II) chloride 124
  tungsten (V) compounds 144
  tungsten (III) compounds 144
  uranium (IV) sulphate 146
  vanadium (IV) acetate 158
  vanadium (II) sulphate 151
Bromide, determination with hypohalites 32
Bromine:
  determination of:
    ammonium salt 49
    antimony 49
    arsenic 49
    ascorbic acid 50
    N,N-dimethyl benzidine 50
    formic acid 49
    glycerol 50
    hydroquinone 50
    hypophosphites 49
    iron (II) 49
    mercury (I) 49
    peroxides 49
    sulphides 49
    sulphites 49
    N,N,N,N-tetramethyl naphthidine 50
    thallium (I) 49
    tin 49
  determination with:
    ascorbic acid 164, 167, 168
    chromium (II) chloride 116
    mercury (I) compounds 133
    phenoxazine 196
    phenyl arsine oxide 195
    sodium arsenite 128
    tin (II) chloride 123
  indicator 49
  references 50
  review of determinations 49
  standardization 49
  standard solution 48
N-Bromosuccinimide:
  determination of:
    allyl alcohol 54
    allyl amine 54
    allylisobutyl-barbituric acid 54
    allylisopropyl-barbituric acid 54
    arsenic (III) salts 54
    arsenite 54
    ascorbic acid 54
    crotonic acid 54
    diallyl barbituric acid 54
    hydrazine 54
    hydroquinone 54
    iodide 54
    iodine 54
    iodine numbers of fats and oils 54
    sulphide 54
    tetrahydro-benzoic acid amide 54
    thiocyanate 54
    thiourea 54
  indicator 53
  references 54
  review of determinations 55
  standardization 53
  standard solution 53
n-Butanol, determination with potassium permanganate 7

SUBJECT INDEX

Cadmium, determination with:
  chloramine-T 44
  periodic acid and its salts 68
Calcium, determination with:
  copper (III) compounds 16
  vanadium (V) compounds 88
Calcium glucoside, determination with lead (IV) acetate 81
Calcium hypochlorite, see Hypohalites 29
Camphor, determination with vanadium (II) sulphate 153
Carbon disulphide, determination with chloramine-T 41, 44
Carbon tetrachloride, determination with chromium salts 116
Carbonyl groups, determination with copper salt 195
Caro acid, determination with:
  sodium arsenite 128
  vanadium (IV) sulphate 159
Catechol, determination of inorganic systems 196
Catechol, determination with lead (IV) acetate 80
Cerium, determination with:
  ascorbic acid 165, 169
  hydroquinone 174, 175, 176
  mercury (I) compounds 133
  periodic acid 68
  potassium hexacyanoferrate (III) 19
  potassium permanganate 4
  sodium arsenite 128
  sodium nitrite 190
  uranium (IV) sulphate 146
  vanadium (II) sulphate 150
  vanadium (IV) sulphate 159
Cerium–iron, determination with uranium (IV) sulphate 146
Cerium (III) salt, determination of hexacyanoferrate 196
Cerium (III) salts, determination with:
  cobalt (III) compounds 100
  copper (I) compounds 138
  hypohalites 33
  mercury (I) nitrate 135
  mercury (I) perchlorate 135
  molybdenum (V) compounds 141
  molybdenum (III) compounds 141
Cerium-vanadate, determination with hydroquinone 175
Cerium–vanadium–chromium, determination with hydroquinone 175
Cerium–vanadate–iron, determination with uranium (IV) sulphate 146
Cerium (IV) salt–vanadate, determination with molybdenum (V) compounds 141
Cerium (IV) salt–vanadate (V)–iron (III) salt, determination with molybdenum (V) compounds 141
Chloral hydrazine, determination with vanadium (V) compounds 89
Chloramine, determination with ascorbic acid 167
Chloramine-B, determination with ascorbic acid 168
Chloramine-T:
  determination of:
    acetaldehyde 43
    aldehydes 43, 44
    aluminium 44
    antimony 39, 43, 44
    arsenic 39, 43
    ascorbic acid 43
    benzaldehyde 43
    bromate 43
    cadmium 44
    carbon disulphide 41, 44
    chlorate 43
    cyanide 42
    2,4-dinitrophenylhydrazine 44
    dixanthate 42
    ethylene glycol 44
    furfurole 43
    glucose 44
    glycerol 44
    hexacyanoferrate (II) 41, 44
    hexacyanoferrate (III) 43
    hydrazine 41, 43, 44
    hydrazine—organic derivatives of 44
    hydrogen peroxide 43
    hydrogen sulphide 42
    hydroquinone 42
    p-hydroxy-benzaldehyde 43
    hydroxy-quinoline 42
    hypophosporous acid 42
    iodide 42, 43
    iron 40, 43, 44
    isonicotinylhydrazide 4
    lactic acid 44
    lead (IV) oxide 43

SUBJECT INDEX

Chloramine-T—*Continued*
  determination of;
    magnesium 44
    mannitol 44
    mercury monovalent 43
    nitrite 43
    nitro-furfurane 44
    p-nitrophenylhydrazine 44
    oxalic acid 44
    oxine 44
    periodate 43
    permanganate 43
    phenols 44
    phenylhydrazine 44
    polythionates 44
    potassium dichromate 43
    quinhydrone 42
    salicylaldehyde 43
    salicylic acid 44
    selenium (IV) oxide 43
    semicarbazide 44
    sodium formate 43
    sodium hydrogen sulphite 42
    sodium sulphide 42
    sugars 44
    sulphide 42
    sulphide—organic compounds 44
    sulphite 42, 44
    thallium 42, 44
    thiobarbituric acid—derivatives of 44
    thiocyanate 42, 43, 44
    thiosemicarbazide 44
    thiosulphate 41, 42
    thiourea 44
    tin 40, 43
    titanium 41, 44
    urea 44
    uric acid 44
    vanadium 41, 44
    vanillin 43
    xanthates 44
    zinc 44
  indicator 38
  references 45
  review of determinations 39
  standardization 38
  standard solution 38
Chlorate:
  determination of:
    inorganic systems 194
    iodide 194
  determination with:
    ascorbic acid 168
    chloramine 43
    chromium (II) salts 114
    mercury (I) compounds 135
    organic dyes 196
    vanadium (II) sulphate 151
Chlorine, determination with:
  ascorbic acid 164, 167
  phenoxazine 190
  phenyl arsine oxide 195
  potassium iodide 195
  sodium arsenite 128
Chlorine in water, determination with organic dyes 196
Chlorite, determination with:
  arsenate 128
  ascorbic acid 164, 167
  sodium arsenite 128
N-Chlorobenzamide 194
o-Chlorobenzol semicarbazone, determination with vanadium (V) compounds 89
Chloroform, determination with chromium (II) salts 116
Chlorotungstate, determination with tungsten (V) (III) compounds 144
Chromate, determination with:
  ascorbic acid 167, 168
  chromium (II) salts 111
  vanadium (IV) sulphate 157
Chromic acid, determination with vanadium (V) acetate 158, 159
Chromium, determination with:
  hydrogen peroxide 103
  hydroquinone 174, 175
  iodine monochloride 59
  manganese (III) compounds 11
  potassium hexacyanoferrate (III) 20
  potassium manganate 194
  potassium permanganate 4
  tin (II) chloride 122
  vanadium (V) compounds 89
  vanadium (II) sulphate 150
  vanadium (IV) sulphate 157
Chromium blue R, determination with vanadium (II) sulphate 152
Chromium (II) salts:
  determination of:
    acetoxime 116
    acetylene compounds 115
    aminoazobenzene 115

SUBJECT INDEX

anthraquinone-2-sulphonic acid 115
anthraquinone-2,7-disulphonic acid 115
antimony 107, 112
arsenic 114
azo-compounds 114
azoxybenzene 115
azoxycompounds 115
bismuth 109, 112
carbon tetrachloride 116
chlorate 114
chloroform 116
chromate 111
cobalt 109
copper 107, 112
copper–antimony 112
copper–gold 112
copper–gold–mercury 112
copper–iron 112
copper–iron–dichromate 112
copper–mercury 112
copper–molybdenum 112
copper–selenium 112
copper–silver 112
copper–titanium 112
copper–tungsten 112
copper–vanadium 112
diacetyldioxime 116
diazosalts 115
dicarboxylic acid 115
dichromate 111
gold 108
hexacyanoferrate (III) 114
hydrazobenzene 115
hydrogen peroxide 114
indigoide 116
iron 109, 112
iron–chromium 113
iron–chromium–uranium 114
iron–molybdenum 113
iron–molybdenum–chromium 113
iron salts 114
iron–titanium 112
iron–tungsten 113
iron–tungsten–molybdenum 113
iron–vanadium 113
iron–vanadium–chromium 113
mercury 108
mercury–bismuth 114
mercury–iron 114

molybdenum 109
molybdenum–chromium 113
molybdenum–chromium–vanadium 113
molybdenum–iron–copper 113
molybdenum–titanium 113
molybdenum–tungsten 113
molybdenum–vanadium 113
nitrate 114
nitrocompounds 114, 115
o-nitrophenol 115
m-nitrophenol 115
p-nitrophenol 115
nitrosocompounds 114, 115
o-nitrosodimethylaniline 115
p-nitrosodimethylaniline 115
1-nitroso-2-naphthol 115
3-nitroso-2-naphthol 115
p-nitrosophenol 115
organic dyes 115
osmium 114
oxygen 114
permanganate 114
picric acid 115
picrolonates 116
picrolonic acid 116
plutonium 114
potassium disulphate 114
potassium periodate 114
p-quinone 115
quinones 114
quinone imine dyes 116
rhenium 114
ruthenium 114
selenite 114
selenium–tellurium 114
silver 108
sodium peroxide 114
sugar 115
tetracene 115
thallium 111
tin 112
tin–antimony 112
tin–bismuth 112
tin–copper 112
tin–copper–antimony 112
tin–copper–bismuth 112
tin–iron 112
tin–iron–bismuth 112
titanium 111
trinitrotoluene 115
triphenyl methane 116

Chromium (II) salts—*Continued*
determination of:
    tungsten 110
    tungsten–chromium 114
    uranium 110
    uranium–vanadium 114
    uranium–vanadium–iron–chromium 113
    vanadate 111
    vanadium–iron 114
    vanadium–titanium 114
determination with:
    mercury (I) compounds 135
    7-oxiphenthiazone-2 196
    indicator 106
    references 116
    review of determinations 107
    standardization 106
    standard solution 104
    titration of mixtures 112
Chromoxan brown 5R, determination with vanadium (II) sulphate 152
Cinnamic acid, determination with potassium permanganate 7
Citric acid, determination with:
    lead (IV) acetate 81
    manganese (III) compounds 12
    potassium permanganate 7
    vanadium (V) compounds 90
Cobalt (III) compounds:
determination of:
    cerium salts 100
    hexacyanoferrate 100
    iron ions 100
determination with:
    ascorbic acid 169
    chromium (II) salts 109
    molybdenum (III) compounds 142
    periodic acid 69
    potassium hexacyanoferrate (III) 20
    vanadium (V) compounds 88
    vanadium (IV) sulphate 159
    indicator 99
    references 100
    review of determinations 100
    standardization 99
    standard solution 99
Cobalt carbonyl, determination with methylene blue 196

Congo red, determination with chromium (II) salts 116
Copper, determination with:
    ascorbic acid 165
    chromium (II) chloride solution 107
    chromium (II) salts 107, 112
    hydrazine sulphate 182
    hydroquinone 175
    mercury (I) compounds 134
    molybdenum (III) compounds 141
    periodic acid 69
    tin (II) chloride 121
    tungsten (V) (III) compounds 144
    vanadium (V) compounds 87
    vanadium (II) sulphate 149
    vanadium (IV) sulphate 157
Copper–antimony, determination with chromium (II) salts 112
Copper (I) compounds:
determination of:
    cerium salts 138
    dichromate 138
    gold salts 138
    hexacyanoferrate 138
    iodate 138
    iridium compounds 138
    permanganate 138
    platinum 138
    vanadate 138
    indicator 138
    references 138
    review of determinations 138
    standardization 137
    standard solution 137
Copper (III) compounds:
determination of:
    acetanilide 17
    acetoacetic acid 17
    amino acids 16
    aniline 17
    antimony 15
    arsenic 15
    calcium 16
    cyanide 16
    diacetyl 17
    4,4-dihydroxidiphenyl 16
    dimethylglyoxime 17
    ethylenediamine 17
    ethyleneglycol 17
    formaldehyde 17
    fumaric acid 17
    glycerol 17

SUBJECT INDEX

hexacyanoferrate (II) 16
hydroquinone 17
inositol 17
iodates 16
iodides 16
lactose 17
maleic acid 17
mannitol 17
mannose 17
oleic acid 17
oxine 17
phenyl hydrazine 17
phloroglucinol 17
proteins 16, 17
pyrocatechol 17
pyrogallol 17
quinone 17
resorcinol 17
starch 17
sugar in blood 16
tartrate 16, 17
thallium 16
thiosulphates 16
urine 16
vitamin B 17
differential percuprimetric titration 16
indicator 15
references 17
review of determinations 15
standardization 15
standard solution 14
Copper–gold, determination with chromium (II) salts 112
Copper–gold–mercury, determination with chromium (II) salts 112
Copper–iron, determination with chromium (II) salts 112
Copper–iron–dichromate, determination with chromium (II) salts 112
Copper–iron–mercury, determination with chromium (II) salts 112
Copper–molybdenum, determination with chromium (II) salts 112
Copper (II) salt:
  determination of:
    carbonyl groups 195
    hydrazine derivatives 195
    hydroxylamine 195
    reducing sugars 195
  determination with:
    iron salts 96

mercury (I) nitrate 135
mercury (I) perchlorate 135
potassium iodide 195
thiosulphate 196
tungsten (V) (III) compounds 144
Copper–selenium, determination with chromium (II) salts 112
Copper–silver, determination with chromium (II) salts 112
Copper (II) sulphate, determination of:
  cystein 195
  molybdenum 195
  tungsten 195
Copper–titanium, determination with chromium (II) salts 112
Copper–tungsten, determination with chromium (II) salts 112
Copper–vanadium, determination with chromium (II) salts 112
Crotonic acid, determination with N-bromosuccinimide 54
Crystal violet, determination with chromium (II) salts 116
Cupferron, determination with molybdenum (III) compounds 142
Cyanide, determination with:
  chloramine-T 42
  copper (III) compounds 16
  hypohalites (hypobromite) 32
  iodine monochloride 59
  potassium permanganate 6, 7
Cyclohexanols, determination with vanadium (V) compounds 90
Cystein, determination with:
  copper (II) sulphate 195
  lead (IV) acetate 80
  potassium hexacyanoferrate (III) 24
Cystein chloride, determination with periodic acid 70

Diacetyl, determination with copper (III) compounds 17
Diacetyldioxime, determination with chromium (II) salts 116
  copper (III) compounds 17 *see* Dimethylglyoxime
Diallyl-barbituric acid, determination with N-bromosuccinimide 54
Diamines, aromatic, determination with sodium nitrite 188

α-Diamines, determination with periodic acid 73
Diantimonate, determination of molybdenum 195
Diazo phenols, determination with vanadium (II) sulphate 151
Diazosalts, determination with chromium (II) salts 115
Dibromohydroxyquinoline, determination with vanadium (V) compounds 90
Dibromooxinates, determination with vanadium (V) compounds 90
Dicarboxylic acid, determination with chromium (II) salts 115
Dichloramine B, see Chloramine B
Dichloramine T, see Chloramine T
2,6-Dichloro phenol indophenol:
determination of:
ascorbic acid 195
determination with:
ascorbic acid 169
Dichromate, determination with:
chromium salts 111
copper (I) compounds 138
hydrazine sulphate 182
hydroquinone 174
hypohalites 31
mercury (I) compounds 135
molybdenum (V), (III) compounds 141
Sodium arsenite 128
tungsten (V) compounds 144
uranium (IV) sulphate 146
Dichromate–iron, determination with:
molybdenum (V) compounds 141
uranium (IV) sulphate 146
4,4-Dihydroxydiphenyl, determination with copper (III) compounds 16
α-Diketones, determination with periodic acid 73
Dimercaptopropanol, determination with iodine monochloride 63
4-Dimethylamino-1,5-dimethyl-2-phenylpyrazolone (amidopyrine), determination with potassium permanganate 7
N,N-Dimethyl benzidine, determination with bromine 50
Dimethylglyoxime, determination with copper (III) compounds 17

Dinitrobenzene, determination with vanadium (II) sulphate 153
2,6-dinitro phenol indophenol, determination with ascorbic acid 169
2,4-Dinitrophenylhydrazine, determination with:
chloramine-T 37
vanadium (II) sulphate 153
Dinitroresorcinol, determination with tin (II) chloride 125
Dithionate, determination with vanadium (V) compounds 89
Dithionite, determination with indigo 195
Dixanthate, determination with chloramine-T 42

Enoles, determination with iodine monochloride 63
Eosin, determination with chromium (II) salts 116
Eriochromazurol, determination with vanadium (II) sulphate 152
Erythritol, determination with potassium permanganate 7
Ethylenediamine, determination with copper (III) compounds 17
Ethylene glycol, determination with:
chloramine-T 44
copper (III) compounds 17
lead (IV) acetate 80
periodic acid 74
Ethylsalicylate, determination with iodine monochloride 63

Ferrocene, determination of iron thiocyanate complex 196
Ferrocyanide, determination with periodic acid 70
Fiat Microfilm PBL 191
Formaldehyde, determination with:
copper (III) compounds 17
hypohalites 34
mercury iodide 195
potassium hexacyanoferrate (III) 25
potassium permanganate 7
vanadium sulphate 153
Formate, determination with:
periodic acid 69
potassium permanganate 6
vanadium (V) compounds 96

SUBJECT INDEX

Formic acid, determination with:
  bromine 49
  lead (IV) acetate 80
  potassium manganate 194
  potassium permanganate 7
Free iodine, determination with ascorbic acid 168
Fumaric acid, determination with:
  copper (III) compounds 17
  potassium permanganate 7
Furfurole, determination with chloramine-T 43

Glucose, determination with:
  chloramine-T 44
  vanadium (II) sulphate 153
Glycerol, determination with:
  bromine 50
  chloramine-T 44
  copper (III) compounds 17
  hypobromites 34
  hypohalites 34
  hypochlorite 34
  lead (IV) acetate 80, 81
  periodic acid 74
  potassium permanganate 7
  vanadium (V) compounds 90
Glycol bond, determination with lead (IV) acetate 80
Glycols, vicinal, determination with lead (IV) acetate 80
Glyoxal, determination with periodic acid 73
Gold, determination with:
  ascorbic acid 166, 169
  chromium (II) salts 108
  copper (I) compounds 138
  hydrazine sulphate 182
  hydroquinone 175
  mercury (I) compounds 133
  molybdenum (III) sulphate 158
  potassium iodide 195
  sulphite 196
  tin (II) chloride 123
  vanadium (IV) sulphate 158
Gold chloride, determination of rifamycine (antibiotics) 195
  tocopherol 195

Halide, determination with hydroquinone 177

Halogen compounds, determination with:
  hydrazine sulphate 181
  hydroquinone 173
Heavy metals, determination with iodine monochloride 63
Hemin, determination with potassium hexacyanoferrate (III) 24
Hemoglobin, determination with potassium hexacyanoferrate (III) 24
Hexacyanoferrate (III), determination with:
  ascorbic acid 167
  cerium salt 196
  chromium (II) salts 114
  chloramine 41, 43, 44
  copper (I) compounds 16
  copper (III) compounds 100
  hydrazine sulphate 182
  hydroquinone 171
  iodine monochloride 59
  mercury (I) compounds 135
  potassium permanganate 5
  sodium arsenite 128
  sodium nitrite 190
  tin (II) chloride 124
  tungsten (V) compounds 144
  uranium (IV) sulphate 146
  vanadium (IV) sulphate 156
Hydrazide, determination with sodium nitrite 190
Hydrazide of isonicotinic acid, determination with vanadium (V) compounds 90
Hydrazine, determination with:
  ascorbic acid 169
  N-bromosuccinimide 54
  chloramine-T 41, 43, 44
  iodine monochloride 61
  mercury (I) 135
  periodic acid 69, 70
  potassium hexacyanoferrate (III) 23
  potassium permanganate 6, 7
  potassium peroxidisulphate 194
  sodium nitrite 189
  vanadium (V) compounds 89
Hydrazine, derivatives of, determination with:
  copper (II) salt 195
  manganese (III) compounds 12
  sodium nitrite 189
  vanadium (V) compounds 89

Hydrazides of the isonicotinic acid, determination with:
  manganese (III) compounds 12
  periodic acid 69
Hydrazine sulphate:
  determination of:
    bromate 181
    copper 182
    dichromate 182
    gold 182
    halogen compounds 181
    hexacyanoferrate 182
    iodate 181
    lead oxides 183
    manganese 183
    mercury 182
    molybdate 183
    nitrite 183
    periodate 181
    selenate 183
    selenium 182
    silver 182
    thallium 182
    vanadate 182
  determination with:
    periodic acid 69
    vanadium (IV) sulphate 157
  indicator 181
  indirect determinations 182
  references 183
  review of determinations 181
  standard solution 181
Hydrazobenzene, determination with chromium (II) salts 115
Hydrogen atoms, active, determination with periodic acid 73
Hydrogen peroxide:
  determination of:
    ammonium salts 103
    arsenic 103
    chromium 103
    hypobromites 103
    sulphides 103
    sulphites 103
    thiosulphites 103
  determination with:
    ascorbic acid 164
    chloramine-T 43
    chromium (II) salts 114
    mercury (I) compounds 135
    organic dyes 196
    periodic acid 70
    potassium manganate 194
    potassium permanganate 6
    sodium nitrite 190
    vanadium (V) compounds 89
    vanadium (IV) sulphate 157
  indicator 102
  references 103
  review of determinations 102
  standardization 101
  standard solution 101
Hydrogen sulphide, determination with:
  chloramine-T 42
  periodic acid 70
  vanadium (V) compounds 89
Hydrogen sulphite, determination with
  periodic acid 70
  vanadium (V) compounds 89
Hydroquinone and similar reducing agents:
  determination of:
    p-aminophenol 177
    azides 176
    cerium 174, 175, 176
    chromium 174, 175
    copper 175
    gold 175
    halide 177
    halogen compounds 173
    hexacyanoferrate 173
    iridium 174
    lead 174
    lead, higher oxides of 176
    manganese 176
    manganese, higher oxides of 176
    metol 177
    peroxidisulphate 176
    p-phenylenediamine 177
    quinhydrone 177
    ruthenium 175
    silver 175
    sulphate 177
    thallium 173
    vanadium 174, 175
  determination with:
    bromine 80
    N-bromosuccinimide 54
    chloramine-T 42
    copper (III) compounds 17
    hypohalites 34

SUBJECT INDEX

iodine monochloride 61
lead (IV) acetate 80
manganese (III) compounds 12
periodic acid 69, 70
potassium hexacyanoferrate (III) 25
vanadium (V) compounds 90
indicator 173
indirect determination 176
references 177
review of determinations 173
standardization 172
standard solution 172
α-Hydroxyacids, determination with periodic acid 73
α-Hydroxy aldehydes, determination with periodic acid 73
p-Hydroxy-benzaldehyde, determination with chloramine 43
Hydroxylamine, determination with:
ascorbic acid 169
copper (II) salt 195
hypohalites 34
iodine monochloride 61
iron salts 96
mercury (I) compounds 135
potassium hexacyanoferrate (III) 23
vanadium (V) compounds 90
vanadium (II) sulphate 151
Hydroxylammonium sulphate, determination of 196
Hydroxy quinoline, determination with chloramine-T 42
Hydroxy-triphenylmethane dyes, determination with vanadium (II) sulphate 151
Hypobromite (see Hypohalites), determination with:
ascorbic acid 164, 167, 169
hydrogen peroxide 103
mercury (I) compounds 133
Hypochlorite (see Hypohalites), determination with:
ascorbic acid 164, 167, 169
mercury (I) compounds 133, 135
organic dyes 196
potassium iodide 196
Hypohalites (Hypobromite, Hypochlorite):
determination of:
acetone 34

amino acids 34
ammonium salts 31
antimony 31
arsenic 31
benzaldehyde 34
benzene-sulphinic acid 34
bromide 32
cerium (III) salts 33
chromium (III) salts 33
cyanide 32
formaldehyde 34
glycerol 34
hydroquinone 34
hydroxylamine 34
iodide 32
iron 33
isopropanol 34
lead 33
mercury 33
nitrites 33
peroxides 33
phosphites 33
selenites 33
selenium (IV) salts 33
silver 33
sugar 34
sulphide 32
sulphite 32
sulphonic acids, aromatic 34
tellurium (IV) salts 33
thallium 33
thioacetamide 34
thiocyanate 32
3-thioketo-5-keto-6-benzyl-1,2,4-triazine 34
thiosulphate 32
tin 31
urea 34
determination with:
sodium arsenite 128
indicator 31
references 34
review of determinations 31
standardization 30
standard solution 30
Hypophosphite, determination with:
bromine 49
potassium permanganate 7
vanadium (V) compounds 89
Hypophosphorous acid, determination with chloramine-T 42

IG Analysenvorschriften 125, 192
Indigo:
  determination of:
    dithionite 195
  determination with:
    chromium (II) salts 116
    vanadium (V) compounds 90
Indigoide, determination with chromium (II) salts 116
Indigosols, determination with iron salts 96
Inorganic compounds, determination with manganese (III) compounds 11
Inorganic substances, determination with potassium peroxidisulphate 194
Inorganic systems, determination with:
  arsenic chloride 196
  catechol 196
  chlorate 194
Inositol, determination with copper (III) compounds 17
Iodate, determination with:
  ascorbic acid 167, 168
  copper (I) compounds 138
  copper (III) compounds 16
  hydrazine sulphate 181
  molybdenum (III), (V) compounds 141
  potassium iodide 196
  potassium permanganate 7
  thiosulphate 196
  tin (II) chloride 124
Iodide, determination with:
  ascorbic acid 168
  N-bromosuccinimide 54
  chloramine-T 43, 42
  chlorate 194
  copper (III) compounds 16
  hypohalites 32
  iodine monochloride 59
  manganese (III) compounds 11
  periodic acid 69, 70
  potassium hexacyanoferrate (III) 25
  potassium permanganate 7
  vanadium (V) compounds 88
Iodimetric determinations, mercury (I) compounds 134
Iodine, determination with:
  ascorbic acid 167
  N-bromosuccinimide 54
  chromium (II) chloride solution 116
  mercury (I) compounds 134
  thiosulphate 196
  tin (II) chloride 123
Iodine, free, determination with ascorbic acid 168
Iodine monochloride:
  determination of:
    amidoguanodinium chloride 63
    amidopyrine 63
    anaesthesine 63
    analgine 63
    aniline 63
    anthranilic acid 63
    antifebrine 63
    antimony 57, 63
    antipyrine 63
    arsenic 57, 63
    ascorbic acid 60
    barbituric acid, derivatives of 63
    bismuthon 63
    chromium 59
    cyanide 59
    dimercaptopropanol 63
    enoles 63
    ethylsalicylate 63
    heavy metals 63
    hexacyanoferrate (II) 59
    hydrazine 61
    hydroquinone 61
    hydroxylamine 61
    iodide 59
    iron 58, 63
    mercaptobenzothiazol 62
    mercury salts 63
    methionine 63
    methylsalicylate 63
    metol 62
    novocaine 63
    olefines 63
    oleic acid 63
    oxine 63
    phenolphthaleine 63
    phenols 63
    phenylenediamine 62
    phenylhydrazine 61
    4-phenylsemicarbazone chloride 63
    rivanol 63
    rutine 63
    salicylaldoxime 63
    semicarbazide chloride 63

SUBJECT INDEX

sphaerophysine 63
sulphite 69, 63
sulphonamides 63
thiocyanate 59
thiosemicarbazide 62
thiosemicarbazone of *p*-acetamidobenzaldehyde 63
thiosulphate 60
thiourea 62
tibon 63
tin 57
titanium 58
determination with:
  ascorbic acid 169
  chromium (II) chloride solution 116
  indicator 3
  references 64
  standardization 56
  standard solution 56
Iodine numbers of fats and oils, determination with *N*-bromosuccinimide 54
Iridium, determination with:
  ascorbic acid 166
  hydroquinone 174
  copper (I) compounds 138
  molybdenum (III) compounds 141
Iron, determination with:
  ascorbic acid 163
  bromine 49
  chloramine-T 40, 43, 44
  chromium (II) chloride solution 116
  chromium (II) salts 109, 112
  hypohalites 33
  iodine monochloride 58, 63
  manganese (III) compounds 11
  mercury (I) compounds 133
  molybdenum (III), (V) compounds 141
  periodic acid 68, 69, 70
  potassium hexacyanoferrate (III) 22
  thiosulphate 196
  tin (II) chloride 121
  tungsten (V) compounds 144
  uranium (IV) sulphate 146
  vanadium (V) compounds 87
  vanadium sulphate 149
Iron-cerium, determination with ascorbic acid 169
Iron-chromium, determination with chromium (II) salts 113

Iron-chromium-uranium, determination with chromium (II) salts 114
Iron-copper, determination with tungsten (III) compounds 144
Iron ions, determination with cobalt (III) compounds 100
Iron-molybdenum, determination with chromium (II) salts 113
Iron-molybdenum-chromium, determination with chromium (II) salts 113
Iron salts:
  determination of:
    antimony 96
    arsenic 96
    ascorbic acid 96
    copper salts 96
    hydroxylamine 96
    indigosols 96
    molybdenum 95
    niobium compounds 96
    perrhenate 96
    phenylhydroxylamine 96
    phytine 96
    platinum salts 96
    tin 95
    titanium 94
    tungsten 96
    uranium 95, 96
    vanadium 95
  determination with:
    chromium (II) salts 114
    organic dyes 196
    perbenzoic acid 195
  indicator 94
  references 96
  review of determinations 94
  standardization 94
  standard solution 94
Iron (III) salt-dichromate, determination with molybdenum (V) compounds 141
Iron (III) salt-vanadate, determination with molybdenum (V) compounds 141
Iron-thiocyanate-copper, determination with ferrocene 196
Iron-titanium, determination with chromium (II) salts 112
Iron-tungsten, determination with chromium (II) salts 113

Iron–tungsten–molybdenum, determination with chromium (II) salts 113
Iron–vanadate, determination with tungsten (V) compounds 144
Iron–vanadium, determination with chromium (II) salts 113
Iron–vanadium–chromium, determination with chromium (II) salts 113
Isacene, determination with potassium hexacyanoferrate (III) 25
Isatine, determination with chromium (II) salts 116
Isoamylalcohol, determination with potassium permanganate 7
Isonicotinic acid hydrazide, determination with:
potassium hexacyanoferrate (III) 23
vanadium (V) compounds 90
Isonicotinic acid hydrazide, determination with:
chloramine-T 44
manganese (III) compounds 12
sodium nitrite 190
Isopropanol, determination with hypohalites 34

Ketals, determination with potassium hexacyanoferrate (III) 25
α-Keto acids, determination with periodic acid 73
α-Ketoaldehydes, determination with periodic acid 73
α-Ketols, determination with periodic acid 73
Ketones, determination with:
periodic acid 73
sodium nitrite 189
vanadium (V) compounds 90

Lactic acid, determination with:
chloramine-T 44
vanadium (V) compounds 90
Lactose, determination with copper (III) compounds 17
Lead, determination with:
ascorbic acid 167
hydroquinone 174
hypohalites 33
periodic acid 70
potassium permanganate 5
Lead (IV) acetate:
determination of:
1,4-anhydro erythritol 80
1,4-anhydro-mannitol 80
ascorbic acid 80
benzylmercaptan 80
calcium glucoside 81
catechol 80
citric acid 81
cystein 80
ethylene glycol 80
formic acid 80
glycol bond 80
glycerol 80, 81
hydroquinone 80
malic acid 81
mandelic acid 79, 81
mannitol 79, 80
methyl-2,6-anhydro-α-D-altropyranoside 80
methyl-α-D-mannofuranoside 80
molybdenum 79
ruthenium 79
tartaric acid 79, 80
tartrate 81
tetrachloro hydroquinone 80
thioglycolic acid 80
thiosemicarbazide 80
thiourea 80
vicinal glycols 80
uranium 79
determination with:
ascorbic acid 168
vanadium (IV) acetate 158, 159
indicator 78
procedure 80
references 81
review of determinations 79
standardization 78
standard solution 78
Lead iodide, determination with periodic acid 70
Lead (IV) oxide, determination with:
chloramine-T 43
hydrazine sulphate 183
periodic acid 69
Lead oxides, higher, determination with hydroquinone 176

SUBJECT INDEX

Magnesium, determination with chloramine-T 44
Malachite green, determination with chromium (II) salts 116
Maleic acid, determination with copper (III) compounds 17
Malic acid, determination with:
  lead (IV) acetate 81
  potassium permanganate 7
  vanadium (V) compounds 90
Malonic acid, determination with:
  manganese (III) compounds 12
  vanadium (V) compounds 90
Mandelic acid, determination with:
  lead (IV) acetate 79, 81
  potassium permanganate 7
Manganese, determination with:
  hydrazine sulphate 183
  hydroquinone 176
  mercury compounds 133
  molybdenum (III) compounds 142
  potassium hexacyanoferrate (III) 20
  potassium iodide 196
  sodium arsenite 128
  vanadium (V) compounds 88
  vanadium (IV) sulphate 159
Manganese (III) compounds:
  determination of:
    amines 12
    antimony 11
    arsenic 11
    ascorbic acid 12
    chromium 11
    citric acid 12
    hydrazine—derivatives 12
    hydrazides of the isonicotinic acid 12
    hydroquinone 12
    inorganic compounds 11
    iodides 11
    iron 11
    isonicotinic acid, hydrazides of 12
    malonic acid 12
    molybdenum 11
    nitrites 11
    organic compounds 11
    oxalic acid 12
    peroxides 11
    phenols p-substituted 12
    salicylic acid 12

tartaric acid 12
thiocyanates 11
tin 11
titanium 11
tungsten 11
uranium 11
vanadium 11
vanadyl 11
indicator 11
references 12
review of determinations 11
standardization 11
standard solution 10
Manganese, higher oxides, determination with hydroquinone 176
Manganese oxide, determination with periodic acid 69
Manganese salts, determination with potassium manganate 194
Mannitol, determination with:
  chloramine-T 44
  copper (III) compounds 17
  lead (IV) acetate 79, 80
Mannose, determination with copper (III) compounds 17
Mercaptans, determination with potassium hexacyanoferrate (III) 25
Mercaptobenzothiazol, determination with iodine monochloride 62
Mercury, determination with:
  ascorbic acid 165
  bromine 49
  chloramine-T 43
  chromium salts 108
  hydrazine sulphate 182
  hypohalites 33
  periodic acid 69, 70
  potassium permanganate 5
  vanadium (V) compounds 85
  vanadium (IV) sulphate 157
Mercury–bismuth, determination with chromium salts 114
Mercury iodide:
  determination of:
    formaldehyde 195
    photographic developers 195
    sugars, reducing 195
  determination with:
    periodic acid 70
Mercury–iron, determination with chromium (II) salts 114

Q

SUBJECT INDEX

Mercury (I) nitrate and mercury (I) perchlorate:
  determination of:
    arsenic salts 135
    bromate 133
    bromine 133
    cerium salts 133, 135
    chlorate 135
    chromium (III) salts 135
    copper salts 134, 135
    dichromate 135
    gold 133
    hexacyanoferrate 135
    hydrazine 135
    hydrogen peroxide 135
    hydroxylamine 135
    hypobromite 133
    hypochlorite 133, 135
    iodine 134
    iron 133
    manganese 133
    mercury salts 135
    molybdenum 134
    permanganate 135
    persulphate 135
    vanadate 135
  indicator 132
  iodimetric determinations 134
  references 135
  review of determinations 133
  standardization 132
  standard solution 132
Mercury (I) salts determination with iodine monochloride 63
Methanol, determination with potassium permanganate 7
Methionine, determination with iodine monochloride 63
p-Methoxybenzal-semicarbazone, determination with vanadium (V) compounds 89
Methyl-2,6-anhydro-α-D-altropyranoside, determination with lead (IV) acetate 80
Methylene blue
  determination of:
    cobalt carbonyl 196
    tin salts 196
  determination with:
    chromium (II) salts 116

Methylethylketone semicarbazone, determination with vanadium (V) compounds 89
N-Methyl-p-aminophenol, see Metol
Methyl-α-D-mannofuranoside, determination with lead (IV) acetate 80
Methyl orange, determination with chromium (II) salts 116
Methyl red, determination with chromium (II) salts 116
Methylsalicylate, determination with iodine monochloride 63
Methyl violet, determination with chromium (II) salts 116
Metol, determination with:
  hydroquinone 177
  iodine monochloride 61
  vanadium (V) compounds 90
Molybdate, determination with hydrazine sulphate 183
Molybdenum, determination with:
  chromium (II) salts 109
  copper sulphate 195
  iron (III) salts 95
  lead (IV) acetate 74
  manganese (III) compounds 11
  mercury compounds 134
  potassium hexacyanoferrate (III) 21
  tin (II) chloride 123
  vanadium (V) compounds 85, 88
Molybdenum (III) compounds
  determination of:
    bromate 141
    cerium salts 141
    cobalt 142
    copper 141
    cupferron 142
    dichromate 141
    gold 141
    iodate 141
    iridium salts 141
    iron 141
    manganese 142
    nitro compounds 141
    nitroso compounds 142
    picric acid 142
    vanadate 141
  indicator 141
  references 142
  review of determinations 141
  standardization 140
  standard solution 140

SUBJECT INDEX

Molybdenum (V) compounds:
  determination of:
    bromate 141
    cerium salts 141
    cobalt 142
    dichromate 141
    iodate 141
    iron 141
    manganese 142
    vanadate 141
  indicator 141
  reference 142
  review of determinations 141
  standardization 140
  standard solution 140
Molybdenum–chromium, determination with chromium (II) salts 113
Molybdenum–chromium–vanadium, determination with chromium (II) salts 113
Molybdenum–iron–copper, determination with chromium (II) salts 113
Molybdenum–tervalent, determination with diantimonate 195
Molybdenum–titanium, determination with chromium (II) salts 113
Molybdenum–tungsten, determination with chromium (II) salts 113
Molybdenum–vanadium, determination with chromium (II) salts 113
Morphine, determination with potassium hexacyanoferrate (III) 25

Naphtho-chromazurine, determination with vanadium (II) sulphate 152
α-Naphthylamine, determination with tin (II) chloride 125
Nickel, determination with vanadium (V) compounds 88
Nickel (IV) dimethylglyoximate, determination with tin (II) chloride 125
Niobium compounds, determination with iron (III) salts 96
p-Nitraniline, determination with:
  tin (II) chloride 125
  vanadium (II) sulphate 153
Nitrate, determination with:
  chromium (II) salts 114
  tin (II) chloride 124
  vanadium (II) sulphate 151

Nitrite, determination with:
  chloramine-T 43
  hydrazine sulphate 183
  hypohalites 33
  lead (IV) acetate 79
  manganese (III) compounds 11
  potassium hexacyanoferrate (III) 25
Nitrite number, see Sodium nitrite 187
Nitrobenzaldehyde, determination with vanadium (II) sulphate 153
Nitrobenzene, determination with vanadium (II) sulphate 153
o-Nitro benzoic acid, determination with vanadium (II) sulphate 153
Nitro compounds, determination with:
  chromium (II) salts 114, 115
  molybdenum (III) compounds 141
  sodium nitrite 188
  vanadium (II) sulphate 153
Nitro-furfurane, determination with chloramine-T 44
m-Nitrophenol, determination with chromium (II) salts 115
o-Nitrophenol, determination with chromium (II) salts 115
p-Nitrophenol, determination with chromium (II) salts 115
p-Nitrophenylhydrazine determination with:
  chloramine-T 44
  lead (IV) acetate 79
  vanadium (II) sulphate 153
Nitroguanidine, determination with vanadium (II) sulphate 153
Nitroso-compounds, determination with:
  chromium (II) salts 114, 115
  molybdenum (III) 142
  sodium nitrite 188
o-Nitrosodimethylaniline, determination with chromium (II) salts 115
p-Nitrosodimethylaniline, determination with chromium (II) salts 115
Nitrosodiphenylamine, determination with tin (II) chloride 125
α-Nitroso-β-naphthol, determination with tin (II) chloride 125
1-Nitroso-2-naphthol, determination with chromium (II) salts 115
3-Nitroso-2-naphthol, determination with chromium (II) salts 115

p-Nitrosoquinoline-iodomethylate determination with potassium hexacyanoferrate (III) 24
2-Nitro-*m*-xylene, determination with vanadium (II) sulphate 153
Novocaine, determination with iodine monochloride 63

Olefines, determination with iodine monochloride 63
Oleic acid, determination with:
 copper (III) compounds 17
 iodine monochloride 63
Organic aromatic compounds, determination with potassium chlorate 194
Organic compounds, determination with:
 manganese (III) compounds 11
 tin (II) chloride 125
Organic dyes:
 determination of:
  chlorate 196
  chlorine in water 196
  hydrogen peroxide 196
  hypochlorite 196
  iron salts 196
  vanadate 196
 determination with:
  chromium (II) salts 115
Organic pharmaceutical preparates determination with vanadium (V) compounds 90
Osmium, determination with:
 chromium (II) salts 114
 vanadium (II) compounds 88
1,3,4-Oxadiazole, determination with sodium nitrite 190
Oxalic acid, determination with:
 chloramine-T 44
 manganese (III) compounds 12
 vanadium (V) compounds 90
Oxine, determination with:
 chloramine-T 44
 copper (III) compounds 17
 iodine monochloride 63
7-Oxiphentiazone-2, determination of:
 chromium (II) salts 196
 tin 196
 titanium 196

o-Oxybenzalsemicarbazone, determination with vanadium (V) compounds 89
Oxygen determination with:
 ascorbic acid 168
 chromium (II) salts 114

Palladium, determination with:
 tin (II) chloride 123
 vanadium (V) compounds 88
 vanadium (II) sulphate 150
Penicillin, determination with:
 periodic acid 70
 potassium hexacyanoferrate (III) 24
Penthathionate, determination with vanadium (V) compounds 89
Perbenzoic acid, determination of:
 antimony 195
 arsenic 195
 iron salts 195
 tin 195
Perborate, determination with vanadium (II) sulphate 151
Periodate, determination with:
 chloramine-T 43
 hydrazine sulphate 181
 sodium arsenite 128
Periodic acid and its salts:
 determination of:
  allyl isothiocyanate 71
  antimony 68, 69, 70
  arsenic 68, 69, 70
  ascorbic acid 69, 70
  cadmium 68
  cerium 68
  cobalt 69
  copper 69
  cystein chloride 70
  ferrocyanide 70
  formates 69
  hydrazide of isonicotinic acid 69
  hydrazine 69, 70
  hydrazine sulphate 69
  hydrogen peroxide 70
  hydrogen sulphite 70
  hydroquinone 69, 70
  iodide 69, 70
  iron 68, 69

lead 70
lead iodide 70
lead oxide 69
manganese oxide 69
mercury 69, 70
mercury iodide 70
penicillin 70
phenylhydrazine 69, 70
potassium antimony tartrate 70
semicarbazide 69
silver iodide 70
sulphide 69
sulphite 70
tellurium 68
tetrathionate 70
thallium 70
thiocyanate 69, 70
thiosemicarbazide 69
thiosine amine 71
thiosulphate 70
thiourea 71
tin 69, 70
vanadium 71
zinc 69
determination with:
  vanadium (IV) sulphate 159
indicator 68
oxidation of organic substances 71
  active hydrogen atoms 73
  acyl derivatives of α-amino alcohols 73
  alcohols, simple 73
  aldehydes 73
  amino acids 73
  amino alcohols 72, 73
  amino alcohols, acyl derivatives of 72, 73
  α-diamines 73
  α-diketones 73
  ethylene glycol 74
  glycerol 74
  glyoxal 73
  hydrogen atoms, active 73
  α-hydroxy acids 73
  α-hydroxy aldehydes 73
  α-keto acids 73
  α-ketoaldehydes 73
  α-ketols 73
  ketones 73
  polyalcohols 73
  polyhydroxycompounds 72

polyhydroxy dicarboxylic acids 73
polyhydroxy monocarboxylic acids 73
1,2-propylene glycol 74
sugars 73
references 74
review of determinations 68
splitting of α-dihydroxy compounds 71
standardization 67
Permanganate, determination with
  ascorbic acid 164, 167, 168
  chloramine-T 43
  copper (I) compounds 138
  chromium salts 114
  hypohalites 31
  mercury (I) compounds 135
  sodium arsenite 128
  sodium nitrite 190
  tin (II) chloride 125
  tungsten (III) compounds 144
  uranium (IV) sulphate 146
  vanadium (IV) acetate 159
  vanadium (IV) sulphate 158
Permanganate–iron, determination with uranium (IV) sulphate 146
Permanganate–vanadate–iron, determination with uranium (IV) sulphate 146
Peroxide, determination with:
  ascorbic acid 169
  bromine 49
  hydrogen peroxide 103
  hypohalites 33
  manganese (III) compounds 11
  potassium hexacyanoferrate (III) 22
  vanadium (II) sulphate 151
Peroxodisulphate, determination with
  ascorbic acid 164
  hydroquinone 176
  vanadium (II) sulphate 151
Perrhenate, determination with iron salts 96
Persulphate, determination with:
  mercury (I) compounds 135
  vanadium (V) compounds 89
Pharmaceutical preparates, organic determination with vanadium (V) compounds 90
Phenolphthaleine, determination with iodine monochloride 63

Phenols, determination with:
  chloramine-T 44
  iodine monochloride 63
  potassium permanganate 7
  sodium nitrite 189
  vanadium (II) sulphate 153
Phenols p-substituted, determination with manganese (III) compounds 12
Phenoxazine, determination of:
  bromine 196
  chlorine 196
Phenyl arsine oxide, determination of:
  bromine 195
  chlorine 195
Phenylenediamine, determination with iodine monochloride 62
p-Phenylenediamine:
  determination of:
    gold (III) salts 177
  determination with:
    vanadium (V) compounds 90
Phenylhydrazine, determination with
  chloramine-T 44
  copper (III) compounds 17
  iodine monochloride 61
  periodic acid 69, 70
Phenylhydrazine chloride, determination with lead (IV) acetate 79
Phenylhydroxylamine, determination with iron salts 96
4-Phenylsemicarbazone chloride, determination with iodine monochloride 63
Phloroglucinol, determination with copper (III) compounds 17
Phosphate, determination with vanadium (V) compounds 89
Phosphite, determination with:
  hypohalites 33
  potassium permanganate 6, 7
  vanadium (V) compounds 89
Photographic developers,, determination with mercury iodide 195
Phthalocyanine, determination with vanadium (V) compounds 90
Phytine, determination with iron salts 96
Picric acid, determination with:
  chromium (II) salts 115
  molybdenum (III) compounds 142
  vanadium (II) sulphate 153

Picrolonates, determination with chromium (II) salts 116
Picrolonic acid, determination with chromium (II) salts 116
Platinum, determination with:
  ascorbic acid 166
  copper (I) compounds 138
  tin (II) chloride 123
  vanadium (V) compounds 88
Platinum salts, determination with iron salts 96
Plutonium, determination with chromium (II) salts 114
Polyalchols, determination with periodic acid 73
Polyhydroxy compounds, determination with periodic acid 72
Polyhydroxy dicarboxylic acid, determination with periodic acid 73
Polyhydroxy monocarboxylic acid, determination with periodic acid 73
Polythionates, determination with chloramine-T 44
Potassium antimony tartrate, determination with periodic acid 70
Potassium chlorate, determination of:
  organic aromatic compounds 194
  tin salts 194
  vanadium 194
Potassium dichromate, determination with chloramine-T 43
Potassium disulphate, determination with chromium (II) salts 114
Potassium hexacyanoferrate (III):
  determination of:
    acetals 25
    adrenalin 24
    antimony 19
    arsenic 19
    ascorbic acid 25
    cerium 19
    chromium 20
    cobalt 20
    cystein 24
    formaldehyde 25
    hemin 24
    hemoglobin 24
    hydrazine 23
    hydroquinone 25
    hydroxylamine 23
    iodide 25

SUBJECT INDEX

iron 22
isonicotinic acid hydrazide 23
ketals 25
manganese 20
mercaptans 25
molybdenum 21
morphine 25
nitrite 25
p-nitrosoquinoline-iodomethylate 24
penicillin 24
peroxide 22
quinidine-iodoethylate 24
quinoline-iodomethylate 24
selenium 21
sugars, reducing 24
sulphides 22
sulphur compounds 22
thallium 21
thiourea 24
tin 19
titanium 22
uranium 22
uric acid 25
vanadium 22
variamine blue 25
vinyl ether 25
indicator 19
references 25
review of determinations 19
standardization 19
standard solution 18
Potassium iodide, determination of:
arsenate 195
chlorine 195
copper salts 195
gold 195
hypochlorite 195
iodate 195
manganese 195
selenate 195
tellurate 195
Potassium manganate, determination of:
antimony 194
arsenic 194
chromium 194
formic acid 194
hydrogen peroxide 194
manganese salts 194
tellurium 194
thallium 194

Potassium periodate, determination with chromium (II) salts 114
Potassium permanganate in alkaline solution:
determination of:
acetaldehyde 7
acetone 7
antimony 3
arsenic 3
barium peroxide 6
benzaldehyde 7
benzylalcohol 7
n-butanol 7
cinnamic acid 7
citric acid 7
cerium 4
chromium 4
cyanide 6, 7
4-dimethylamino-1,5-dimethyl-2-phenylpyrazolone (amidopyrine) 7
erythritol 7
formaldehyde 7
formate 6
formic acid 7
fumaric acid 7
glycerol 7
hexacyanoferrate (II) 5
hydrazine 6, 7
hydrogen peroxide 6
hypophosphite 7
iodate 7
iodide 7
isoamylalcohol 7
lead 5
malic acid 7
mandelic acid 7
mercury 5
methanol 7
phenol 7
phosphite 6, 7
reducing substances in natural and waste water 8
ruthenium 5
salicylic acid 7
selenium 4
sodium peroxide 6
sulphide 6
sulphite 6
tartaric acid 7
tellurium 4
thallium 5

Potassium permanganate in alkaline solution—*Continued*
determination of:
thiocyanate 7
thiosulphate 6, 7
tin 5
vanadium 4
indicator 3
references 8
review of determinations 3
standardization 3
standard solution 3
Potassium peroxodisulphate, determination of:
hydrazine 194
inorganic substances 194
1,2-Propylene glycol, determination with periodic acid 74
Proteins, determination with copper (III) compounds 16, 17
Pyrazol anthrone, determination with vanadium (II) sulphate 152
Pyrocatechol, determination with copper (III) compounds 17
Pyrogallol, determination with copper (III) compounds 17

Quinhydrone, determination with:
chloramine-T 42
hydroquinone 177
Quinidine-iodoethylate, determination with potassium hexacyanoferrate (III) 24
Quinoline-iodomethylate, determination with potassium hexacyanoferrate (III) 24
Quinone, determination with:
chromium (II) salts 114
copper (III) compounds 17
thiosulphate 196
p-Quinone, determination with chromium (II) salts 115
Quinone imine dyes, determination with chromium (II) salts 116

Reducing substances in natural and waste water, determination with potassium permanganate 8
Reducing sugars, determination with potassium hexacyanoferrate (III) 24

Resazurine, determination with ascorbic acid 169
Resorcinol, determination with copper (III) compounds 17
Resorufine, determination with ascorbic acid 169
Rhenium, determination with:
chromium (II) salts 114
tin (II) chloride 123
Rhodamine, determination with chromium (II) salts 116
Rifamycin, determination with:
ascorbic acid 169
gold chloride 195
Rivanol, determination with iodine, monochloride 63
Rosaniline, determination with chromium (II) salts 116
Ruthenium, determination with:
chromium (II) salts 114
hydroquinone 175
lead (IV) acetate 79
potassium permanganate 5
Rutine, determination with iodine monochloride 63

Safranine, determination with chromium (II) salts 116
Salicalaldehyde, determination with chloramine-T 43
Salicylaldoxime, determination with iodine monochloride 63
Salicylic acid, determination with:
chloramine-T 44
manganese (III) compounds 12
potassium permanganate 7
Selenate, determination with:
hydrazine sulphate 183
potassium iodide 196
Selenite, determination with:
chromium (II) salts 114
hypohalites 33
Selenium, determination with:
ascorbic acid 166
hydrazine sulphate 182
potassium hexacyanoferrate (III) 21
potassium permanganate 4
Selenium (IV) oxide, determination with:
chloramine-T 43
vanadium (V) compounds 89

SUBJECT INDEX

Selenium (IV) salts, determination with hypohalites 33
Selenium-tellurium, determination with chromium (II) salts 114
Semicarbazide, determination with:
  ascorbic acid 169
  chloramine-T 44
  periodic acid 69
  sodium nitrite 190
Semicarbazide chloride, determination with:
  iodine monochloride 63
  vanadium (V) compounds 89
Silicic acid, determination with tin (II) chloride 125
Silver, determination with:
  ascorbic acid 165, 169
  chromium (II) salts 108
  hydrazine sulphate 182
  hydroquinone 175
  hypohalites 33
  vanadium (II) sulphate 149
  vanadium (IV) sulphate 158
Silver compounds, determination with vanadium (V) compounds 89
Silver iodide, determination with periodic acid 70
Sodium arsenite:
  determination of:
    bromine 128
    caro acid 128
    cerium 128
    chlorine 128
    chlorites 128
    dichromate 128
    hexacyanoferrate (III) 128
    hypohalites 128
    iodine 128
    manganese 128
    periodate 128
    permanganate 128
  indicator 127
  references 129
  review of determinations 128
  standard solution 127
Sodium formate, determination with chloramine-T 43
Sodium hydrogen sulphite, determination with chloramine-T 42
Sodium nitrite:
  determination of:
    aliphatic acids 190
    amines, aromatic, primary 186
    amines, aromatic, secondary 188
    amines, aromatic, tertiary 188
    diamines, aromatic 188
    hydrazides 190
    hydrazine 189
    hydrazine derivatives 189
    ketones, cyclic 189
    nitrite number 187
    nitrocompounds 188
    nitrosocompounds 188
    phenols 189
    semicarbazide 190
    thiosemicarbazide 190
  indicator 186
  references 191
  review of determinations 186
  standardization 185
  standard solution 185
Sodium peroxide, determination with:
  chromium (II) salts 114
  potassium permanganate 6
Sodium sulphide, determination with chloramine-T 42
Sphaerophysine, determination with iodine monochloride 63
Starch, determination with copper (III) compounds 14
Sugar, determination with:
  chloramine-T 44
  chromium (II) salts 115
  hypohalites 34
  periodic acid 73
Sugar in blood, determination with copper (III) compounds 16
Sugars, reducing, determination with:
  mercury iodide 195
  potassium hexacyanoferrate (III) 24
Sulphate, determination with hydroquinone 177
Sulphide, determination with:
  bromine 49
  N-bromosuccinimide 54
  chloramine-T 42
  hydrogen peroxide 103
  hypohalites 32
  potassium hexacyanoferrate (III) 24
  potassium permanganate 6
  periodic acid 59
  vanadium (V) compounds 88

Sulphide organic compounds, determination with chloramine-T 44
Sulphinic acid, aromatic, determination with sodium nitrite 190
Sulphite:
  determination of:
    gold salts 196
  determination with:
    bromine 49
    chloramine-T 42, 44
    hydrogen peroxide 103
    hypohalites 32
    iodine monochloride 60, 63
    periodic acid 70
    potassium permanganate 6
    vanadium (V) compounds 88
Sulphonamides, determination with iodine monochloride 63
Sulphonic acid, determination with hypohalites 34
Sulphur compounds determination with potassium hexacyanoferrate (III) 22

Tannin, determination with vanadium (V) compounds 90
Tartaric acid, determination with:
  lead (IV) acetate 79, 80
  manganese (III) compounds 12
  potassium permanganate 7
  vanadium (V) compounds 90
Tartrate, determination with:
  copper (III) compounds 16, 17
  lead (IV) acetate 81
Tellurate, determination with:
  potassium iodide 196
  uranium (IV) sulphate 146
Tellurium, determination with:
  hypohalites 33
  periodic acid 68
  potassium manganate 194
  potassium permanganate 4
Tetracene, determination with chromium (II) salts 115
Tetrachlorohydroquinone, determination with lead (IV) acetate 80
Tetrahydro-benzoic acid amide, determination with N-bromosuccinimide 54
N,N,N,N-Tetramethyl naphthidine, determination with bromine 50

Tetrathionate, determination with:
  periodic acid 70
  vanadium (V) compounds 89
Thallium, determination with:
  ascorbic acid 169
  bromine 49
  chloramine-T 42, 44
  chromium (II) salts 111
  copper (III) compounds 16
  hydrazine sulphate 182
  hydroquinone 173
  hypothalites 33
  periodic acid 70
  potassium hexacyanoferrate (III) 21
  potassium manganate 194
  potassium permanganate 5
  thiosulphate 196
  vanadium (II) sulphate 150
Thiazine, determination with ascorbic acid 169
Thiazone, determination with ascorbic acid 169
Thioacetamide, determination with hypohalites 34
Thiobarbituric acid derivatives, determination with chloramine-T 44
Thiocyanate, determination with:
  N-bromosuccinimide 54
  chloramine-T 42, 43, 44
  hypohalites 32
  iodine monochloride 59
  manganese (III) compounds 11
  periodic acid 69, 70
  potassium permanganate 7
  vanadium (V) compounds 88
Thioglycolic acid, determination with lead (IV) acetate 80
3-Thioketo-5-keto-6-benzyl-1,2,4-triazine, determination with hypohalites 34
Thiosemicarbazide, determination with:
  chloramine-T 44
  iodine monochloride 62
  lead (IV) acetate 80
  periodic acid 69
  sodium nitrite 190
Thiosemicarbazone of p-acetoamidobenzaldehyde, determination with iodine monochloride 63
Thiosine amine, determination with periodic acid 71

Thiosulphate:
  determination of:
    bromate 196
    copper salts 196
    iodate 196
    iodine 196
    iron 196
    quinone 196
    thallium 196
  determination with:
    chloramine-T 41, 42
    copper (III) compounds 16
    hypohalites 32
    iodine monochloride 60
    periodic acid 70
    potassium permanganate 6, 7
    vanadium (V) compounds 88
Thiosulphite, determination with hydrogen peroxide 103
Thiourea, determination with:
  N-bromosuccinimide 54
  chloramine-T 44
  iodine monochloride 62
  lead (IV) acetate 80
  periodic acid 71
  potassium hexacyanoferrate (III) 24
Tibon, determination with iodine monochloride 63
Tin, determination with:
  ascorbic acid 167
  bromine 49
  chloramine-T 40
  chromium (II) salts 112
  hypohalites 31
  iodine monochloride 57
  iron salts 95
  manganese (III) compounds 11
  methylene blue 196
  7-oxiphenthiazone-2 196
  perbenzoic acid 195
  periodic acid 69, 70
  potassium chlorate 194
  potassium hexacyanoferrate (III) 19
  potassium permanganate 5
  vanadium (II) sulphate 151
Tin (II) chloride:
  determination of:
    antimony 123
    bromate 124
    bromine 123
    chromium 122
    copper 121
    dinitroresorcinol 125
    gold 123
    hexacyanoferrate 124
    iodate 124
    iodine 123
    iron 121
    mercury 122
    molybdenum 123
    α-naphthylamine 125
    nickel (IV) dimethylglyoximate 125
    p-nitraniline 125
    nitrate 124
    nitrosodiphenylamine 125
    α-nitroso-β-naphthol 125
    organic compounds 125
    palladium 123
    permanganate 125
    platinum 123
    rhenium 123
    silicic acid 125
    vanadium 122
    variamine blue 125
  indicator 121
  procedure 120
  references 125
  review of determinations 121
  standardization 121
  standard solution 120
Tin–antimony, determination with chromium (II) salts 112
Tin–bismuth, determination with chromium (II) salts 112
Tin–copper, determination with chromium (II) salts 112
Tin–copper–antimony, determination with chromium (II) salts 112
Tin–copper–bismuth, determination with chromium (II) salts 112
Tin–iron, determination with chromium (II) salts 112
Tin–iron–bismuth, determination with chromium (II) salts 112
Titanium, determination with:
  chloramine-T 41, 44
  chromium (II) chloride 116
  chromium salts 111
  iodine monochloride 58
  iron salts 94

Titanium, determination with—*Continued*
  manganese (III) compounds 11
  7-oxiphenthiazone-2 196
  potassium hexacyanoferrate (III) 22
  vanadium (V) compounds 85, 88
  vanadium (II) sulphate 150
  vanadium (IV) sulphate 158
Tocopherol, determination with gold chloride 195
Trinitrotoluene, determination with chromium (II) salts 115
Triphenyl methane, determination with chromium (II) salts 116
Triphenylmethane dyes, determination with vanadium (II) sulphate 151
Trithionate, determination with vanadium (V) compounds 89
Tungstate, determination with tungsten (III) compounds 144
Tungsten (III) compounds, determination of:
  bromate 144
  cerium salts 144
  copper 144
  iron–copper 144
  permanganate 144
  tungstate 144
Tungsten (V) compounds:
  determination of:
    bromate 144
    cerium salts 144
    copper 144
    dichromate 144
    hexacyanoferrate (III) 144
    iron 144
    iron–vanadate 144
    vanadate 144
  determination with:
    ascorbic acid 169
    chromium (II) salts 110
    iron salts 96
    manganese (III) compounds 11
    vanadium (V) compounds 86
  indicator 143
  reference 144
  review of determinations 144
  standardization 143
  standard solution 143
Tungsten–chromium, determination with chromium (II) salts 114

Uranium (IV) sulphate:
  determination of:
    bromate 146
    cerium 146
    cerium–iron 146
    dichromate 146
    dichromate–iron 146
    hexacyanoferrate (III) 146
    iron 46
    permanganate 146
    permanganate–iron 146
    permanganate–vanadate–iron 146
    tellurate 146
    vanadate 146
    vanadate–iron 146
    vanadate–cerium–iron 146
  indicator 145
  references 146
  review of determinations 146
  standardization 145
  standard solution 145
Uranium, determination with:
  chromium (II) salts 10
  iron salts 95, 96
  lead (IV) acetate 79
  manganese (III) compounds 11
  peroxide solution 103
  potassium hexacyanoferrate (III) 22
  vanadium (V) compounds 86
  vanadium (II) sulphate 151
Uranium (III, IV), determination with lead (IV) acetate 79
Uranium–vanadium–iron–chromium, determination with chromium (II) salts 113
Uranium–vanadium, determination with chromium (II) salts 114
Urea, determination with:
  chloramine-T 44
  hypohalites 34
Uric acid, determination with:
  chloramine-T 44
  potassium hexacyanoferrate (III) 25
Urine, determination with copper (III) compounds 16

Vanadate, determination with:
  ascorbic acid 168
  chromium (II) salts 111
  copper (I) compounds 138

SUBJECT INDEX

hydrazine sulphate 182
mercury compounds 135
molybdenum (III), (V) compounds 141
organic dyes 196
tungsten (V) compounds 144
uranium (IV) sulphate 146
Vanadate–cerium–iron, determination with uranium (IV) sulphate 146
Vanadate–iron, determination with uranium (IV) sulphate 146
Vanadate–iron (III) salt, determination with molybdenum (V) compounds 141
Vanadium, determination with:
  ascorbic acid 164, 169
  chloramine-T 41, 44
  hydroquinone 174, 175
  iron salts 95
  manganese (III) compounds 11
  periodic acid 71
  potassium chlorate 194
  potassium hexacyanoferrate (III) 22
  potassium permanganate 4
  sodium nitrite 190
  tin (II) chloride 122
  vanadium (V) compounds 84
  vanadium (II) sulphate 149
  vanadium (IV) sulphate 159
Vanadium (IV) acetate:
  determination of:
    bromate 158
    chromic acid 158
    lead acetate 158
    permanganate 158, 159
  indicator 156
  references 159
  review of determinations 156
  standardization 156
  standard solution 155
Vanadium–chromium–molybdenum, determination with ascorbic acid 169
Vanadium (V) compounds:
  determination of:
    amidopyrine 90
    α-amino acids 90
    amino guanidonium chloride 89
    analgine 90
    antimony 88
    arsenic 84
    benzalazine 89

benzalsemicarbazone 89
calcium 88, 89
citric acid 90
chloralhydrazine 89
o-chlorobenzalsemicarbazone 89
chromium 89
cobalt 88
copper 87
dibromohydroxyquinoline 90
dibromooxinates 90
dithionate 89
formate 90
glycerol 90
hexacyanoferrate 89
hydrazide of isonicotinic acid 90
hydrazine 89
hydrazine derivatives 89
hydrogen peroxide 89
hydroquinone 90
hydroxylamine 90
hypophosphite 89
indigo 90
iodide 88
iron 87
isonicotinic acid hydrazide of 90
lactic acid 90
lead 88
malic acid 90
malonic acid 90
manganese 88
mercury 85
metol 90
p-methoxybenzalsemicarbazone 89
methylethylketone semicarbazone 89
molybdenum 85, 88
nickel 88
organic pharmaceutical preparates 90
osmium 88
oxalic acid 90
o-oxybenzalsemicarbazone 89
palladium 88
penthathionate 89
persulphate 89
pharmaceutical preparates—organic 90
p-phenylenediamine 90
phosphate 89
phosphite 89
phthalocyanine 90

Vanadium (V) compounds:
  determination of—*Continued*
    platinum 88
    selenium (IV) oxide 89
    semicarbazide chloride 89
    silver 89
    sulphide 88
    sulphite 88
    tannin 90
    tartaric acid 90
    tetrathionate 89
    thallium 85
    thiocyanate 88
    thiosulphate 88
    titanium 85, 88
    trithionate 89
    tungsten 86
    uranium 86
    vanadium 84
    zinc 88
  oxidation of:
    alcohols 90
    cyclohexanol 90
    glycerol 90
    ketones 90
  indicator 84
  references 90
  review of determinations 84
  standardization 83
  standard solution 83
Vanadium–iron, determination with:
  ascorbic acid 169
  chromium (II) salts 114
Vanadium (II) sulphate:
  determination of:
    acetaldehyde 153
    acetone 153
    alcohols 153
    amines 153
    anthraquinone 152
    antimony 150
    azobenzene 151
    azo dyes 151
    benzaldehyde 153
    benzophenone 153
    bismuth 150
    bromate 151
    camphor 153
    cerium 150
    chlorate 151
    chromium 150
    copper 149
    diazo phenols 151
    dinitrobenzene 153
    2,4-dinitrophenylhydrazine 153
    formaldehyde 153
    glucose 153
    hydrocarbons, aromatic 153
    hydroxylamine 151
    hydroxy-triphenylmethane dyes 151
    iron 149
    p-nitraniline 153
    nitrate 151
    nitrobenzaldehyde 153
    nitrobenzene 153
    o-nitro benzoic acid 153
    nitro compounds 153
    p-nitrophenyl hydrazine 153
    2-nitro-m-xylene 153
    nitroguanidine 153
    palladium 150
    perborate 151
    peroxide 151
    peroxodisulphate 151
    phenols 153
    picric acid 153
    pyrazol anthrone 152
    silver 149
    thallium 150
    tin 151
    titanium 150
    triphenylmethane dyes 151
    uranium 151
    vanadium 149
    vanillin 153
  indicator 149
  references 153
  review of determinations 149
  standardization 148
  standard solution 147
Vanadium (IV) sulphate:
  determination of:
    antimonite 157
    arsenite 157
    bismuth 158
    caro's acid 159
    cerium 159
    chromate 157
    chromium 157
    cobalt (III) 159
    copper 157
    gold 158
    hexacyanoferrate (III) 156

hydrazine sulphate 157
hydrogen peroxide 157
manganese 159
mercury 157
periodic acid 159
permanganate 158, 159
silver 158
titanium 158
vanadium 159
indicator 156
references 159
review of determinations 156
standardization 156
standard solution 155
Vanadium–titanium, determination with chromium (II) salts 114
Vanadyl, determination with manganese (III) compounds 11
Vanillin, determination with:
 chloramine-T 43
 vanadium (II) sulphate 153

Variamine blue, determination with:
 potassium hexacyanoferrate (III) 25
 tin (II) chloride 125
Vicinal glycols, determination with lead (IV) acetate 80
Vinyl ether, determination with potassium hexacyanoferrate (III) 25
Vitamin B, determination with copper (III) compounds 17

Xanthates, determination with chloramine-T 44

Zinc, determination with:
 ascorbic acid 167
 chloramine-T 44
 periodic acid 69
 vanadium (V) compounds 88

OTHER TITLES IN THE SERIES
ANALYTICAL CHEMISTRY

- Vol. 1 WEISZ—*Microanalysis by the Ring Oven Technique*
- Vol. 2 CROUTHAMEL—*Applied Gamma-Ray Spectrometry*
- Vol. 3 VICKERY—*The Analytical Chemistry of the Rare Earths*
- Vol. 4 HEADRIDGE—*Photometric Titrations*
- Vol. 5 BUSEV—*The Analytical Chemistry of Indium*
- Vol. 6 ELWELL and GIDLEY—*Atomic Absorption Spectrophotometry*
- Vol. 7 ERDEY—*Gravimetric Analysis, Parts I–III*
- Vol. 8 CRITCHFIELD—*Organic Functional Group Analysis*
- Vol. 9 MOSES—*Analytical Chemistry of the Actinide Elements*
- Vol. 10 RYABCHIKOV and GOL'BRAIKH—*The Analytical Chemistry of Thorium*
- Vol. 11 CALI—*Trace Analysis for Semiconductor Materials*
- Vol. 12 ZUMAN—*Organic Polarographic Analysis*
- Vol. 13 RECHNITZ—*Controlled-Potential Analysis*
- Vol. 14 MILNER—*Analysis of Petroleum for Trace Elements*
- Vol. 15 ALIMARIN and PETRIKOVA—*Inorganic Ultramicroanalysis*
- Vol. 16 MOSHIER—*Analytical Chemistry of Niobium and Tantalum*
- Vol. 17 JEFFERY and KIPPING—*Gas Analysis by Gas Chromatography*
- Vol. 18 NIELSEN—*Kinetics of Precipitation*
- Vol. 19 CALEY—*Analysis of Ancient Metals*
- Vol. 20 MOSES—*Nuclear Techniques in Analytical Chemistry*
- Vol. 21 PUNGOR—*Oscillometry and Conductometry*

10-6-66

OHIO UNIVERSITY LIBRARY

Please return this book as soon as you have finished with it. In order to avoid a fine it must be returned by the latest date stamped below.

JAN 11 1980

JAN 14 1980

QTR. LOAN

SEP 13 1987

SEP 11 1987

CF   20M-10/64